Conference Proceedings of the Society for Experimental Mechanics Series

Series Editor

Kristin B. Zimmerman, Ph.D.
Society for Experimental Mechanics, Inc.,
Bethel, CT, USA

More information about this series at http://www.springer.com/series/8922

Allison M. Beese • Alan T. Zehnder • Shuman Xia
Editors

Fracture, Fatigue, Failure and Damage Evolution, Volume 8

Proceedings of the 2015 Annual Conference on Experimental and Applied Mechanics

Springer

Editors
Allison M. Beese
Materials Engineering Department
Pennsylvania State University
Pennsylvania, USA

Alan T. Zehnder
Cornell University
Ithaca, New York, USA

Shuman Xia
Georgia Institute of Technology
Atlanta, Georgia, USA

ISSN 2191-5644 ISSN 2191-5652 (electronic)
Conference Proceedings of the Society for Experimental Mechanics Series
ISBN 978-3-319-21610-2 ISBN 978-3-319-21611-9 (eBook)
DOI 10.1007/978-3-319-21611-9

Library of Congress Control Number: 2015951232

Springer Cham Heidelberg New York Dordrecht London

Printed on acid-free paper

Springer International Publishing AG Switzerland is part of Springer Science+Business Media (www.springer.com)

Preface

Fracture, Fatigue, Failure, and Damage Evolution represents one of nine volumes of technical papers presented at the SEM 2015 SEM Annual Conference & Exposition on Experimental and Applied Mechanics organized by the Society for Experimental Mechanics and held in Costa Mesa, CA, June 8–11, 2015. The complete Proceedings also include volumes on: *Dynamic Behavior of Materials*; *Challenges in Mechanics of Time-Dependent Materials*; *Advancement of Optical Methods in Experimental Mechanics*; *Experimental and Applied Mechanics*; *MEMS and Nanotechnology; Mechanics of Biological Systems and Materials; Mechanics of Composite & Multifunctional Materials and Residual Stress, Thermomechanics & Infrared Imaging, Hybrid Techniques and Inverse Problems.*

Each collection presents early findings from experimental and computational investigations on an important area within Experimental Mechanics, Fracture and Fatigue being one of these areas.

Fatigue and fracture are two of the most critical considerations in engineering design. Understanding and characterizing fatigue and fracture has remained as one of the primary focus areas of experimental mechanics for several decades. Advances in experimental techniques, such as digital image correlation, acoustic emissions, and electron microscopy, have allowed for deeper study of phenomena related to fatigue and fracture. This volume contains the results of investigations of several aspects of fatigue and fracture such as microstructural effects, the behavior of interfaces, the behavior of different and/or complex materials such as composites, and environmental and loading effects. The collection of experimental mechanics research included here represents another step toward solving the long-term challenges associated with fatigue and fracture.

Pennsylvania, PA — Allison M. Beese
Ithaca, NY — Alan T. Zehnder
Atlanta, GA — Shuman Xia

Contents

Chapter 1
Reflection-Mode Digital Gradient Sensing Technique for Experimental Fracture Mechanics

Amith S. Jain and Hareesh V. Tippur

Abstract In this work, the reflection-mode Digital Gradient Sensing (r-DGS) method is extended for visualizing and quantifying crack-tip deformations in solids under quasi-static and dynamic loading conditions. The r-DGS technique employs digital image correlation principles to quantify two orthogonal surface slopes simultaneously in specularly reflective solids by quantifying small deflections of light rays. For the first time, r-DGS has been implemented here to study both mode-I and mixed-mode (I/II) problems and quantify fracture parameters. Under dynamic loading conditions, r-DGS is implemented in conjunction with high-speed digital photography to map surface slopes in edge cracked plates subjected to one-point impact. The measured surface slopes have been used to successfully evaluate stress intensity factor histories by pairing measurements with the corresponding asymptotic crack tip field descriptions using overdeterministic least-squares analyses.

Keywords Optical metrology • Surface slopes • Stress wave propagation • High-speed photography • Fracture mechanics

1.1 Introduction

Measurement of stresses, strains, and deformations is at the heart of experimental mechanics studies. Over the years, experimental mechanics community has developed numerous measurement techniques to study stress concentration problems. In this spirit, a new technique called Digital Gradient Sensing (DGS) based on 2-D digital image correlation, has been recently introduced for studying transparent [1, 2] and reflective [3] solids by quantifying small angular deflections of light rays caused by defects or deformation of solids. In the former works, the method has been successfully demonstrated by measuring two orthogonal stress gradients near stress concentrations in solids whereas the latter has been demonstrated via full-field orthogonal surface slope fields in flexural problems. The purpose of the current work is to extend the reflection-mode DGS (or, simply r-DGS) method to quasi-static and dynamic fracture mechanics. Visualization of time-resolved deformations near mode-I and mixed-mode (mode I/II) cracks for quantifying fracture parameters are among the goals of this work.

1.2 Optical Setup and Working Principle

The schematic of the optical setup used for reflection-mode DGS (or, r-DGS) is shown in Fig. 1.1. It consists of a specularly reflective planar object (cracked specimen), a beam splitter, a randomly speckled/textured planar target and a digital camera. The speckle pattern on the target is prepared by decorating a planar surface with alternative black/white mists of paint.

The optical arrangement is such that the digital camera is used to record speckles on the target plane via the reflective specimen and the beam splitter. In the undeformed or the reference state of the specimen, a generic point P on the target is photographed through a point O on the specimen surface. When the specimen deforms, O moves to O′ on the target due to an out-of-plane displacement w in the z-direction (the optical axis). This results in the camera photographing a neighboring point Q on the target in the deformed state of the specimen. By correlating these two images corresponding to the reference and deformed states, the local displacement components δ_x and δ_y in the x- and y-directions, respectively, can be evaluated

A.S. Jain • H.V. Tippur (✉)
Department of Mechanical Engineering, Auburn University, Auburn, AL 36849, USA
e-mail: tippuhv@auburn.edu

A.M. Beese et al. (eds.), *Fracture, Fatigue, Failure and Damage Evolution, Volume 8*, Conference Proceedings of the Society for Experimental Mechanics Series, DOI 10.1007/978-3-319-21611-9_1

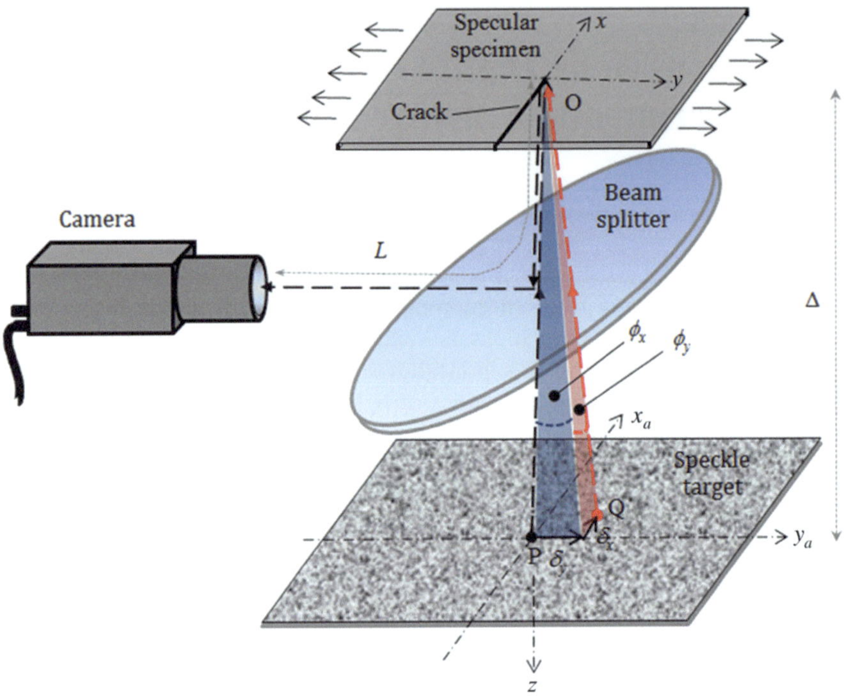

Fig. 1.1 Schematic depicting the experimental setup and working principle of reflection-mode DGS (r-DGS) method

in the whole field. Using this information, the surface slope components in, say, the *x–z* and *y–z* planes can be evaluated as [3],

$$\phi_{x;y} \approx \tan\phi_{x;y} = 2\frac{\partial w}{\partial x;\, y} = \frac{\delta_{x;y}}{\Delta} \tag{1.1}$$

where $\phi_{x;y}$ represent angular deflections of light rays. Here Δ denotes the distance between the specimen and target planes and controls the sensitivity of the method.

1.3 Quasi-Static Mixed-Mode Experiments

Quasi-static 3-point bend experiments were performed on edge notched PMMA specimens using r-DGS. A $130 \times 60 \times 8.9$ mm^3 specimen was used. An initial crack of length 12 mm cut at 45° to the edge as shown in Fig. 1.2. The specimen resting on two anvils (120 mm span) was loaded in a displacement control mode (cross-head speed = 0.005 mm/s). One of the two 130 mm × 60 mm faces of the specimen was deposited with a thin aluminum film to make the surface specularly reflective. A beam splitter was positioned between the specimen and the camera at an angle of 45° to the optical axis of the camera. A speckle target plate was placed at 45° to the beam splitter. The normal distance between the specimen surface and the speckle plane was ($\Delta =$) 65 mm. A digital SLR camera fitted with 28–300 mm focal length macro lens was placed in front of the specimen at a distance (*L*) of ~1150 mm. A reference image was recorded with a camera resolution of 1504 × 1000 pixels in no-load/reference state. During the experiment, speckle images were recorded at different load levels.

Due to deformations in the crack-tip vicinity, the reflected light rays carry surface slope information relative to the reference state. Using 2-D DIC, the angular deflection fields ϕ_x and ϕ_y were obtained by correlating images corresponding to the deformed state with the one from the reference state. The images were divided into 15 × 15 pixel sub-images and correlated with a 5 pixel overlap during image correlation.

Figure 1.3 shows the resulting surface slopes near the crack-tip at a select load level. The contours represent surface slopes $(\partial w/\partial x)$ and $(\partial w/\partial y)$ where *w* is the displacement in the out-of-plane (in the *z*-) direction. These contours show qualitative similarities with the reflection-mode coherent gradient sensing counterparts where the same field quantities are measured but one component at a time.

The availability of two orthogonal fields offers flexibility to evaluate slopes in a rotated coordinate system aligned with the crack orientation, denoted by the coordinates *x′* and *y′* in Fig. 1.2. Accordingly, in this research, the orthogonal angular

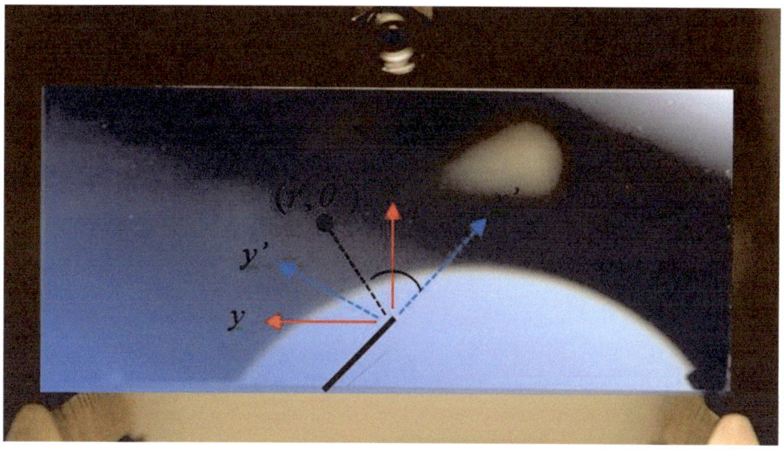

Fig. 1.2 Specimen geometry for crack tip deformation measurement for mixed-mode-I/II loading

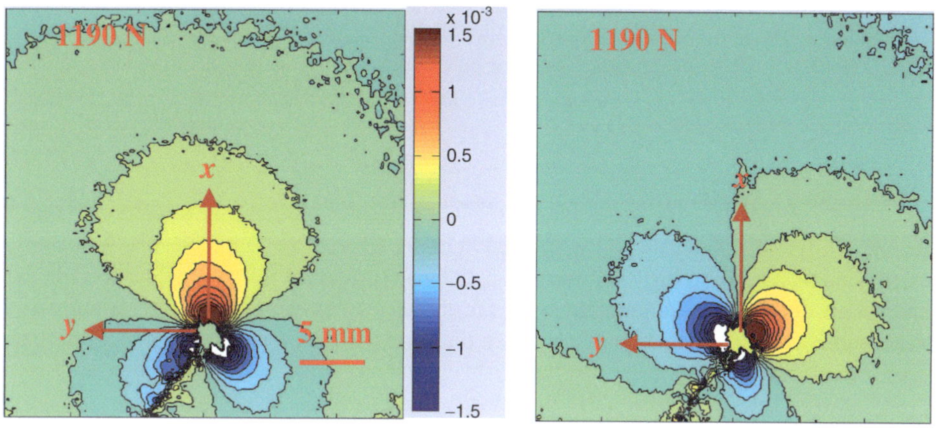

Fig. 1.3 Contours representing surface slopes in the mixed-mode crack tip vicinity. Contour increments $= 15 \times 10^{-5}$ rad

deflections of light rays and hence the respective surface slopes in the local x' and y' directions were evaluated using transformations,

$$\begin{Bmatrix} \frac{\phi'_x}{2} \\ \frac{\phi'_y}{2} \end{Bmatrix} = \begin{Bmatrix} \frac{\partial w}{\partial x'} \\ \frac{\partial w}{\partial y'} \end{Bmatrix} = \begin{pmatrix} \cos\alpha & \sin\alpha \\ -\sin\alpha & \cos\alpha \end{pmatrix} \begin{Bmatrix} \frac{\partial w}{\partial x} \\ \frac{\partial w}{\partial y} \end{Bmatrix} \tag{1.2}$$

where α (=45° in this case) is the crack orientation angle.

The asymptotic expansion for in-plane surface slopes for mixed-mode loading condition is obtained by superposing the mode-I and mode-II fields [4],

$$\begin{aligned} \frac{\phi'_x}{2} = \frac{\partial w}{\partial x'} = \frac{-\nu B}{2E} & \left[\begin{array}{l} \left\{ \sum_{N=1}^{\infty} \left(A_N \left(\frac{N}{2} - 1 \right) r^{\left(\left(\frac{N}{2} \right) - 2 \right)} \cos \left(\frac{N}{2} - 2 \right) \theta' \right) \right\} + \\ \left\{ \sum_{N=1}^{\infty} \left(B_N \left(\frac{N}{2} - 1 \right) r^{\left(\left(\frac{N}{2} \right) - 2 \right)} \sin \left(\frac{N}{2} - 2 \right) \theta' \right) \right\} \end{array} \right] \\ \frac{\phi'_y}{2} = \frac{\partial w}{\partial y'} = \frac{-\nu B}{2E} & \left[\begin{array}{l} \left\{ \sum_{N=1}^{\infty} \left(A_N \left(\frac{N}{2} - 1 \right) r^{\left(\left(\frac{N}{2} \right) - 2 \right)} \sin \left(\frac{N}{2} - 2 \right) \theta' \right) \right\} + \\ \left\{ \sum_{N=1}^{\infty} \left(B_N \left(\frac{N}{2} - 1 \right) r^{\left(\left(\frac{N}{2} \right) - 2 \right)} \cos \left(\frac{N}{2} - 2 \right) \theta' \right) \right\} \end{array} \right] \end{aligned} \tag{1.3}$$

where (r, θ') are polar coordinates centered at the crack-tip in the local coordinates x' and y' with $\theta' = \theta + \alpha$, E the elastic modulus, ν the Poisson's ratio, B the nominal specimen thickness, $A_1; B_1 = \sqrt{2/\pi}\,(K_I; K_{II})$ with K_I and K_{II} being the mode-I and mode-II stress intensity factors, respectively. Solutions based on N up to 3 were used to extract the stress intensity

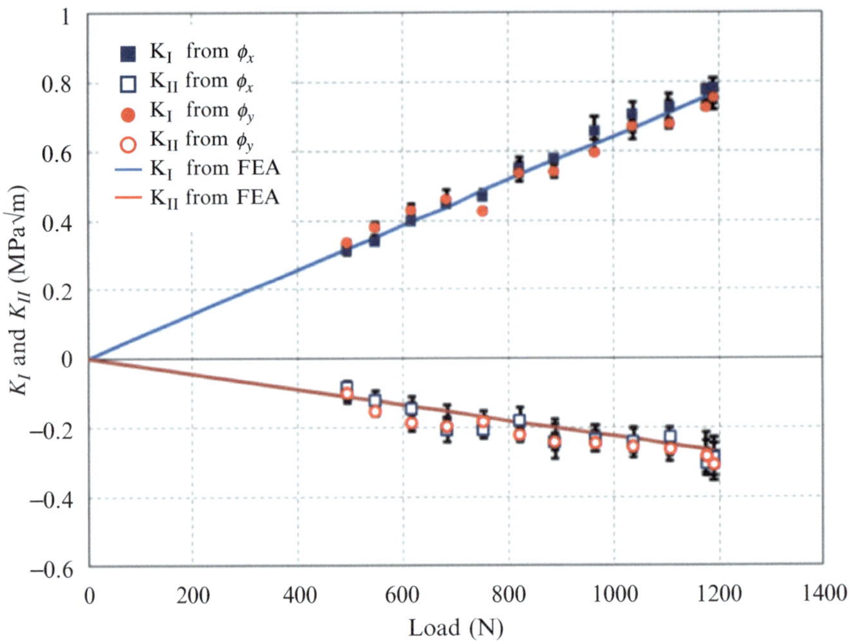

Fig. 1.4 Comparison of mixed-mode stress intensity factors from r-DGS and finite element simulation

factors. The results thus obtained are plotted in Fig. 1.4 for different load levels. A finite element analysis for the problem was carried out using ABAQUS™ software package to complement mixed-mode stress intensity factors obtained experimentally and details of the same are suppressed here for brevity. Figure 1.4 shows the variation of these computations overlaid on the experimental results. A good match between the two is evident from the plot, suggesting the feasibility of r-DGS for studying mixed-mode fracture problems.

1.4 Dynamic Mixed-Mode Experiments

The crack-tip deformation measurements under mixed-mode (I/II) conditions for an edge notched PMMA specimen ($130 \times 60 \times 8.9$ mm^3) during stress wave propagation were performed using a long-bar impactor used in conjunction with high-speed photography and r-DGS method. The schematic of the experimental setup is as shown in Fig. 1.5. The loading device consisted of an aluminum 7075-T6 long-bar (25.4 mm diameter and 2 m long) with a cylindrical impacting head, a gas-gun and a high-speed digital camera. The long-bar was aligned co-axially with the barrel of the gas-gun housing a 305 mm long, 25.4 mm diameter aluminum striker. The crack-tip deformations were photographed using a Cordin 550 high-speed digital camera equipped with 32 CCD sensors and two high-energy flash lamps to illuminate the target plate. A beam splitter positioned between the lens and the specimen at 45° angle was used to view the speckle pattern on the target via the reflective face of the specimen. The cylindrical tip of the long-bar was registered against the notch-free edge of the specimen. To achieve mixed-mode loading an eccentricity of 20 mm with respect to the crack line and the axis of the long-bar (see, Fig. 1.6) was used. As in the quasi-static experiments, one of the faces of the specimen (130 mm × 60 mm) was made specularly reflective by sputter coating it with aluminum. In these experiments, the distance between the target plate and specimen was (Δ=) 102 mm (and the distance between the camera lens and the specimen was ~715 mm).

A set of 32 reference images (one image from each CCD sensor of the high-speed camera) prior to loading was captured by operating the camera at 150,000 frames per second. Next the camera was triggered as the striker contacted the long-bar. A second set of 32 speckle images was captured at the same framing rate while the specimen was experiencing transient loading. The deformed and reference image pairs recorded by the same sensor of the high-speed camera were paired and correlated to obtain orthogonal displacements δ_x and δ_y on the target plane (see, Fig. 1.1). These were subsequently converted into surface slopes ($\partial w/\partial x$) and ($\partial w/\partial y$) on the specimen plane. Figures 1.7 show r-DGS contours near a dynamically loaded mixed-mode crack-tip for a time instant just before crack initiation. The contour plots of surfaces slopes in Fig. 1.8 represent the ones for the post-initiation regime following crack kinking from its initial orientation.

The stress intensity factors in the pre-crack initiation phase was performed in the global coordinate system (x and y) defined at the crack-tip aligned with the loading direction and the specimen edges (Fig. 1.2). The stress intensity factors for a

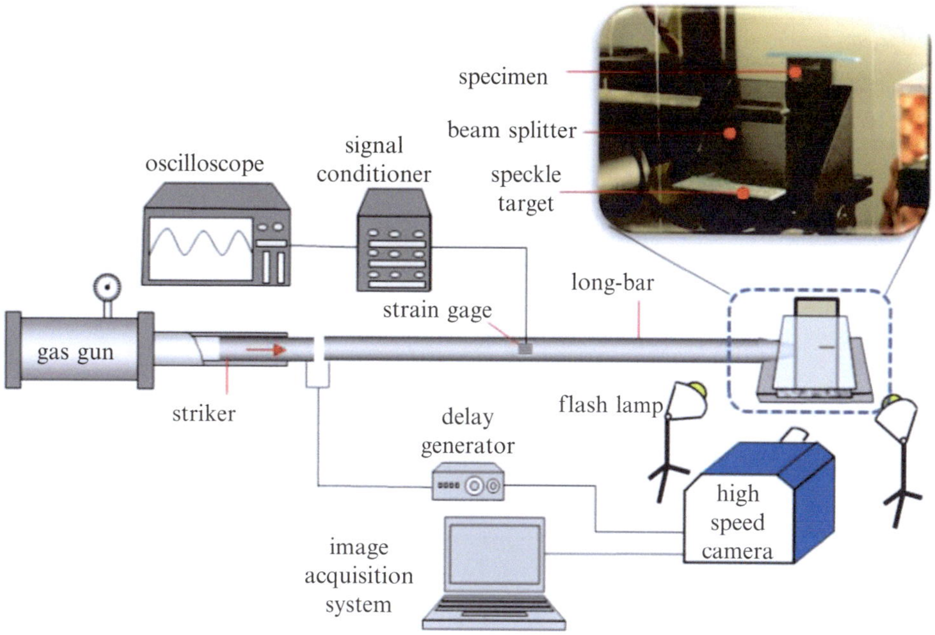

Fig. 1.5 Schematic of the experimental setup used in the dynamic mode-I fracture study. *Inset* is the photograph showing the close-up view of the optical arrangement in front of the sample

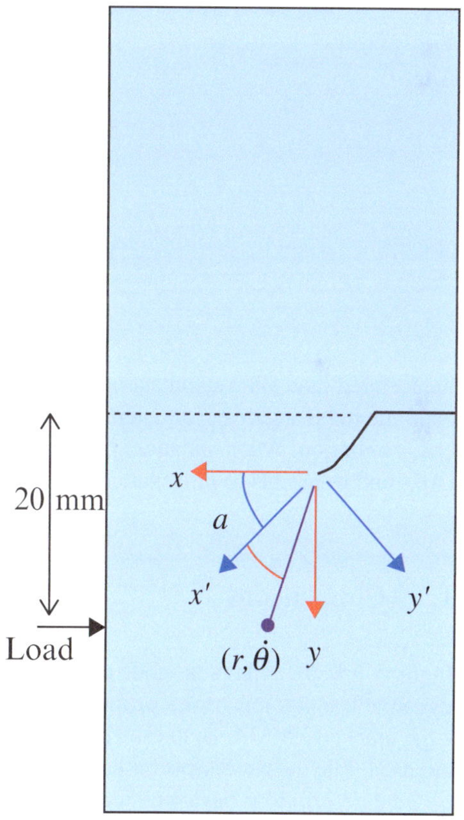

Fig. 1.6 Specimen loading configuration and crack tip coordinate system used in dynamic fracture experiments. Global and local coordinates for mixed-mode (mode-I/II) crack propagation. Specimen size = 130 mm × 60 mm

stationary but dynamically loaded crack were evaluated using the mixed-mode asymptotic equations [4] given in (1.3). The details of the analysis are suppressed here for brevity. In the post-crack initiation regime, the data analysis utilized crack tip field equations for a steadily growing mixed-mode crack that takes into account the instantaneous crack speed.

The results from the analysis are shown in Fig. 1.9. Initially both mode-I and -II stress intensity factors are negative. Though unintuitive, the former is caused when the notch (initially ~300 μm wide) closes due to eccentric loading above the crack line (relative to the initial notch direction) as shown in Fig. 1.6. With the passage of time, upon reflection of compressive stress waves as tensile waves from the free edges (particularly from the cracked edge) of the sample, the crack

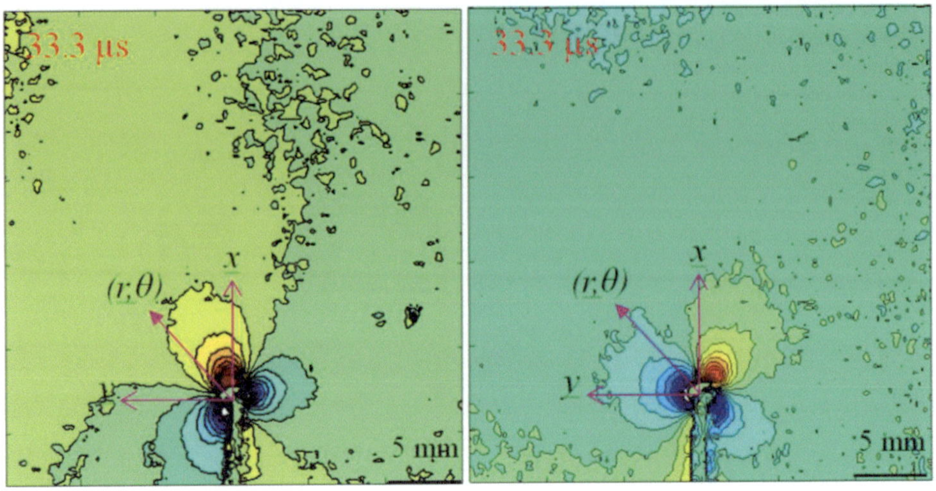

Fig. 1.7 Experimental ϕ_x (*left*) and (*right*) ϕ_y contours near the crack-tip before crack initiation. Contours are plotted every 20×10^{-5} rad

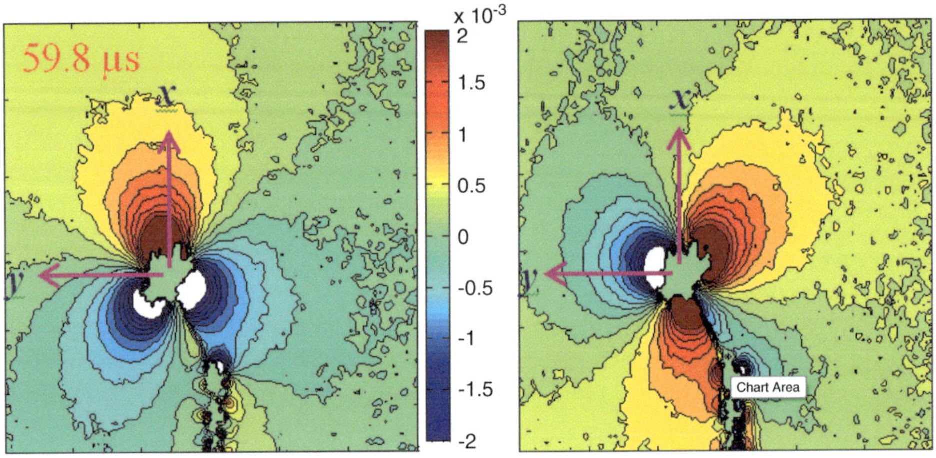

Fig. 1.8 Experimental ϕ_x (*left*) and ϕ_y (*right*) contours near the crack-tip in global *x*, *y* coordinates. Contours increments $= 20 \times 10^{-5}$ rad

flanks open (and slide simultaneously) causing mode-I stress intensity factor to turn positive. Mode-II stress intensity factor continue to be negative, consistent with the loading configuration used. Further, its magnitude increases monotonically until crack initiation. After initiation, however, the magnitude of mode-II stress intensity factors decreases to zero as the crack kinks and re-orients to propagate in an increasingly dominant mode-I condition.

1.5 Conclusions

In this work, the reflection-mode digital gradient sensing (r-DGS) method has been successfully extended to study static and dynamic fracture mechanics problems. First, the method has been demonstrated here for mixed-mode (I/II) problems using edge cracked PMMA samples. Three-point bend configurations with inclined edge cracks in otherwise symmetric loading are used. The measurements based on an overdetermininistic least-squares analysis of optical data in conjunction with crack tip field descriptions have produced reliable fracture parameter estimates.

Next, the r-DGS method has been extended to study mixed-mode dynamic fracture problems under impact loading conditions. The orthogonal deformations near the tip of dynamically loaded stationary cracks as well as transiently propagating cracks have been mapped by coupling r-DGS method with high-speed digital photography. The results presented include kinked crack growth relative to the initial crack orientation. The evolution of both mode-I and mixed-mode (I/II) stress intensity factor histories have been analyzed using measured data in conjunction with plane stress elasto-dynamic steady-state crack tip field descriptions.

Acknowledgments The partial support for this research through grants from the National Science Foundation (grant #1232821) and Department of Defense (grant # W31P4Q-14-C-0049) are gratefully acknowledged.

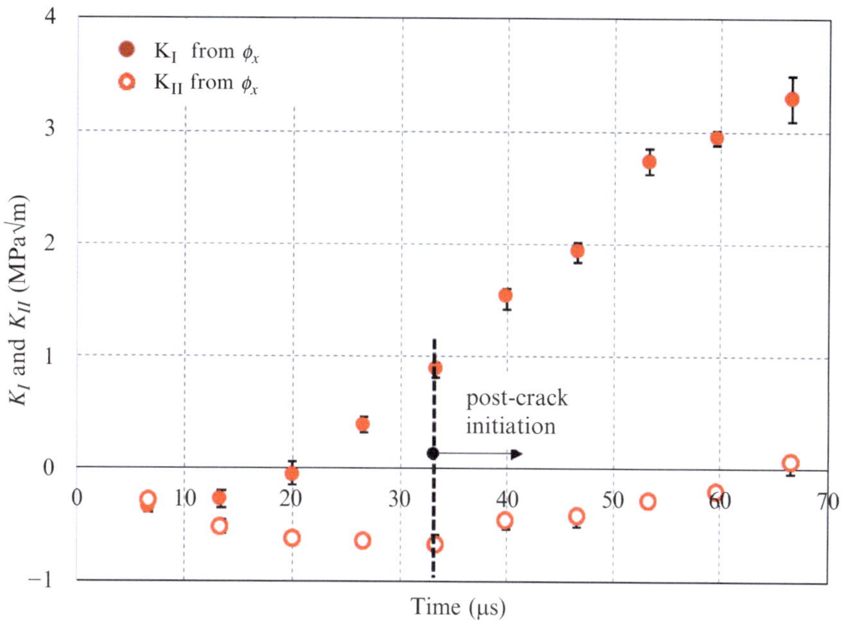

Fig. 1.9 Dynamic mixed-mode stress intensity factor histories from regression analysis of surface slope data (*symbols*)

References

1. Periasamy, C., Tippur, H.V.: A full-field digital gradient sensing method for evaluating stress gradients in transparent solids. Appl. Optics. **51**(12), 2088–2097 (2012)
2. Periasamy, C., Tippur, H.V.: Measurement of orthogonal stress gradients due to impact load on a transparent sheet using digital gradient sensing method. Exp. Mech. **53**(1), 97–111 (2013)
3. Periasamy, C., Tippur, H.V.: A full-field reflection-mode digital gradient sensing method for measuring orthogonal slopes and curvatures of thin structures. Meas. Sci. Technol. **24**, 025202 (2013)
4. Kirugulige, M.S., Tippur, H.V.: Mixed-mode dynamic crack growth in functionally graded glass-filled epoxy. Exp. Mech. **46**(2), 269–281 (2006)

Chapter 2
Experimental and Computational Investigation of Out-of-Plane Low Velocity Impact Behavior of CFRP Composite Plates

O.T. Topac, B. Tasdemir, B. Gozluklu, E. Gurses, and D. Coker

Abstract Strength of composite materials under transverse loading has remained a major weakness despite numerous advancements in composite technologies. Most frequent and critical result of this characteristic is internal delamination damage, which is undetectable and lead to major strength reduction in the structure. This condition is usually encountered in low-velocity impact situations which frequently occur during the maintenance of aircraft. Past studies have successfully developed experimental and analysis methods for accurately predicting impact force history and damage footprint based on the comparison with post-impact results. However, there is almost no experimental work on the progression sequence of damage during impact in the literature. This paper focuses on experimental and computational investigation of the damage initiation and growth process during low-velocity impact of $[0_7/90_4]_s$ and $[90_7/0_4]_s$ cross-ply CFRP laminates. In the experiments, through-the-thickness direction is tracked using ultra-high speed camera and DIC technique to record damage progression and dynamic strain fields. In the numerical part of the study 3-D explicit, finite element analysis is conducted to model matrix crack initiation and propagation. The finite element results are then compared with experiments in terms of failure modes and sequence.

Keywords Delamination • Matrix cracking • Transverse impact • High-speed camera • Experimental validation

2.1 Introduction

Due to the advances in composite technologies, carbon-fiber reinforced polymer (CFRP) composite laminates are widely used in a large variety of engineering applications. Specifically, in aerospace industry, composite material application has increased up to 500 % in the past 20 years [1]. In spite of greater strength of composites, they are much weaker under loading in through-the-thickness (transverse) direction. This results in a sudden and catastrophic brittle failure. This kind of loading is encountered in impact situations, which frequently occur during ground operations and maintenance of aircraft. Among impact conditions with different velocities, low-velocity impact (LVI) has a special importance since it creates insidious delamination damage inside the laminate. It occurs frequently and usually without any visible damage from the outside of structure. Hence, they can stay unrepaired for the remaining life of the aircraft. In aerospace industry, this phenomenon is considered as one of the most critical weaknesses of composite materials. For damage tolerant design, element level LVI tests are generally relied upon, which usually requires hundreds of specimens for different test and specimen configurations. However, it is impossible to conduct impact tests for each condition. Due to the abundant requirement of tests and uncertainty of the process, the designed damage tolerant structures are generally thicker and heavier than required.

O.T. Topac
Department of Aerospace Engineering, Middle East Technical University, Ankara 06800, Turkey

Helicopter Group, Turkish Aerospace Industries, Ankara 06980, Turkey
e-mail: tanaytopac@gmail.com

B. Tasdemir • E. Gurses
Department of Aerospace Engineering, Middle East Technical University, Ankara 06800, Turkey

B. Gozluklu
Helicopter Group, Turkish Aerospace Industries, Ankara 06980, Turkey

D. Coker (✉)
Department of Aerospace Engineering, Middle East Technical University, Ankara 06800, Turkey

METU Center for Wind Energy, Middle East Technical University, Ankara 06800, Turkey
e-mail: coker@metu.edu.tr

A.M. Beese et al. (eds.), *Fracture, Fatigue, Failure and Damage Evolution, Volume 8*, Conference Proceedings of the Society for Experimental Mechanics Series, DOI 10.1007/978-3-319-21611-9_2

In the literature, low-velocity impact response of composite laminates is investigated in terms of load and displacement histories of the laminate and resulting damages. One of the pioneering low velocity impact experiments on laminated composites was conducted by Cantwell et al. where the mass was dropped from a simple drop-weight tower and resulting damage was successfully observed by non-destructive inspections [2]. In their experimental and numerical study, Choi et al. investigated impact-induced damage mechanisms [3]. Beam-like 2D composite plates with several cross-ply stacking sequences were impacted by 2D impactors to achieve a damage scheme uniform in width direction. Resulting damages obtained for $[0_n/90_m/0_n]$ and $[0_n/90_m/0_n/90_m/0_n]$ laminates showed coupled matrix cracking and delamination, the characteristic impact-induced damage. In a computational effort, a finite element analysis of a simple 2D [0/90/0] beam done where matrix cracking and delamination damage interaction were simulated [4]. More recently, effects of dispersed stacking sequence were investigated in a combined experimental and numerical study, in which element level impact experiments were conducted on composite laminates according to ASTM D7136 standards. In numerical side, finite element analyses were conducted with matrix, fiber and delamination damage considerations to completely simulate experiments [5, 6]. A good agreement was obtained for post-impact results of experiments and analyses in terms of resulting damage, force, displacement and energy histories. However, to this date, almost no effort has been made to get a complete understanding and computational analysis of matrix and delamination damage progression sequence under impact conditions.

This study aims to observe damage progression sequence in carbon-epoxy composite laminated plates and verify the damage modeling of finite element analyses (FEA). Low velocity impact experiments are conducted on two types of [0/90] cross-ply specimens. Resulting matrix cracking and delamination process is captured by a high speed camera (1M fps). In computations, the experiment conditions are simulated in ABAQUS/Explicit and matrix damage evolution is investigated. In those, intralaminar matrix cracking damage is modeled via continuum damage mechanics (CDM) based failure approach proposed by Christensen [7] and compared with real-time results from high-speed camera.

2.2 Experimental Method

2.2.1 Experimental Setup

The experimental setup consists of an external structure for drop weight-tower, a base fixture to clamp the composite specimen and a guiding rail to provide an aimed drop of the impactor. CAD drawing and photograph of the setup with high speed camera and lighting are shown in Fig. 2.1. At the base fixture, both ends of the test specimen are fixed between two thick aluminum plates. In order to prevent possible sliding, the specimen is drilled and bolted from its ends to the base structure. Guiding rail is provided by a hollow aluminum tube with a rectangular cross-section vertically fixed above the center of the specimen.

Impact specimens, are made of Cytec T300/895 CFRP material with individual ply thickness of 0.16 mm. The size of the specimen is 85 mm × 15 mm which provides plane strain condition. Two different stacking sequences of $[0_7/90_4]_s$ and $[90_7/0_4]_s$ are manufactured where the thickness of the specimens are equal to 3.5 mm. The impactor, seen in Fig. 2.2a, is made of SAE 304 stainless steel material and designed to have variable mass of 0.365, 0.788 and 1.21 kg. It is designed with semi-cylindrical cross section to create line load impact conditions in the width direction of the specimen. The progression of damage during impact event is captured by Photron SA5 high speed camera, capable of capturing images at rates up to one million frame per second (fps).

a b

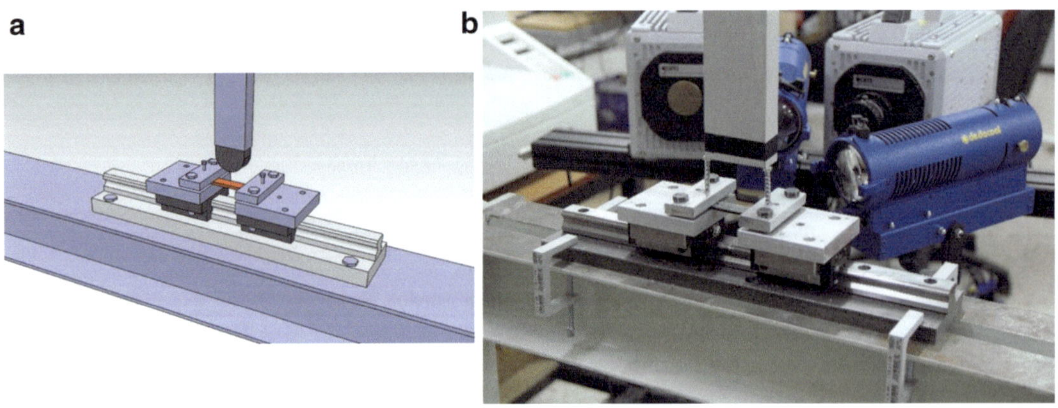

Fig. 2.1 Experimental set-up for line load impact testing; (**a**) CAD drawing and (**b**) photograph of setup with high speed camera and lighting

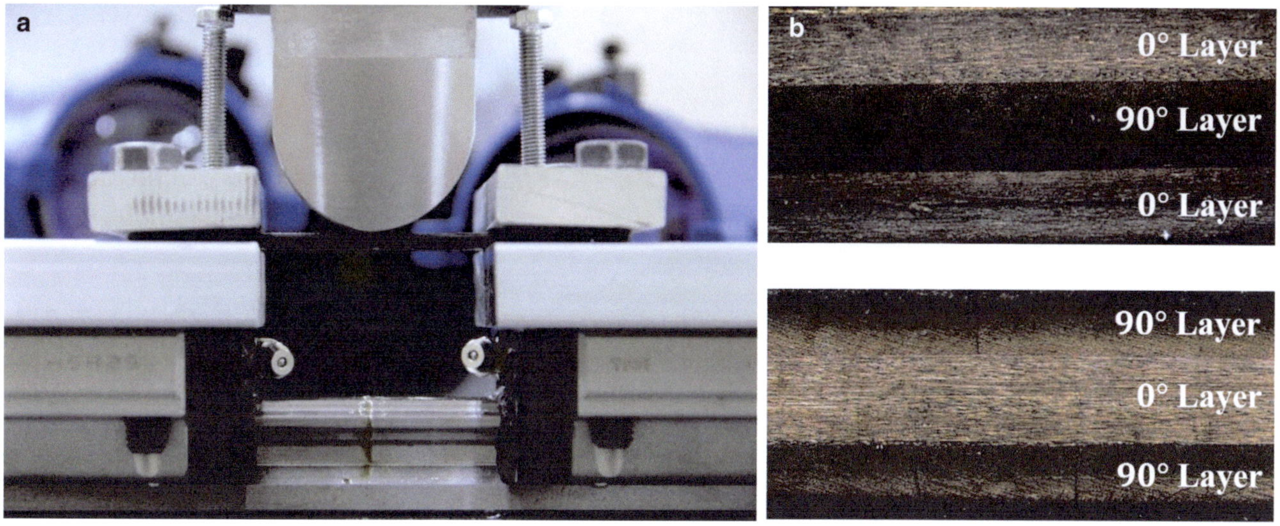

Fig. 2.2 (**a**) Close-up side-view of the specimen with the impactor and grips, (**b**) micrograph images of two stacking sequences used in this paper: $[0_7/90_4]_s$ and $[90_7/0_4]_s$

2.2.2 *Experimental Procedure*

Firstly, side sections of composite specimens are polished by Minitech-233 polishing machine in order to achieve clear visualization of their side section. Before conducting the tests, polished specimens were scanned with Huvitz HDS-5800 Digital Microscope from the edge section to reveal manufacturing flaws which were in the acceptable limits. Images are captured with lens having zoom rate of 50×.

The experiments are conducted as drop-weight impact events. Two sets of specimens with different stacking sequences are tested, each set having four individual composite beams. The high-speed camera is focused on at the edge section of the composite with frame rates from 15,000 to 210,000 fps are used to provide adequate compromise between capture rate of crack progression and picture resolution. Steel impactor of 0.365 kg is symmetrically placed inside the guiding rail and dropped from 1 m height resulting in 3.58 J kinetic energy at the instant of impact.

A free fall of impactor is assumed and friction between impactor and guiding rail is neglected in analyses and evaluation of tests. Image acquisition module of high-speed camera is triggered just before impact and the whole impact event is successfully captured. After first impact, bounce-back of the impactor is prevented to avoid any further damage after the first contact. An illustrative picture just before impact is given in Fig. 2.2a. After the experiments, resulting damage scheme is optically examined with microscope.

2.3 Computational Method

In the computational part of this study, simplified experimental impact case is modeled as seen in Fig. 2.3 and simulated in ABAQUS/Explicit finite element software. For the geometrical model, only the region of laminate between gripping tabs is modeled which results in a 45 × 15 mm plate. Rest of the plate beyond the tabs is lumped in the clamped boundary conditions. Material properties of T300/985 carbon/epoxy composite are obtained from manufacturer specification sheet and previous studies [8, 9] which are provided in Table 2.1. The impactor is modeled as an analytical rigid body with lumped mass of 0.365 kg at its centroid. An initial vertical velocity of 4.3 m/s that corresponds to a free fall from 1 m is defined for the impactor.

In computational analysis, only matrix cracking damage is considered and compared with experiments. The composite laminate is meshed with solid C3D8R (Reduced integration, eight noded hexagonal) elements as in Fig. 2.3 where each layer is modeled with one element in thickness direction, and in longer side of the beam, elements are biased to have a finer mesh near the impacting region. The total number of elements is 580,800. Contact condition between impactor and composite laminate is defined with general contact algorithm of ABAQUS. Normal behavior is defined as hard contact with penalty algorithm and tangential Coulomb friction between impactor and composite is defined with a friction coefficient of $\mu = 0.3$ based on the study of [6, 10].

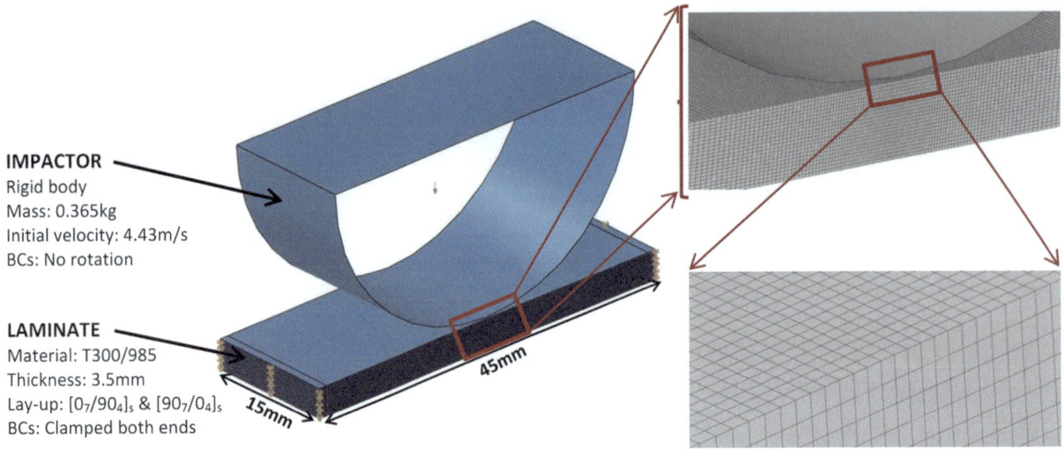

Fig. 2.3 Finite element model and mesh of the impact analysis

Table 2.1 Cytec T300/985 ply properties

Density	1600 kg/m^3
Elastic properties	$E1 = 106$ GPa; $E2 = E3 = 7.8$ GPa; $\nu_{12} = \nu_{13} = 0.30$; $\nu_{23} = 0.44$; $G_{12} = G_{13} = 10.5$ GPa; $G_{23} = 2.7$ GPa
Strength properties	$S_{22}^{+} = 50$ MPa; $S_{22}^{-} = 170$ MPa; $S_{12} = 170$ MPa; $S_{13} = S_{23} = 95$ MPa

In computational analysis, only matrix cracking damage is considered and compared with experiments. Intraply damage in composite materials is commonly studied by continuum damage mechanics (CDM), where several load interaction criteria exist to predict initiation of crack under normal and shear loading of the plies. In this study, Christensen CDM criterion with only matrix cracking consideration is applied using Autodesk Simulation Composite Analysis 2015 plugin. Christensen failure criterion is chosen due to its high order transverse stress terms ($\sigma_{33}^{m}, \sigma_{13}^{m}, \sigma_{23}^{m}$) in crack prediction, which are dominant in transverse impact situations. Christensen criterion considers tensile and compressive failure strengths of a ply in normal orthotropic directions $\left(S_{11}^{+}, S_{11}^{-}, S_{22}^{+}, S_{22}^{-}\right)$ and absolute values of longitudinal and transverse shear strengths (S_{12}, S_{23}). It assumes that matrix crack initiates when the interaction equation becomes equal to or greater than one:

$$\left(\frac{1}{S_{22}^{+}} - \frac{1}{S_{22}^{-}}\right)(\sigma_{22} + \sigma_{33}) + \frac{1}{S_{22}^{+}S_{22}^{-}}(\sigma_{22} + \sigma_{33})^2 + \frac{1}{S_{23}^{2}}\left(\sigma_{23}^{2} + \sigma_{22}\sigma_{33}\right) + \frac{1}{S_{12}^{2}}\left(\sigma_{12}^{2} + \sigma_{13}^{2}\right) \geq 1$$

2.4 Results

2.4.1 Experimental Results

2.4.1.1 Impact Testing of $[0_7 / 90_4]_s$ Unidirectional CFRP Laminates

Series of photographs are taken with high speed camera from the side of the specimen. While observable damage occurred at the side section, no visible damage is produced on the contact region between impactor and the specimen. In the results, 0 μs represents the impact instant. In Fig. 2.4, series of high speed camera images taken at 15,000 fps, 66.7 μs interframe time are presented. The photograph in Fig. 2.4a shows the impactor and specimen just before impact and Fig. 2.4b, taken 133 μs later, shows loading of the composite plate with no visible sign of damage. One frame later, shown in Fig. 2.4c complete composite failure is observed. Resulting damage scheme closely resembles the results of Choi et al. on a similar composite laminate [3]. Delaminations between the 0° and 90° plies are observed consisting of a delamination of the upper interface under the impactor and a delamination of the lower interface away from the impact region. Matrix cracks appear in the 90° plies connecting the two delaminations. However, damage progression and its sequence are not captured at this camera speed.

The micrograph images of the front and mirrored back faces of the laminate after the experiment are shown in Fig. 2.5a and b, respectively. The delamination and the matrix cracking patterns are observed to be the same throughout the width of the specimen.

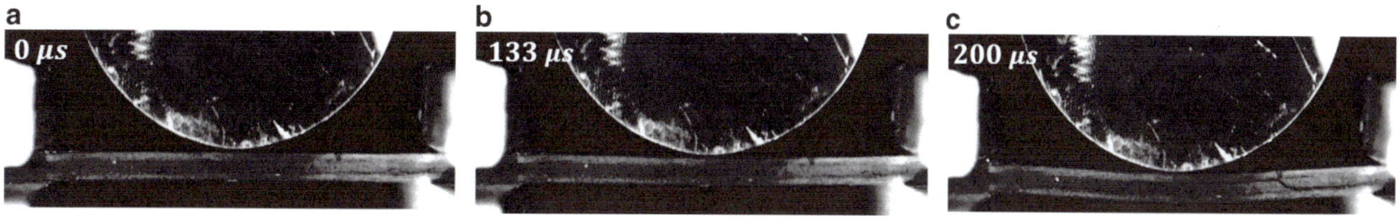

Fig. 2.4 High speed camera images of $[0_7/90_4]_s$ specimen taken at 15,000 fps. (**a**) Just before the impact and (**b**) after impact subjected to loading with no damage, (**c**) after damage initiation

Fig. 2.5 Micrograph images of two faces of damaged specimen showing uniformity of damage

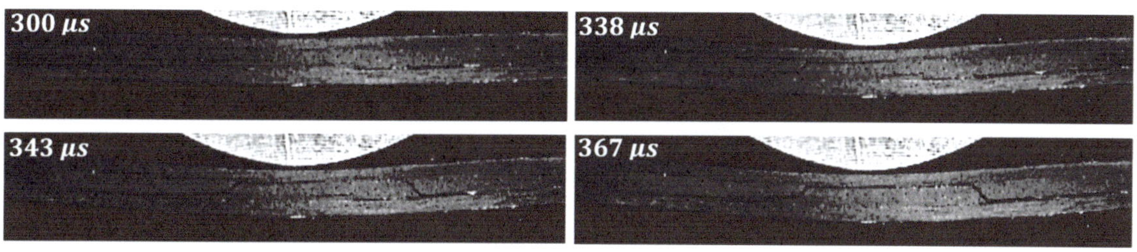

Fig. 2.6 High speed camera images of $[0_7/90_4]_s$ specimen taken at 210,000 fps showing progression of matrix cracking and delamination

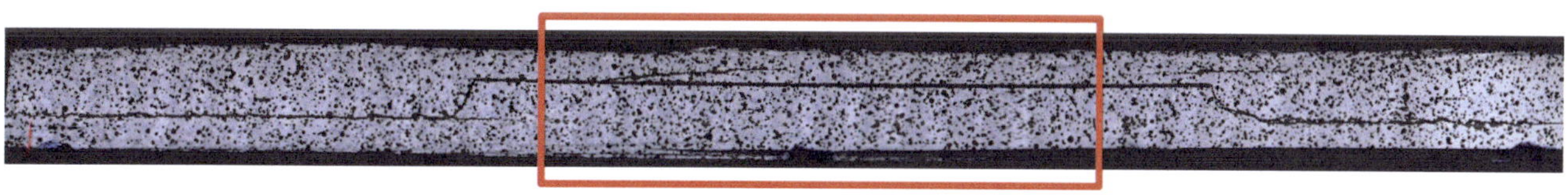

Fig. 2.7 Micrograph of laminate after the impact experiment showing the final failure pattern and the location of the field of view of the high speed camera

To look at the failure sequence at a higher time resolution, an impact test was conducted under the same conditions and the results were captured with the high speed camera at 210,000 fps, 4.76 μs interframe time. In the first frame the composite is loaded and initial delamination in the bottom 0/90 interface can be seen. 38 μs later a matrix crack starts from the lower interface and is followed by a delamination at the upper interface growing towards the impact point. Five microseconds 5 μs later, a second matrix crack initiates at the right hand side from the bottom interface and propagates towards the upper interface. In the last picture, 24 μs later, upper delamination grows from the matrix crack towards the impact point and joins with the upper left delamination front. This illustrates the failure sequence of $[0_7/90_4]_s$ laminates under low velocity impact (Fig. 2.6).

In order to observe the strain fields at the crack front during impact, DIC analysis is performed on the high speed camera images captured at 20,000 fps. The post-impact micrograph is shown in Fig. 2.7 with the final failure pattern similar to the previous cases. Because of the resolution, framing rate and shutter speed limitations, the field of view of the high speed camera is limited to the area just under the impact site, highlighted with red frame in Fig. 2.7. The high speed camera pictures and the Tresca strain fields using DIC method during the impact event are shown in Fig. 2.8a and b, respectively. At 100 μs, initial delamination fronts are observed at upper interface from both sides and lower interface from only the left side. One frame (50 μs) later, the delaminations at the upper interface propagate towards the center from both sides, however the lower delamination is found to have arrested. In the final frame, the upper delaminations from both sides combine to form a single

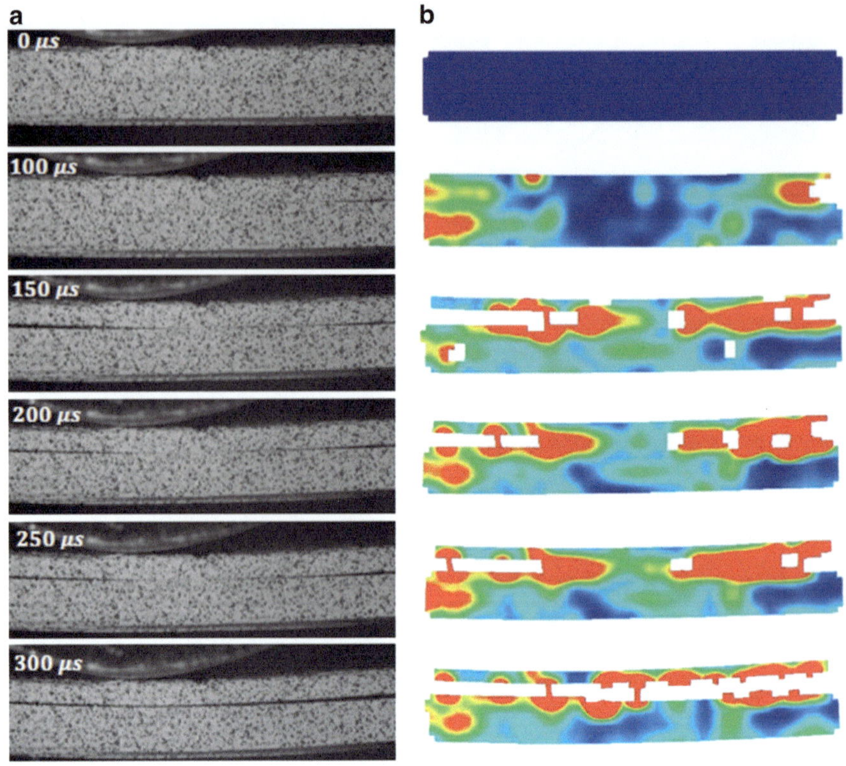

Fig. 2.8 Close-up of the middle section of $[0_7/90_4]_s$ specimen showing stress fields at lower and upper delamination regions. (**a**) High speed camera images, (**b**) Tresca strain contours from Digital Image Correlation (DIC) analysis program ARAMIS

upper delamination. An interesting observation is that between 150 and 250 μs, the delaminations seem to arrest and even show closure of delamination front. However, upon closer examination, the crack surfaces, which have coalesced at 150 μs, are observed to close due to compressive waves from the impactor. At 300 μs, the tensile waves arrive and the delamination opens up.

2.4.1.2 Impact Testing of Unidirectional OE $[90_7 / 0_4]_s$ CFRP Laminates

Second set of impact tests are conducted on $[90_7 / 0_4]_s$ unidirectional CFRP laminates. The images during the impact event were captured at 20,000 fps for DIC analysis; however due to shutter speed limitations blurry images were obtained and could not be processed for DIC. The images of the plate at 0, 200, 300 and 700 μs after impact are shown in Fig. 2.9. At 200 μs crack is initiated from the middle lower layer because of normal tensile stresses on matrix material caused by bending of the beam. Two frames later, at 300 μs, four additional matrix cracks are observed next to other cracks with an average of 3.8 mm distance from each other. At 700 μs, all of the initial matrix cracks are propagated to delamination at lower interface.

2.4.2 Computational Results

Finite element analyses are conducted with Christensen failure criterion to correlate crack location and sequence with experiments. Resulting matrix crack initiation and propagation schemes of $[0_7/90_4]$ and $[90_7/0_4]_s$ configurations are presented in Fig. 2.10a and b respectively.

In $[0_7/90_4]_s$ stacking sequence, represented in Fig. 2.10a, local matrix damage is predicted at the impact location, unlike experimental results that produced no visible surface damage. However at the side section, a similar crack pattern is obtained. Initially, a matrix crack is initiated at bottom close to the center and after 8 μs, it propagated towards the center diagonally reaching to the upper interface. Afterwards, matrix cracking propagated just below and above interfaces resembling a

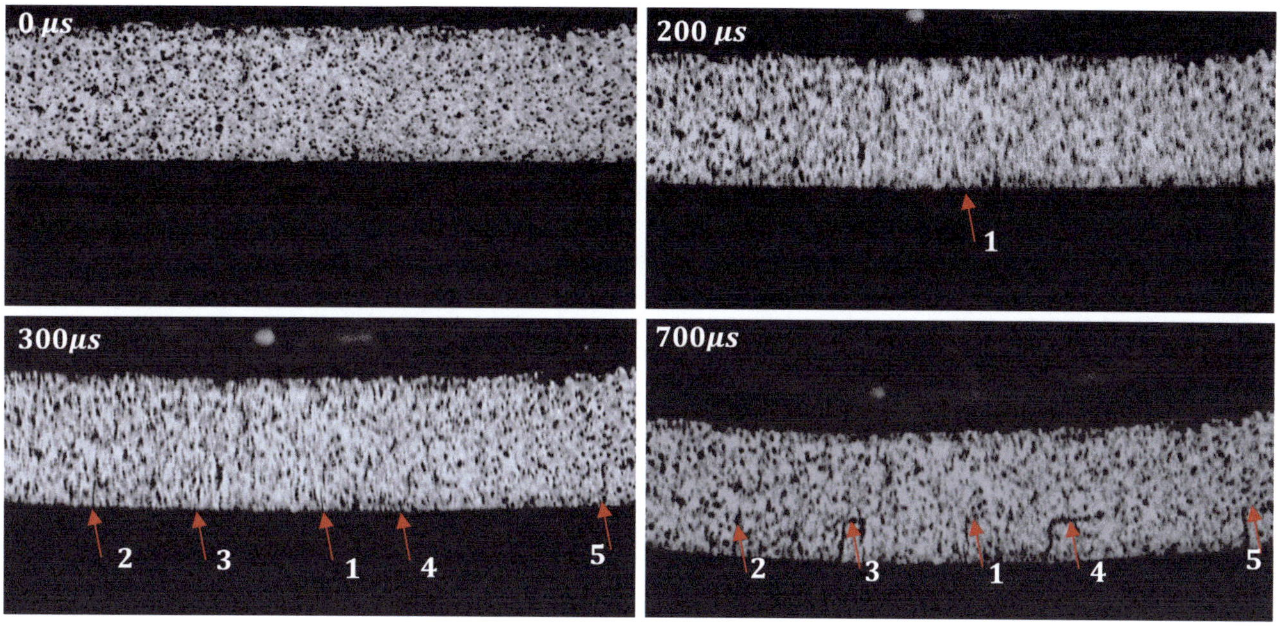

Fig. 2.9 High speed camera images of $[90_7/0_4]_s$ specimen taken in 20,000 fps showing matrix cracking initiation and propagation to delamination

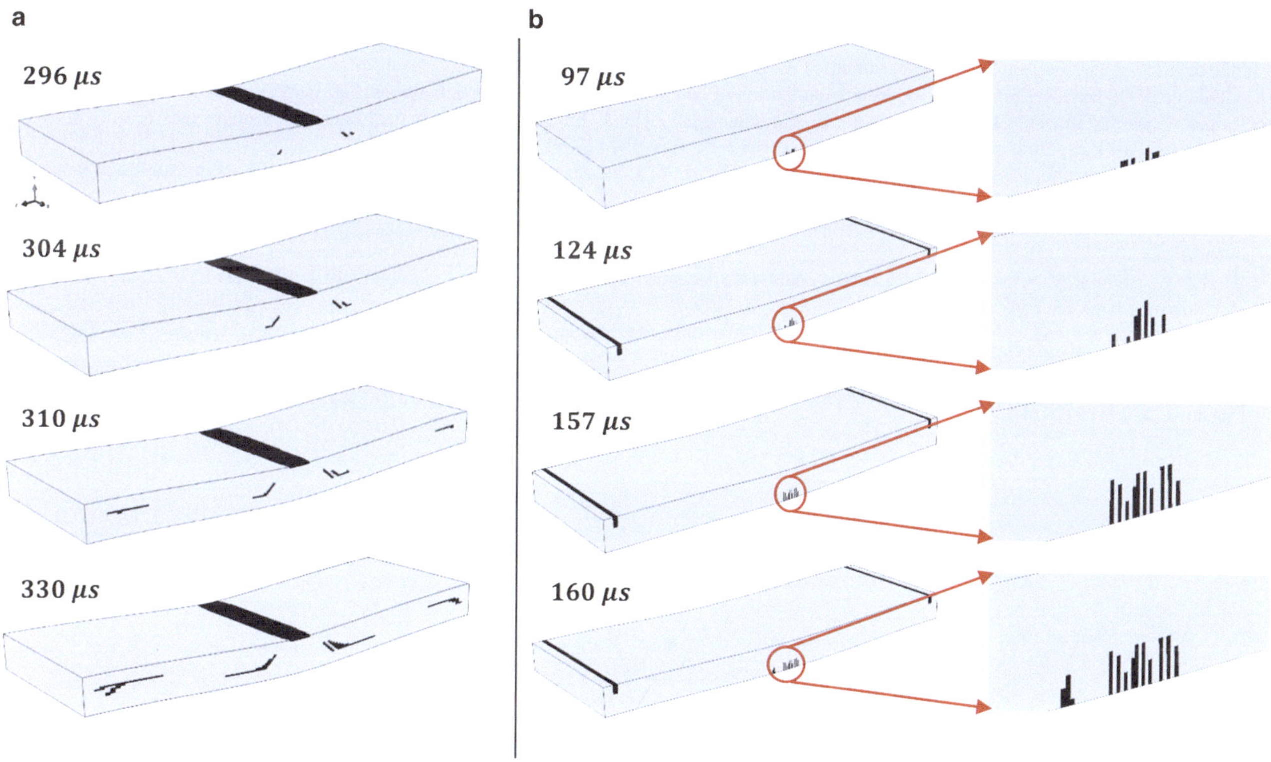

Fig. 2.10 Matrix crack initiation and progression scheme predicted by Christensen criterion for (**a**) $[0_7/90_4]_s$ and (**b**) $[90_7/0_4]_s$ stacking sequences

delamination pattern. While the crack angle of 45° is very close to the experiment results, the distance between two matrix cracks were about two times shorter than experimental results. In $[90_7/0_4]_s$ stacking sequence, represented in Fig. 2.10b, several matrix cracks are initiated in a region of 2 mm at bottom layer progressing perpendicular towards the bottom 0/90 interface. Upon reaching interface at 157 μs, additional matrix cracks are obtained due to lack of delamination damage in the computational model.

2.5 Conclusions

Experimental and computational studies are conducted on $[0_7/90_4]_s$ and $[90_7/0_4]_s$ 2-D beam-like unidirectional CFRP laminates to evaluate the failure progression of composites under out-of-plane impact loading. The line load impact tests on a composite plate was carried out to observe the impact failure phenomena from the edge using a high-speed camera and DIC system. Two different failure modes were captured depending on the stacking sequence. For $[0_7/90_4]_s$ lay-up, the damage initiates as delamination in the bottom interface followed by matrix cracking towards the impact point causing delamination in the upper layer. For $[90_7/0_4]_s$ lay-up, vertical matrix cracks initiate at the lower 90 layers in the line of impact, leading to delamination of the lower interface. Finite element simulations using Christensen matrix failure model was able to capture the general trend of matrix failure initiation and propagation observed in the experiments for both layups. A further study is being carried out to implement Cohesive Zone Method to model subsequent delamination observed in the experiments.

Acknowledgements The authors would like to acknowledge the contributions of METU Center for Wind Energy for allowing access to the experimental facilities. The authors would also like thank to Ayse Begum Erdem, Ali Gezer, and Miray Aydan Arca for their contributions.

References

1. Roeseler, B., Sarh, B., Kismarton, M.: Composite structures-the first 100 years. In: 16th International Conference on Composite Materials, Kyoto, 8–13 July 2007
2. Cantwell, W.J., Curtis, P.T., Morton, J.: An assessment of the impact performance of CFRP reinforced with high-strain carbon fibres. Compos. Sci. Technol. **25**, 133–148 (1986)
3. Choi, H.Y., Downs, R.J., Chang, F.-K.: A new approach toward understanding damage mechanisms and mechanics of laminated composites due to low-velocity impact: Part I—experiments. J. Compos. Mater. **25**, 992–1011 (1991)
4. Geubelle, P.H., Baylor, J.S.: Impact-induced delamination of composites: a 2D simulation. Compos. Part B **29**(5), 589–602 (1998)
5. Lopes, C.S., Seresta, O., Coquet, Y., Gürdal, Z., Camanho, P.P., Thuis, B.: Low-velocity impact damage on dispersed stacking sequence laminates. Part I: experiments. Compos. Sci. Technol. **69**(7–8), 926–936 (2009)
6. Lopes, C.S., Camanho, P.P., Gürdal, Z., Maimí, P., González, E.V.: Low-velocity impact damage on dispersed stacking sequence laminates. Part II: numerical simulations. Compos. Sci. Technol. **69**(7–8), 937–947 (2009)
7. Christensen, R.M.: The numbers of elastic properties and failure parameters for fiber composites. J. Eng. Mater. Technol. **120**(2), 110–113 (1998)
8. Hiel, C.C., Sumich, M., Chappell, D.P.: A curved beam test specimen for determining the interlaminar tensile strength of a laminated composite. J. Compos. Mater. **25**, 854–868 (1991)
9. Cheremisinoff, N.P. (ed.): Handbook of Ceramics and Composites: Mechanical Properties and Specialty Applications, p. 528. CRC Press, New York (1991)
10. Shi, Y., Pinna, C., Soutis, C.: Modelling impact damage in composite laminates: a simulation of intra- and inter-laminar cracking. Compos. Struct. **114**, 10–19 (2014)

Chapter 3
Prediction of Incipient Nano-Scale Rupture for Thermosets in Plane Stress

J.C. Moller, S.A. Barr, T.D. Breitzman, G.S. Kedziora, A.M. Ecker, R.J. Berry, and D. Nepal

Abstract There is limited experimental evidence that fracture nucleation in polymers includes a small number of covalent bond scissions followed by rapid void growth by chemo-mechanical processes. Generalized criteria for predicting such bond scission, then, would help anticipate fracture in polymer matrix composites. Strain states at incipient bond scission for thermoset resins in plane stress are here predicted by atomistic simulation. Several cured epoxy systems were examined, each having a different chain length. For biaxial extension and a portion of the shearing regime, scission occurs at a critical value of the larger principal strain. This value increases with increasing chain length. The corresponding dilatation is largest for biaxial extension and decreases to nearly zero for pure shear. Results are compared with strain invariants at fracture measured from experiments in which polymer matrix composites having various ply stacking sequences were loaded to rupture.

Keywords Plane stress • Strain invariant • Atomistic • Thermoset

3.1 Introduction

Owing to steep gradients in displacement and non-linear material behavior, stress states in the vicinity of a crack tip can be highly varied. It is also well-known that fracture depends upon whether the bulk of the test sample places the crack tip zone in plane strain or plane stress. Unlike an equiaxed multigranular metal, a uniaxial tensile test of a polymer does not necessarily provide results which can be reliably applied to anticipate material response in other loading states.

Published work has shown the results of atomistic simulations of polymer systems in one-directional tensile deformation with simplified boundary conditions and deformation methods. More investigation is therefore necessary to develop a generalized nano-scale characterization of mechanical properties including yield and rupture thresholds. Generalized descriptions are important for informing larger-scale models and for anticipating fracture nucleation in the wide variety

J.C. Moller (✉)
Miami University, Oxford, OH 45056, USA

Universal Technology Corporation, Dayton, OH 45432, USA

Air Force Research Laboratory, WPAFB, Dayton, OH 45433, USA
e-mail: mollerjc@miamioh.edu

S.A. Barr • D. Nepal
Universal Technology Corporation, Dayton, OH 45432, USA

Air Force Research Laboratory, WPAFB, Dayton, OH 45433, USA

R.J. Berry
Air Force Research Laboratory, WPAFB, Dayton, OH 45433, USA

T.D. Breitzman
Exploratory Research Section, Air Force Research Laboratory, WPAFB, Dayton, OH 45433, USA

G.S. Kedziora
Air Force Research Laboratory, WPAFB, Dayton, OH 45433, USA

Engility Corporation, Franklin, OH 45005, USA

A.M. Ecker
Air Force Research Laboratory, WPAFB, Dayton, OH 45433, USA

University of Dayton, Dayton, OH 45469, USA

A.M. Beese et al. (eds.), *Fracture, Fatigue, Failure and Damage Evolution, Volume 8*, Conference Proceedings of the Society for Experimental Mechanics Series, DOI 10.1007/978-3-319-21611-9_3

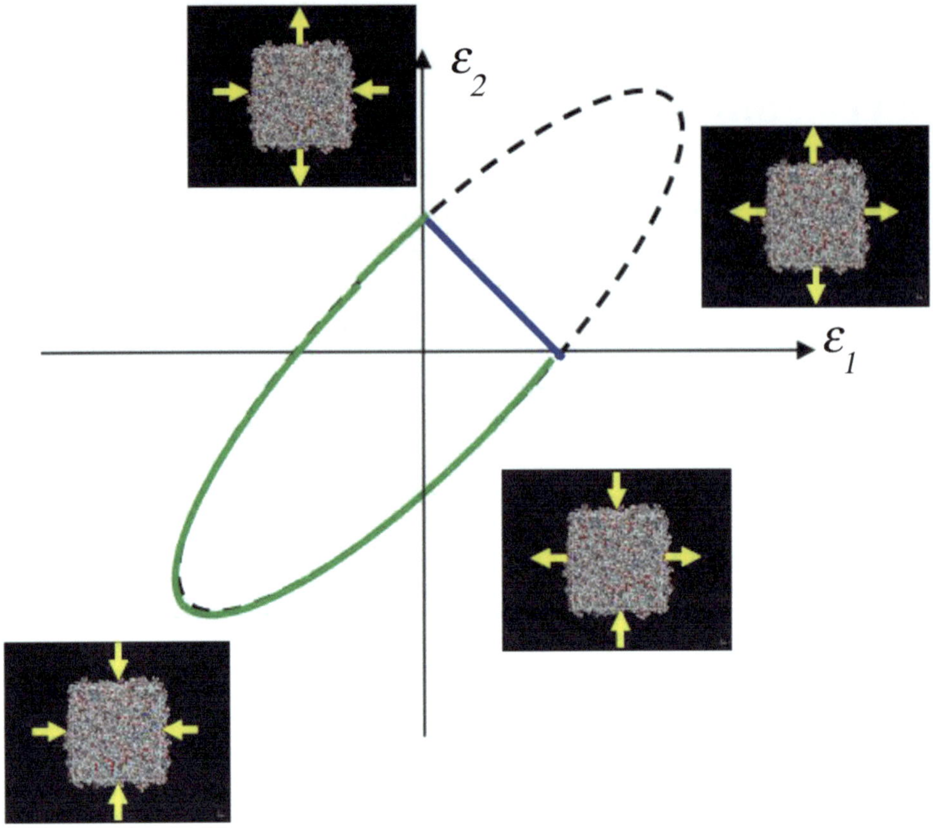

Fig. 3.1 An example of the speculated yield surface for plane stress. When both principal strains are positive, biaxial extension occurs. When the signs of the principal strains differ, various levels of shear occur. The *blue line* represents yield due to void growth. The *green curve* represents yield due to shear

of stress and strain stares that can occur in the loaded composites. As a first important step, a method to strain atomistic in plane stress was developed and used to explore the responses of selected thermoset resins.

One basis for generalizing yield and fracture in polymers is strain-invariant-fracture theory [1]. An example yield surface which has been posed as appropriate for thermosets is shown in Fig. 3.1 where the functions plotted depend upon strain tensor invariants. The invariants are given by

$$J_1 = \varepsilon_1 + \varepsilon_2 + \varepsilon_3 \tag{3.1}$$

$$J_2 = \varepsilon_1\varepsilon_2 + \varepsilon_2\varepsilon_3 + \varepsilon_3\varepsilon_1 \tag{3.2}$$

$$J_3 = \varepsilon_1\varepsilon_2\varepsilon_3 \tag{3.3}$$

where J_1 is often treated as a measure of volumetric dilatation. Further, equivalent strain is given by

$$\varepsilon_{equiv} = \left(\frac{(\varepsilon_1 - \varepsilon_2)^2 + (\varepsilon_2 - \varepsilon_3)^2 + (\varepsilon_3 - \varepsilon_1)^2}{2}\right)^{0.5} \tag{3.4}$$

The first and second invariants can be related as follows.

$$J_2 = \frac{1}{3}\left(J_1^2 - \varepsilon_{equiv}^2\right) \tag{3.5}$$

3.2 Methods

The responses of several cross-linked epoxy systems were investigated. Diglycidyl ether of bisphenol A (DGEBA) having various chain lengths was the resin. Methylene diamine (MDA) was the cross-linker. Each system contained more than 30,000 atoms and had cross-link extents greater than 90 %. Dynamics was simulated in the Large-scale Atomic/Molecular Massively Parallel Simulator (LAMMPS). The consistent valence force field (CVFF) with harmonic potentials and no cross terms was used. A script was developed in which a succession of plane stress states was applied to the system. Stress levels were set to maintain a target effective strain rate (10^{10} s^{-1} for the results presented here), prescribe ratio of in-plane strains, and maintain a selected hydrostatic stress level.

3.3 Results

Over the course of numerous simulations, each system was deformed in a variety of ways ranging from biaxial tension to biaxial compression. Yield was identified from plots of effective stress as a function of effective strain. Yield was defined as the first instance of decline in stress which was not associated with fluctuations having a thermal origin. The locus of points for yield for a short-chain-length system is shown in Fig. 3.2. For biaxial extension strain combinations, the trend matches that shown in Fig. 3.1. For the shear and biaxial compression regimes, trends are not yet discernable. Also provided in Fig. 3.2 are effective stress-effective strain plots for several principal strain combinations. In pure biaxial extension, the system reaches yield and shows softening. In pure shear, the yield strength is less than biaxial extension. Post-yield, the system is perfectly plastic and then strain hardening. In constrained compression ($\varepsilon_1 < 0,\ \ \varepsilon_2 = 0$), yield strength is 50 % larger and there is only strain hardening after yield. In pure biaxial compression, yield strength and strain hardening rate are still larger.

In what follows, the independent variable is shown as the ratio of the principal strains expressed as $\tan^{-1}\ (\varepsilon_2/\varepsilon_1)$. For values up to 90°, the sample is in biaxial tension. From 90° to 180°, the sample is in shear and, from 180° to 225°, the sample is in biaxial compression.

The system stiffness in the elastic region was estimated by calculating the ratio of the effective stress to the effective strain at the yield point. This apparent modulus is plotted as a function of principal strain ratio in Fig. 3.3. For all strain combinations providing biaxial extension, it is essentially constant. In pure shear, it reaches a minimum. It is stiffest in biaxial compression.

For linearly-elastic/plastic materials in plane stress, the value of $(-1)\ \ \varepsilon_3/J_1$ would equal ν (i.e., Poisson's ratio). If the material were isotropic and linearly elastic, the result would be a constant value less than or equal to 0.5. The results (shown in Fig. 3.4) are generally greater than this range. The difference from commonly-found range of values appears to be due to large strains in direction orthogonal to the loading plane. Also, for a portion of the range in shear as well as the range for biaxial compression, the values are significantly larger than for biaxial tension and for shear in which the principal strain having the larger magnitude is positive. The sharply differing values in the vicinity of 135° are due to uncertainty in strain overwhelming the mean value.

For linearly-elastic/plastic materials in plane stress, the value of $(\sigma_1 + \sigma_2)/J_1$ would equal E, the elastic modulus. The results (shown in Fig. 3.5) indicate the system is significantly stiffer in biaxial compression than for biaxial tension. The values found are greater than expected by a factor of 2–4.

The effective stress at yield was approximately constant in biaxial tension and in shear until the principal strain having the larger magnitude was negative. At this point, the effective yield strength progressively rose and was largest for pure biaxial compression (as shown in Fig. 3.6).

The dilatational strain at yield (i.e., $(\varepsilon_1 + \varepsilon_2 + \varepsilon_3)_Y = J_{1Y}$) is a metric which has been used to anticipate yield in biaxial tension. The results are shown in Fig. 3.7. Indeed, the value is essentially constant for biaxial tension and then monotonically declines in the shear regime. The value found in biaxial tension is greater than values measured from experiments for similar materials [1–4].

The equivalent strain at yield was no less than 0.10 and showed no strong trend over the range of principal strain ratio. The results are shown in Fig. 3.8. The values are similar to those found experimentally.

The strain second invariant at yield was also scattered and showed values ranging from –0.0025 to –0.011 (as shown in Fig. 3.9). Measured values have the same sign but are and order of magnitude less than the ones reported here [2–4].

The LAMMPS script for plane stress was enhanced to monitor the lengths of backbone molecular bonds at regular intervals and stop when an incipient bond rupture is identified. For each bond type, bond length was compared with the

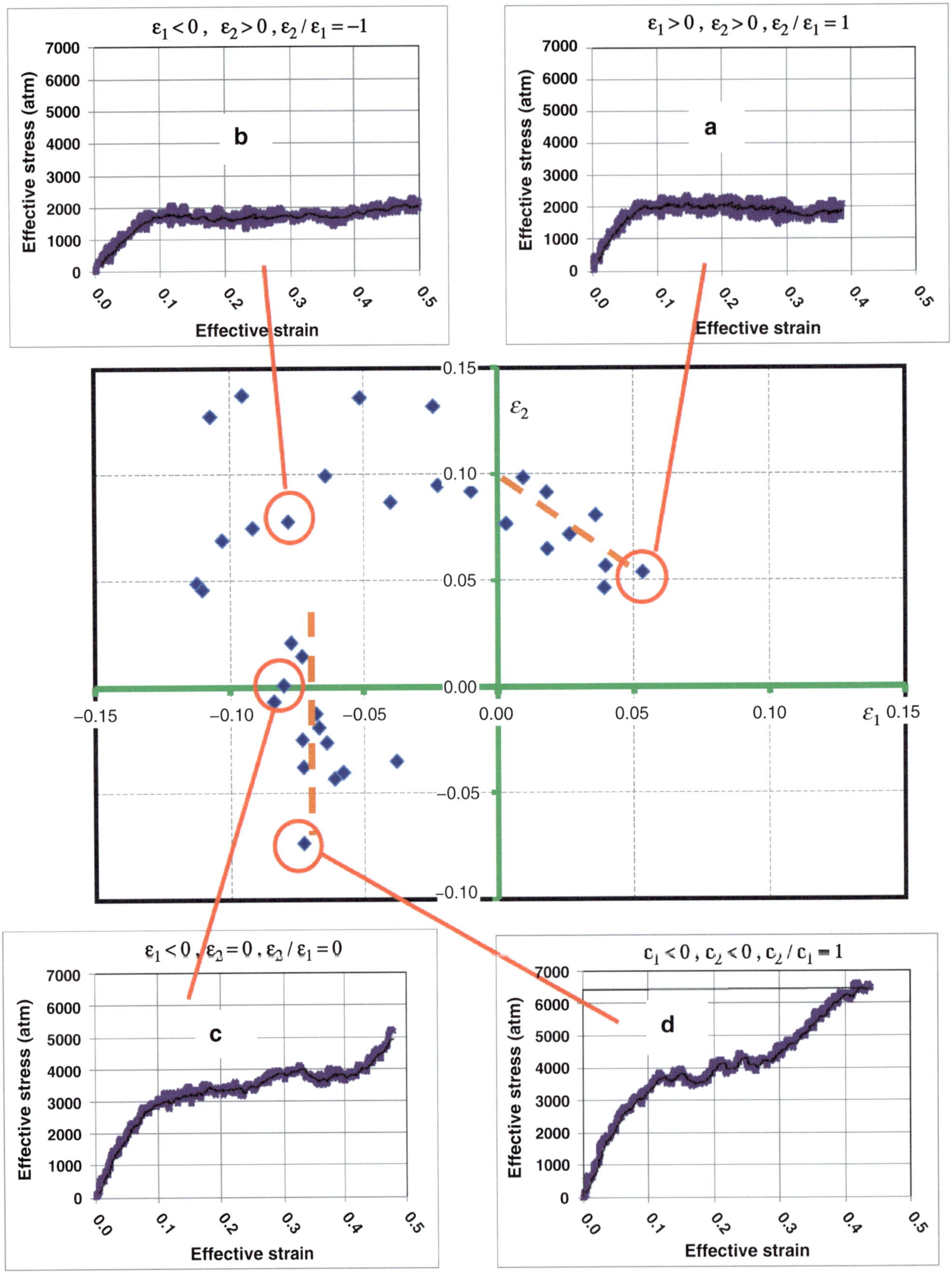

Fig. 3.2 Strain combinations at yield point (*main plot*) along with plots of effective stress versus effective strain for pure biaxial extension (**a**), pure shear (**b**), constrained compression (**c**), and pure biaxial compression (**d**)

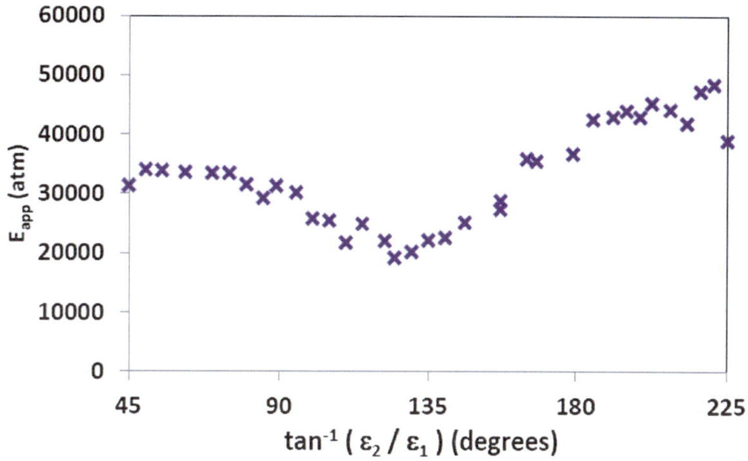

Fig. 3.3 Apparent modulus as a function of strain ratio

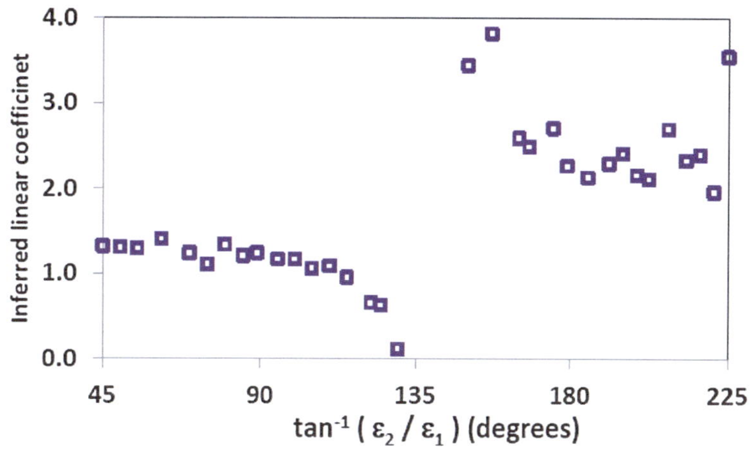

Fig. 3.4 Values of the linear coefficient $(-1)\ \varepsilon_3/J_1$ inferred from the elastic portion of the stress-strain diagrams

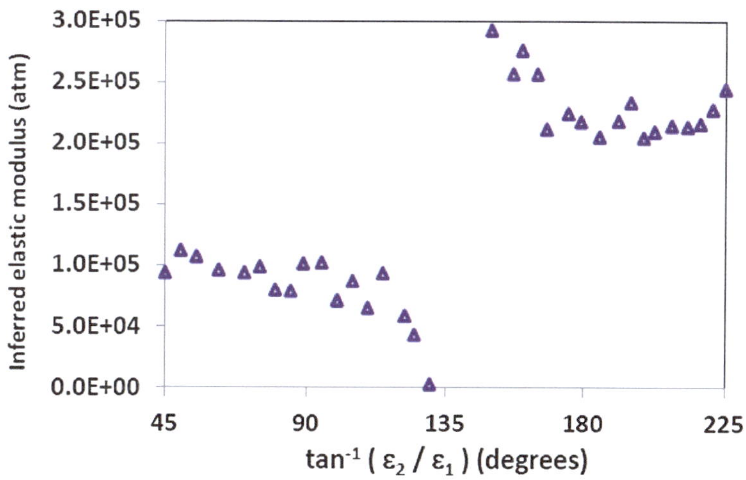

Fig. 3.5 Values of the inferred elastic modulus, $(\sigma_1 + \sigma_2)/J_1$, inferred from the elastic portion of the stress-strain diagrams

length in which the associated energy equaled the energy to rupture the bond. When at least one bond was found to reach or exceed its respective threshold value, the simulation was stopped and the system state archived.

Results have been obtained for some of the possible strain proportions. These are provided in Fig. 3.10. For strain states within the zone delimited by lines 1 and 2, rupture appears to be governed by the magnitude of the positive strain with rupture beginning whenever ε_2 is just greater than 1.0.

The volumetric dilatation at rupture has been considered by others to be an important metric for anticipating rupture in general. The simulation results are provided in Fig. 3.11. For combinationsofstrainin which the magnitude of the compressive strain is greater than that of the tensile strain, the nature of the atomic interaction appears to be distinctly different. While

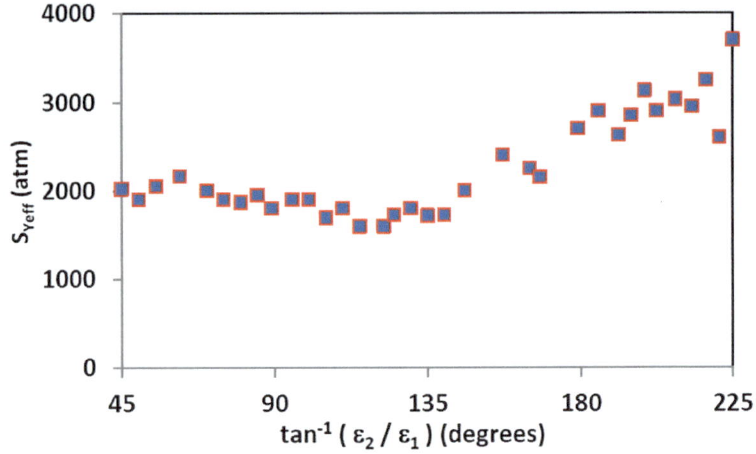

Fig. 3.6 The effective stress at yield as a function of strain ratio

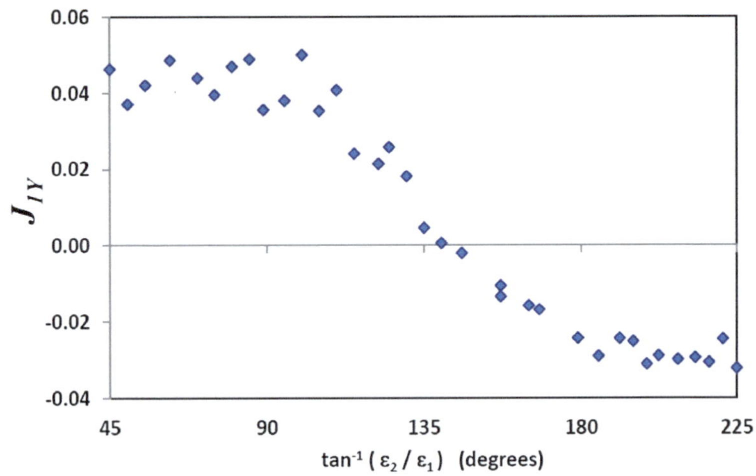

Fig. 3.7 The first strain invariant at yield as a function of in-plane principal strain ratio

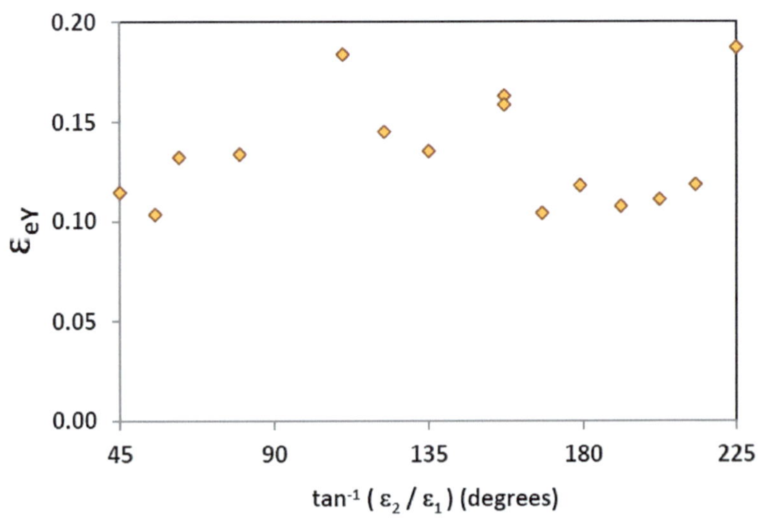

Fig. 3.8 Equivalent strain at yield as a function of in-plane principal strain ratio

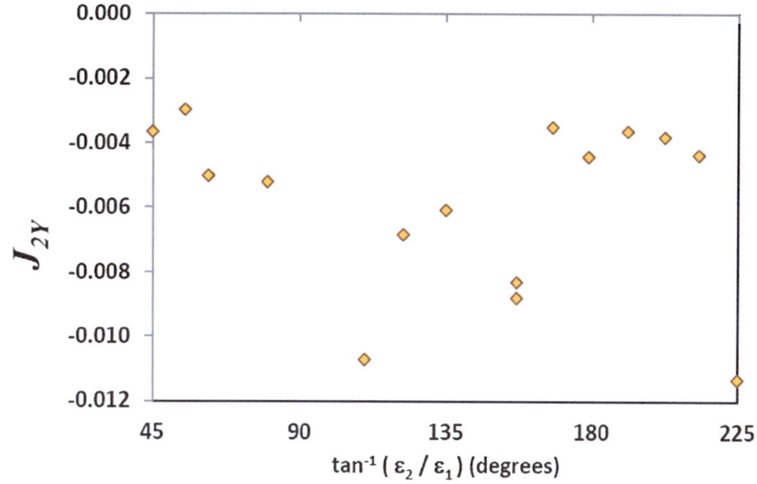

Fig. 3.9 Strain invariant J_2 at yield as a function of in-plane principal strain ratio

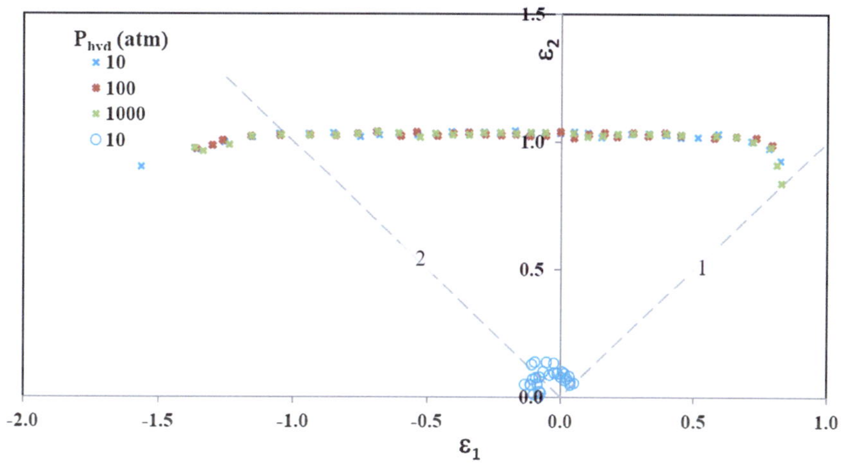

Fig. 3.10 Plot of principal strains at yield (*circles*) and incipient rupture (*x marks*) in plane stress for a short-chain DGEBA/MDA system. The *dashed line* marked 1 denotes biaxial stress with stresses equal in both principal directions. For the *dashed line* marked 2, stresses are equal in magnitude and opposite in sign

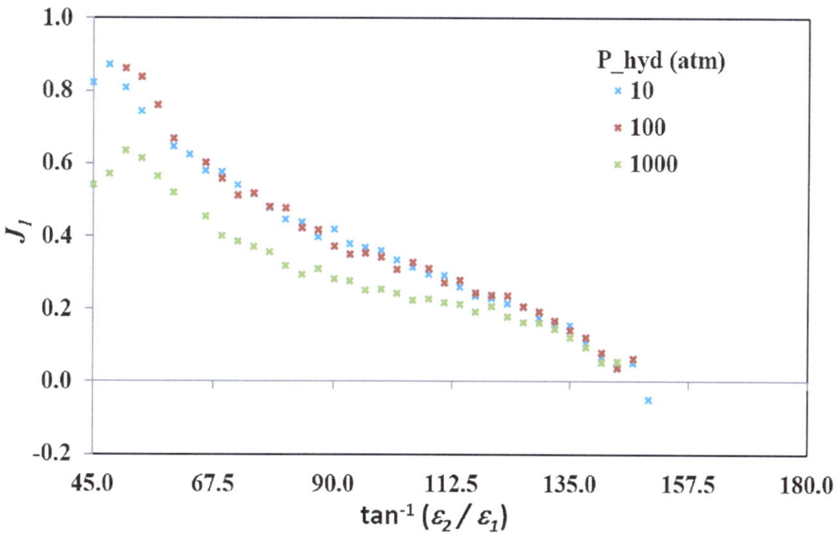

Fig. 3.11 The first invariant of the strain tensor (J_1) as a function of the proportion among principal strains. The amount of dilation decreases with increasing hydrostatic pressure and decreases in the transition from biaxial extension to pure shear

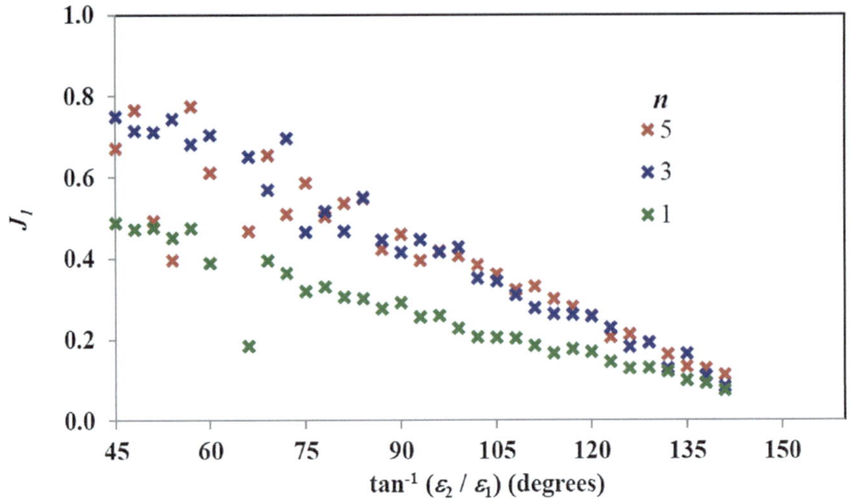

Fig. 3.12 Dilatation at the point of first bond rupture as a function of the principal strain ratio. The result for different resin chain lengths is shown

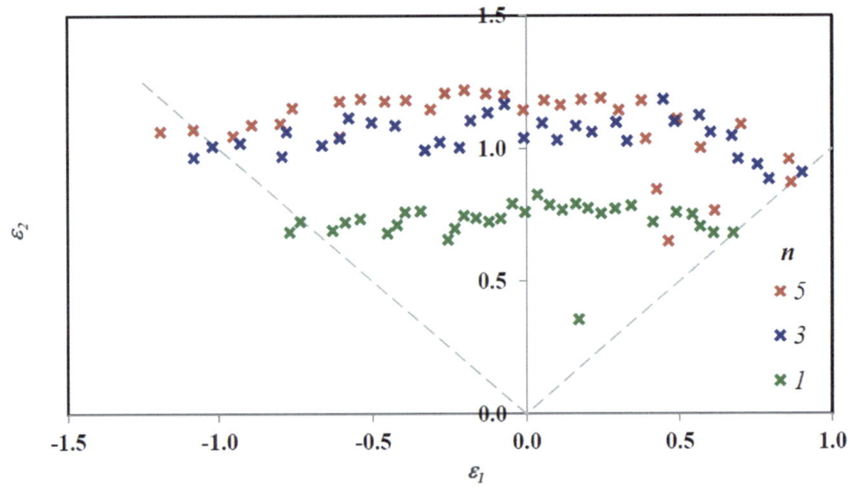

Fig. 3.13 Principal strain combinations at the point of first bond rupture. Results for different resin chain lengths are shown

large positive dilatations at rupture were found for many of the points plotted, dilatation is approximately zero or negative. Run results thus far also indicate rapid molecular rearrangements in that the time step needs to be sharply reduced in order to have realistic computation of atom locations.

The script was used to examine the effect of resin chain length (and therefore cross-link density). The number of repeat groups in the resin chain is indicated by the value of n. Figure 3.12 shows that the amount of volumetric dilatation is smallest for the short chain length but is essentially the same for $n = 3$ and $n = 5$. Figure 3.13 shows a similar trend in terms of the principal strain envelope which defines incipient rupture. Figure 3.14 shows the effective stress at incipient bond rupture. This value can be considered a metric of ultimate strength. The only consistent trend is that strength is generally larger in pure shear ($\tan^{-1}(\varepsilon_1/\varepsilon_2) = 135°$) than it is in biaxial extension ($\tan^{-1}(\varepsilon_1/\varepsilon_2) = 45°$) and there is a continuous variation among intermediate conditions. It also appears the system with the longest chain length is comparatively weak in biaxial extension but is the strongest in pure shear.

3.4 Conclusions

A novel method to deform atomistic representations of molecular systems in plane stress has been developed and applied to covalently-bonded polymer networks. Preliminary results agree with strain-invariant fracture theory in some aspects. Further, strain invariant values at rupture are similar to those inferred from experiments. Once the responses in shear and biaxial compression are completely characterized, results could be used to inform continuum models for thermoset behavior in plane stress.

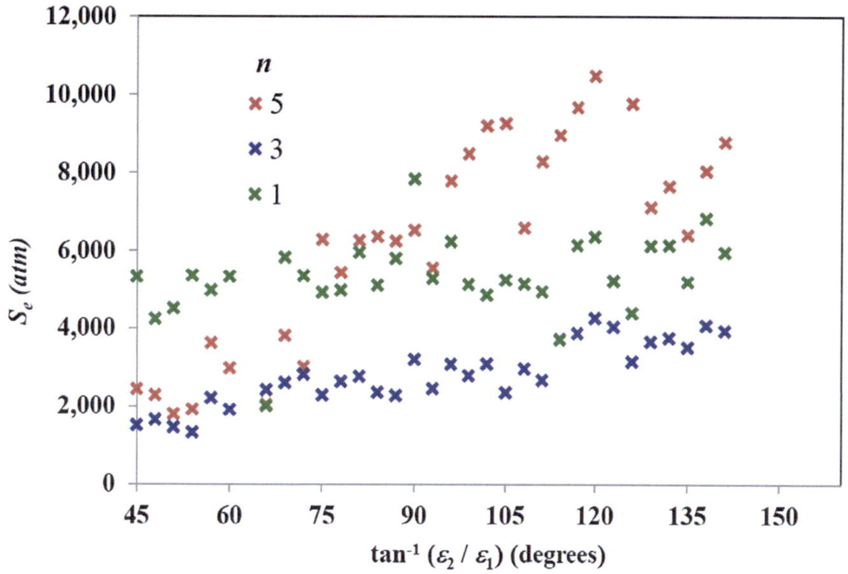

Fig. 3.14 Effective stress at the point of first bond rupture as a function of principal strain ratio. Results for different resin chain lengths are shown

References

1. Gosse, J., Christensen, S.: Strain invariant failure criteria for polymers in composite materials. In: AIAA/ASME/ASCE/AHS/ASC structures, structural dynamics, and materials Conference 2001, Seattle (2001)
2. Herakovich, C.: Mechanics of Fibrous Composites. Wiley, New York (1998)
3. Carlsson, L., Pipes, R.: Experimental Characterization of Advanced Composite Materials. Prentice Hall, Englewood Cliffs (1987)
4. Tsai, S.: Theory of Composites Design. Think Composites, Dayton (1992)

Chapter 4
Effect of Degree of Cure on Damage Development in FRP

Takahiro Ozaki, Tatsuro Kosaka, and Kazuhiro Kusukawa

Abstract We have developed a fiber-optic sensing system for measurement of degree of cure (DOC) of FRP laminates. By applying this system to monitoring of cure process of FRP laminates, molding time cam be optimized. However, it is considered that some part of the composite has incomplete DOC due to non-uniformity of molding temperature when the optimized molding process is conducted. Therefore, the effect of DOC on quality of FRP must be clear to manufacture high quality FRP using the optimized molding process. In the present paper, GFRP cross-ply laminates were manufactured with arbitrary DOC. Tensile tests to investigate the effect of DOC on the damage development behavior of the cross-ply laminates were conducted. DOC of the specimen was monitored by the fiber-optic sensing system developed by our laboratory. Matrix cracks were observed during the tensile tests. From the experimental results, it was found that the lower DOC specimen showed the lower crack initiation stress and the slower development rate of transverse crack in 90° layer. In addition, it appeared that many splitting cracks were generated in 0° layers when the DOC was less than 100 % due to the low toughness of resin. From these results, it was found that DOC governs damage development behavior of FRP laminates strongly.

Keywords FRP laminates • Matrix cracks • Tensile test • Degree of cure • Optical fiber sensors

4.1 Introduction

Fiber reinforced plastics (FRP) are composited of reinforcements such as glass or carbon fibers and resin matrix. FRP have been used in various fields in recent years because they have features of high toughness, light weight and corrosion proof. On the other hand, manufacturing of FRP products take labor and a long time. Moreover, optimization of molding condition is conducted by trial and error. Therefore, the manufacturing cost is very high. This problem is caused by difficulty of evaluation of internal state of materials during molding process. Recently, an in situ measurement method using embeddable fiber optic sensors has been proposed. This technology allows us to measure internal state of FRP during molding process and then to control pressure and temperature automatically.

Matrix cracks occurred in FRP laminates under loading degrades their mechanical properties [1], the behavior of matrix crack development indicate toughness of FRP laminates. In the present paper, the effect of DOC on the matrix crack development of cross-ply GFRP laminates was investigated experimentally. In order to manufacture the specimen with DOC less than 100 %, the heating process was interrupted. The DOC was measured by the Fresnel-based refractive-index optical fiber sensors [2]. The damages were observed by a video camera during tensile tests. The effect of DOC on the damage development were discussed from the experimental results.

4.2 Manufacturing of Specimens

In this study, we used GFRP prepreg (Mitsubishi Rayon GE352G135SB) of 120 μm in thickness for manufacturing cross-ply laminates. Figure 4.1 illustrates stacked laminates and location of embedded sensors for hot-press molding. The 12 prepreg sheets (140 × 140 mm) were laminated on an aluminum plate as the stacking sequence was $[0_3/90_3]_S$. During the lamination

T. Ozaki • T. Kosaka (✉) • K. Kusukawa
School of Systems Engineering, Kochi University of Technology, 185 Miyanokuchi, Tosayamada-cho, Kami-city, Kochi 782-8502, Japan
e-mail: kosaka.tatsuro@kochi-tech.ac.jp

A.M. Beese et al. (eds.), *Fracture, Fatigue, Failure and Damage Evolution, Volume 8*, Conference Proceedings of the Society for Experimental Mechanics Series, DOI 10.1007/978-3-319-21611-9_4

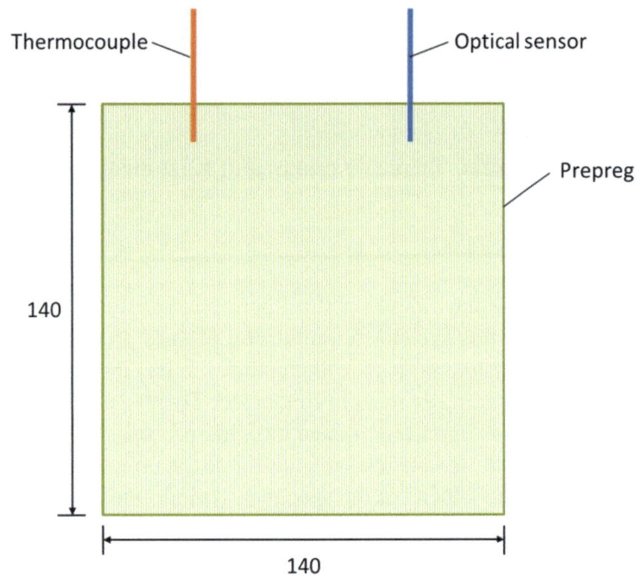

Fig. 4.1 Location of optical fibers and dimensions of the test piece

process, the optical fiber sensor and a thermocouple were embedded between the first and second sheets in order to measure degree of cure and temperature, respectively. The material was covered by another aluminum plate, breathers, and a vacuum film to make vacuum pack for removing voids during molding process.

The Fresnel-based refractive-index fiber optic sensors were used for measurement of degree of cure of FRP during molding. This sensor uses Fresnel's reflection at the flat-end of a glass optical fiber caused by mismatch of refractive index between resin and glass. Refractive index of resin can be calculated from the reflected optical power because the change in refractive index of glass is very small during molding. Degree of cure can be obtained by eliminating temperature effect on refractive index.

The experimental system for measurement of degree of cure of FRP laminates during hot-press molding process is illustrated in Fig. 4.2. The molding pressure was 0.5 MPa for all laminates. The laminates were heated until 137∼138 °C for 58 min and then cooled instantly. After de-molding the laminates form the mold, the test pieces for tensile tests were cut out of the laminates. The pieces where the optical fiber sensors and thermocouples were embedded were heated again at 140 °C for 2½ h in order to cure the laminates perfectly.

Figure 4.3 shows the relationship between degree of cure of the laminate and temperature during hot-press molding and re-heating process. The red line shows degree of cure increased from 0 to 80 % until at 137 °C and from 80 to 82 % during cooling. Therefore, the final degree of cure of the specimen was 82 % and the maximum temperature was 137 °C. The blue line shows that it is confirmed that the degree of cure measured by the embedded optical fiber sensor reached 100 % during re-heating process.

We prepared six specimens. The Table 4.1 shows the maximum molding temperature and degree of cure for each specimen. The two specimens were cured perfectly at 140 °C for 3 h. From the results, it was found that the values of degree of cure were from 82 to 96 % although the maximum temperature was almost same. Therefore, it is very difficult to control degree of cure by temperature control.

4.3 Tensile Test

4.3.1 Testing Method

The dimension of strip specimens used for tensile tests was shown in Fig. 4.4. The length was from 90 to 130 mm, the width was 20 mm, the thickness was from 1.12 to 1.19 mm and the volume fraction of reinforcements was 66.7 %. The aluminum tabs were glued to both surfaces of the specimen and a strain gauge was adhered to center of the specimen. Tensile tests were conducted with a cross-head speed of 0.5 mm/min. by a tensile test machine (Shimadzu Autograph) until the specimens were broken. Load and strain were recorded during the test. The center square area of 20 × 20 mm was recorded by a video

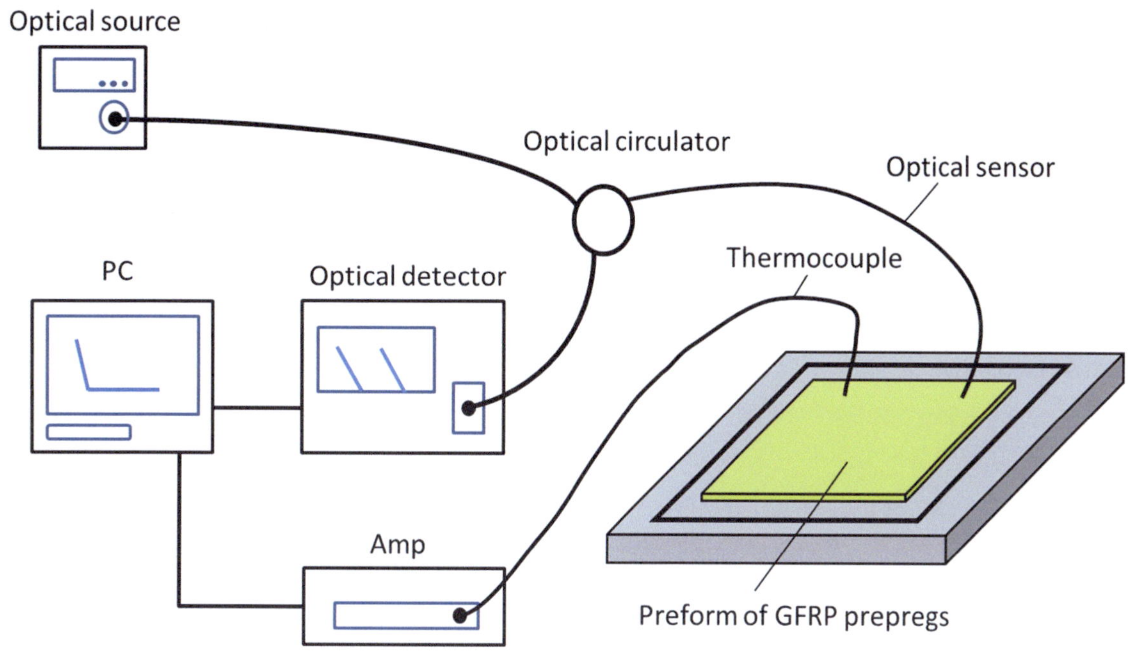

Fig. 4.2 Experimental system for measurement of degree of cure of FRP laminates during hot-press molding process

Fig. 4.3 Relationship between degree of cure and temperature during hot-press molding re-heating process (82 % DOC Specimen)

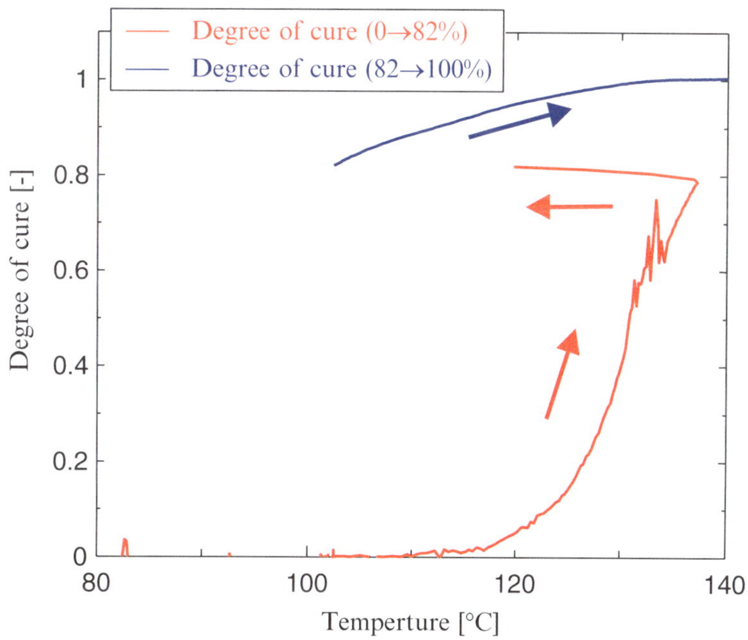

Table 4.1 Degree of cure and maximum temperature for each test specimens

Temperature [°C]	137	137	138	138	140 (for 3 h)	140 (for 3 h)
Degree of cure [%]	82	86	93	96	100	100

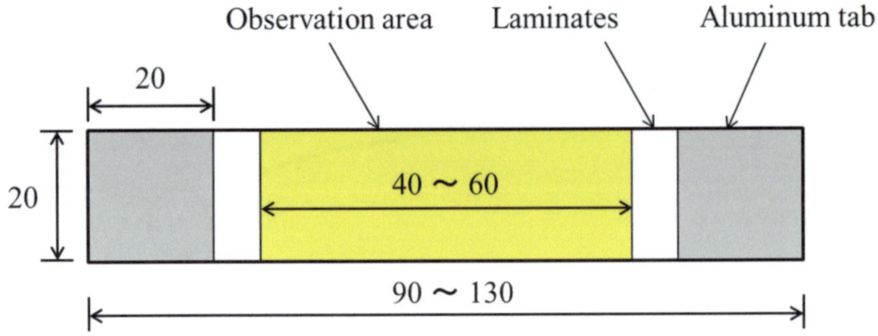

Fig. 4.4 Dimension of the specimens

Table 4.2 Tensile strength and Young's modulus

Degree of cure [%]	Tensile strength [MPa]	Young's modulus [GPa]
82	648	30.4
86	583	27.1
93	631	29.1
96	629	29.7
100	578	27.7
100	563	27.3

camera with a backlight in order to observe initiation and development of cracks. After the test, stress-strain curves were obtained and both the tensile strength and initial Young's modulus was calculated for each specimen. Moreover, number of longitudinal and transverse cracks in 0° and 90° layers were counted from the video record.

4.3.2 *Strength and Initial Young's Modulus*

Tensile strength and initial Young's modulus of all specimens were listed in Table 4.2. From the results, it appeared that the tensile strength of the 82 % specimen was the highest and that of the 100 % specimen was the lowest because elongation of the lower DOC specimen was larger. This fact means that the strength of continuous fibers in 0° layers governed the strength of cross-ply laminates and most reinforcing fibers in 0° layers did not break until the end of test. On the other hand, the initial Young's moduli of all specimens were almost the same as each other. Therefore, it was found that the initial mechanical properties with no damage was not affected by the degree of cure when it is larger than 82 %.

4.3.3 *Stress-Strain Curves and Feature of Damages*

Stress and transverse crack density of 82 and 100 % DOC specimens are plotted against strain in Fig. 4.5. The maximum measured strain was almost 0.01 because the strain gages were broken. From the figure, it was found that the stress-strain curve of the 82 % DOC specimen showed a knee point at 0.003 strain clearly while the curve of the 100 % DOC specimen was linear. The figure shows that the transverse crack initiated at the knee point for the 82 % specimen. Therefore, it was found the nonlinearity of a stress-strain curve of the 82 % DOC specimen resulted from transverse cracks in 90° layer decreased stiffness of the cross-ply laminates.

Figure 4.6 shows photographs of the damaged 82 and 100 % DOC specimens just before break of the specimen. The photograph 4.6a shows many splitting cracks and transverse cracks were generated for the 82 % specimen during tensile tests. On the other hand, only transverse cracks were observed in the photograph 4.6b of the 100 % DOC specimen.

4.3.4 *Development of Cracks*

Relationships between transverse crack density of 90° layer and stress are shown for all specimens in Fig. 4.7. As for the 82 % DOC specimen, transverse cracks initiated at 92.5 MPa, grew with increase of applied stress and then saturated to 20.2/cm at

Fig. 4.5 Stress-strain curves and transverse crack density of 82 % and 100 % DOC specimens

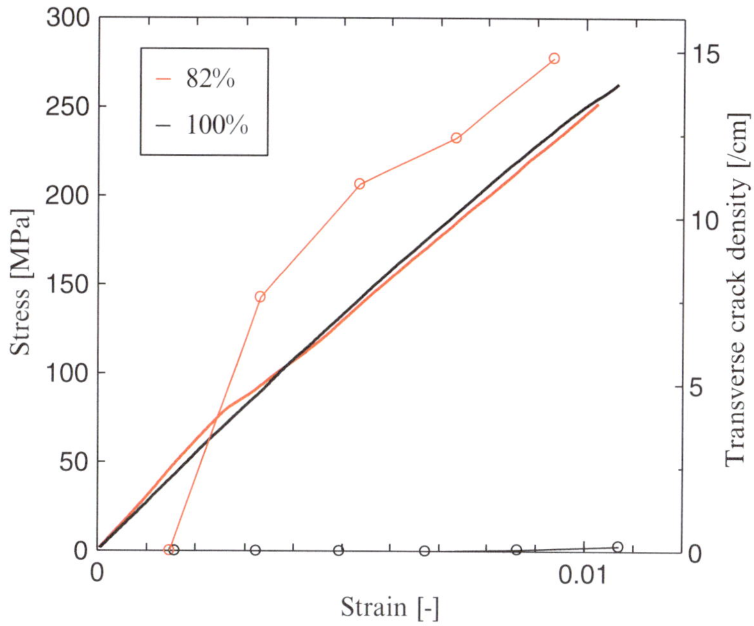

Fig. 4.6 Photographs of damaged specimens

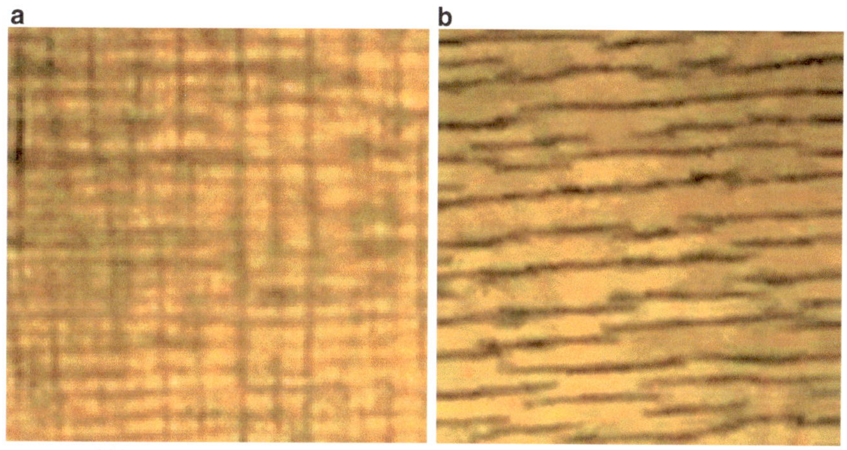

647.5 MPa. The initiation stress of transverse cracks of the 86 %, 93 % and 96 % DOC specimens were 86.5 MPa, 136 MPa and 136.9 MPa, respectively. The saturated transverse crack density of the 86 %, 93 % and 96 % DOC specimens were 15/cm at 583 MPa, 16/cm at 593 MPa and 16/cm at 593 MPa, respectively. In case of 100 % DOC specimens, the initiation stress were 220.4 or 263.3 MPa and the saturated density were 14/cm at 577.9 MPa or 12/cm at 570.6 MPa. From these results, it appeared that the crack initiation stress was larger and the crack density was lower when DOC was higher.

Figure 4.8 shows relationships between splitting crack density of 0° layer and stress. As for the 82 % DOC specimen, splitting cracks initiated at 231 MPa, increased with increase of applied stress and then saturated to 10.3/cm at 648 MPa. The initiation stress of splitting cracks of the 86 %, 93 % and 96 % DOC specimens were 216.2 MPa, 362.6 MPa and 273.7 MPa, respectively. The saturated splitting crack density of the 86 %, 93 % and 96 % DOC specimens were 7.3/cm at 583 MPa, 5.5/cm at 635 MPa and 2.4/cm at 593 MPa, respectively. In case of 100 % DOC specimens, very few splitting cracks were observed. It is known that the splitting occurs by transverse tensile stress in 0° layer of cross-ply laminates due to orthotropic stiffness of FRP laminates. Therefore, this results shows that the toughness of FRP is lower when the degree of cure is lower.

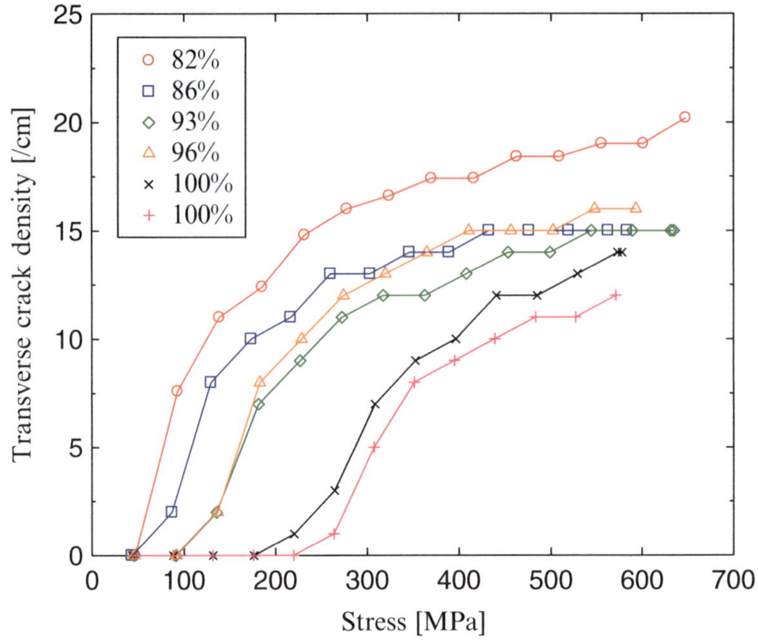

Fig. 4.7 Relationship between transverse crack density and stress

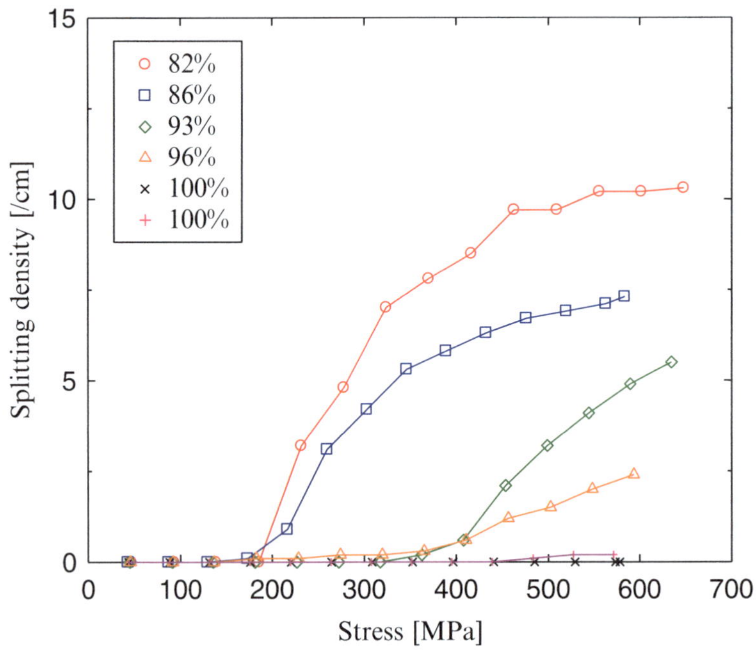

Fig. 4.8 Relationship between splitting crack density and stress

4.4 Summary

In this paper, FRP cross-ply laminates were manufactured with arbitrary degree of cure and tensile test were performed with observation of matrix cracks. From these results, it appeared that the transverse crack initiation stress was larger and the transverse crack density was lower when DOC was higher. Moreover, it appeared that many splitting cracks were generated in 0° layers when the DOC was less than 100 % due to the low toughness of resin. From these results, it was found that DOC governs damage development behavior of FRP laminates strongly.

Acknowledgement This work was supported by Council for Science, Technology and Innovation (CSTI), "Innovative structural materials" of "Cross-ministerial Strategic Innovation Promotion Program (SIP)" (Funding agency: JST).

References

1. Lim, S.G., Hong, C.S.: Prediction of transverse cracking and stiffness reduction in cross-ply laminated composites. J. Compos. Mater. **23**, 695–713 (1989)
2. Kosaka, T., Osaka, K., Sawada, Y.: Cure monitoring of resin by real-time measurement of refractive index using single-mode optical fibers. J. Soc. Mater. Sci. Jpn. **59**(5), 391–397 (2010)

Chapter 5
Stochastic Discrete Damage Simulations of Laminate Composites

Gunjin Yun and Li Zhao

Abstract In this paper, probabilistic failure response and damage patterns in polymer matrix composite laminates was investigated by considering spatially varying strength properties. For this purpose, an efficient random field modeling framework for multiple cross-correlated random fields is proposed whereby different a set of uncorrelated random variables in the Karhunen-Loève (KL) expansion are generated by Latin hypercube sampling technique and transformed to sets of correlated random variables. Discrete Damage Modeling (DDM) is performed by means of Regularized eXtended-Finite Element Method (Rx-FEM). The strength properties represented by spatially varying cross-correlated random fields define the matrix crack insertion patterns, whereas their propagation as well as the delamination growth is governed by cohesive law models with constant fracture toughness properties. One composite laminate a quasi-isotropic carbon/epoxy (Hexply IM7/8552) $[45/90/-45/90]_s$ was modeled by using probabilistic DDM. The effect of statistical parameters such as the correlation length, variance and correlation coefficient between random fields of normal and shear strength within DDM framework was examined for the first time. Significant effects of the statistical parameters on the failure behavior and ultimate component strength was observed, manifesting importance of accurate definitions of the statistical properties for predicting probabilistic failure behavior and damage tolerance of laminate composites.

Keywords Laminate composite • Random field • Karhunen-Loève (KL) expansion method • Discrete damage modeling • Matrix cracking • Delamination

5.1 Introduction

Polymer matrix based composite (PMC) materials have been increasingly used in aircraft structures for their high specific modulus and strength, high damping values, corrosion resistance, and better fatigue life aspect, etc. Since two constituents (i.e. fiber and resin) are combined preserving their original properties, microscopically the PMC materials show geometric heterogeneity, hence being one of the causes that create considerable scatter in mechanical properties. Variability in PMC strength properties is also introduced by manufacturing process, and operating conditions. Therefore, use of statistically-based material properties are recommended according to the military handbook 17 [1].

Probabilistic models of strength and stiffness properties of PMC materials based on experimental test data have been studied in order to statistically characterize FRP composites [2], use them in probabilistic design [3], develop stochastic finite element basis [4–6], predict statistical failure [6, 7], conduct strength analyses in various stochastic analysis studies [8–10], evaluate reliability [11], etc. Sriramula et al. comprehensively overviewed quantification and modeling of uncertainties in FRP composite materials [12]. Onkar et al. studied first-ply failure statistics of laminate composite plates with Tsai-Wu and Hoffman criteria and the first-order perturbation technique under random linear elastic material properties and random transverse loading [13]. Spatial variability of material properties was also considered an important factor that determines the probability of failure according to research by Carbillet et al. [14]. Wu et al. evaluated statistical strength of composite laminates considering two-dimensional random distributions of axial, transverse and shear strengths of the composite lamina [15]. Using the random field theory (RFT) and the stochastic finite element (SFEM), Feng et al. modeled volume fraction as a two-dimensional random field and the fiber angle as a single random variable in order to generate the variance of effective modulus of laminate composite plates [16]. In their research, different correlation lengths were assumed in the normal and tangential directions and random fields were discretized using the spatial average method [16, 17]. Nader et al. conducted Monte Carlo simulation-based probabilistic finite element analyses of woven fabric E-glass/vinyl ester coupons

G. Yun (✉) • L. Zhao
Department of Civil Engineering, The University of Akron, 244 Sumner St. ASEC 210, Akron, OH 44325-3905, USA
e-mail: gy3@uakron.edu

A.M. Beese et al. (eds.), *Fracture, Fatigue, Failure and Damage Evolution, Volume 8*, Conference Proceedings of the Society for Experimental Mechanics Series, DOI 10.1007/978-3-319-21611-9_5

considering spatial variations of elastic properties, strength parameters, and geometric properties [18, 19]. They studied effects of the spatial correlation length, finite element size, probability distribution functions (PDF) models and failure criteria on statistical variations of design properties of the standard-size composite samples. These prior researches certainly have focused on statistical estimation of coupon/component level strengths and moduli. Statistical simulations for initiations of matrix damage are critically important since initial flaws in PMC mostly start from matrix cracking and/or debonding between fibers and matrix and the variability of matrix-dominant properties is larger than that of fiber-dominant properties. To the best knowledge of authors, there has been few research on effects of the spatial variability of strength properties on probabilistic strength and post-cracking behavior of PMC on the basis of discrete damage simulation methods. It is partly because of limited availability of discrete damage simulations tools.

In this paper, a random field modeling framework will be developed for probabilistic mesh-independent discrete damage analyses of PMC materials. Strengths properties will be treated as mutually correlated random fields. The Karhunen-Loève (KL) expansion method will be employed for discretizing spatially varying random fields. For mesh-independent discrete damage simulations, Regularized eXtended Finite Element Method (Rx-FEM) will be used whereby failure criteria are employed. Success of this work will greatly help reproducing statistical and realistic initial crack insertion patterns and assessing *in-situ* strength and post-cracking behavior. In the following Sect. 5.2, random field modeling approach tailored to unidirectional orthotropic ply will be presented. Section 5.3 will present stochastic damage simulation results from the proposed random field modeling approach combined with discrete damage modeling (DDM) and address effects of various probabilistic model parameters. Finally Sect. 5.4 will summarize the results and draw conclusions.

5.2 Random Field Modeling for Fiber-Reinforced Polymer Matrix Laminate

5.2.1 *Cross-Correlated Multivariate Random Field Modeling*

In 2012, Yun proposed an efficient framework for modeling multiple cross-correlated random fields whereby a set of uncorrelated random fields in KLE having independent auto-covariance functions are generated by Latin Hypercube sampling and transformed to a sets of cross-correlated random variables [20]. Vorechovsky also proposed a practical framework for generating cross correlated random fields [21]. However, the method requires all cross correlated fields over the domain to share an identical auto-covariance function [21]. In 2013, Cho et al. proposed two different approaches (i.e. muKL and mcKL) generalizing the KLE for modeling multi-correlated non-stationary stochastic process [22]. This paper employs the method proposed in Ref. [20]. Let's consider multiple n zero-mean and square-integrable random processes,

$$\{RV_1(x,\theta), RV_2(x,\theta), \ldots, RV_n(x,\theta)\}. \tag{5.1}$$

The exponential auto-covariance function of ith random process is expressed as

$$C_i(\mathbf{x}_1, \mathbf{x}_2) = \sigma_i^2 \exp\left(-\frac{|x_1 - x_2|}{l_x} - \frac{|y_1 - y_2|}{l_y}\right) \tag{5.2}$$

where l_x and l_y are the correlation length parameters in x and y directions, respectively; and σ_i^2 is the variance. Then KL expansion of each random field will be

$$RV_i(x,\theta) = \sum_{k=1}^{\infty} \sqrt{\lambda_k^i} f_k^i(x) \{\eta_k^i(\theta)\}. \tag{5.3}$$

where λ_k^i and $f_k^i(x)$ are kth eigenvalue and eigenfunction of the auto-covariance function of the RV_i, respectively and $\{\eta_k^i(\theta)\}$ are uncorrelated Gaussian normal random variables. With a definition $\rho_{km}^{ij} \equiv E\left[\eta_k^i(\theta)\, \eta_m^j(\theta)\right]$, then cross-covariance between RV_i and RV_j will be expressed as

$$C_{ij}(RV_i, RV_j) = \sum_{k=1}^{\infty}\sum_{m=1}^{\infty} \rho_{km}^{ij} \sqrt{\lambda_k^i \lambda_m^j} f_k^i(x_1)\, f_m^j(x_2) \tag{5.4}$$

Then by orthogonality property of the set of eigenfunctions, the correlation coefficient (i.e. ρ_{km}^{ij}) between ith and jth random fields is defined as a projected cross-covariance onto the set of kth and mth mode eigenfunctions.

$$\rho_{km}^{ij} = \frac{1}{\sqrt{\lambda_k^i \lambda_m^j}} \int \int C_{ij} \left(RV_i, RV_j\right) f_k^i (x_1) f_m^j (x_2) \, dx_1 dx_2 \tag{5.5}$$

Since the eigenfunction of each random field is associated with an auto-covariance function and the cross-covariance C_{ij} is populated by the normal variables $\{\eta_k^i(\theta)\}$ and $\{\eta_m^i(\theta)\}$, the cross-correlated n random fields in (5.1) can be realized by a correlation matrix

$$[\boldsymbol{\rho}] = \begin{bmatrix} 1 & \rho^{12} & .. & \rho^{1n} \\ \rho^{21} & 1 & .. & \rho^{2n} \\ . & . & . & . \\ \rho^{n1} & \rho^{n2} & .. & 1 \end{bmatrix}. \tag{5.6}$$

Our strategy to generate ensemble data of cross-correlated random fields is to combine the spectral decomposition of the cross-covariance matrix defined in (5.6) with KL expansion [20]. To this end, uncorrelated normal random variables will be generated by the Latin hypercube sampling technique and transformed to correlated random fields by the following method. An eigenvalue problem for the symmetric and positive definite correlation matrix $[\boldsymbol{\rho}]$ is written as

$$[\boldsymbol{\rho}] \; [\boldsymbol{\Psi}] = [\boldsymbol{\Psi}] \, [\boldsymbol{\Lambda}] \tag{5.7}$$

where $[\boldsymbol{\Psi}]$ and $[\boldsymbol{\Lambda}]$ are the orthogonal eigenvector and the eigenvalue matrices of $[\boldsymbol{\rho}]$. By the orthogonality of the eigenvector matrix $[\boldsymbol{\Psi}]$, the correlation matrix can be decomposed as

$$[\boldsymbol{\rho}] = [\boldsymbol{\Psi}] \, [\boldsymbol{\Lambda}] \, [\boldsymbol{\Psi}]^T = [\mathbf{A}] \, [\mathbf{A}]^T \quad \text{and} \quad [\mathbf{A}] = [\boldsymbol{\Psi}] \, [\boldsymbol{\Lambda}]^{1/2}. \tag{5.8}$$

The transformation matrix $[\mathbf{A}]$ defined in (5.8) can operate on the uncorrelated random variables in order to transform them to cross-correlated random variables. For example, let's write a sample vector of n uncorrelated normal random fields for the kth KL expansion mode as follows

$$\{\boldsymbol{\eta}\}_k = \left[\eta_k^1 \; \eta_k^2 \; .. \, .. \; \eta_k^n \right]^T \tag{5.9}$$

where n is the number of random fields and superscript T indicates the matrix transpose operation. This sample of uncorrelated random fields will be sampled by the Latin hypercube sampling as mentioned previously. Then they are transformed to samples of the cross-correlated random fields as

$$\{\boldsymbol{\xi}\}_k = [\mathbf{A}] \, \{\boldsymbol{\eta}\}_k \tag{5.10}$$

Let's consider three random processes RV_i, RV_j and RV_k. The three random fields at a location $\mathbf{x}$ can be approximated by the KL expansion method as follows

$$\begin{aligned} RV_i (x, \theta) &= \langle RV_i (x, \theta) \rangle + \sum_{m=1}^{M} \sqrt{\lambda_m^i} f_m^i (x) \left\{\xi_m^i (\theta)\right\} \\ RV_j (x, \theta) &= \langle RV_j (x, \theta) \rangle + \sum_{m=1}^{M} \sqrt{\lambda_m^j} f_m^j (x) \left\{\xi_m^j (\theta)\right\} \\ RV_k (x, \theta) &= \langle RV_k (x, \theta) \rangle + \sum_{m=1}^{M} \sqrt{\lambda_m^k} f_m^k (x) \left\{\xi_m^k (\theta)\right\} \end{aligned} \tag{5.11}$$

where $< \cdot >$ is the mean value; and M is the truncation order of KLE series. Because random strength properties must be positive and only Gaussian distribution is assumed in the KLE, the Gaussian distribution was transformed to the lognormal distribution by taking exponential of the random strength samples.

$$RV_k(x,\theta) = e^{RV_k(x,\theta)} \tag{5.12}$$

In this way, realizations of cross-correlated strength properties having lognormal distributions can be generated. The cross-correlated multivariate random strength properties generated at the nodal points of the random field mesh are interpolated into Gauss integration points of the finite element mesh following a generalized mapping technique [23]. Since the random field mesh does not match with the finite element mesh due to the different mesh criteria, the generalized mapping technique is necessary.

5.2.2 *Orthotropic Spatial Auto-Covariance Function of Random Ply Strength Fields*

Variations of strength properties in the transverse direction of unidirectional ply show more fluctuations than in the longitudinal fiber direction because strength properties are influenced by fiber clustering (i.e. geometric heterogeneity), changes of local volume fraction, resin curing extent, etc. Therefore, it can be deduced that strength properties will have shorter spatial correlation length in the transverse direction than in the fiber direction. In this paper, direction-sensitive strength properties are considered by an orthotropic spatial auto-covariance function defined as

$$\begin{aligned}
&C(\mathbf{x}_1, \mathbf{x}_2) = \sigma^2 exp\left(-\frac{|l\cos(\theta_f - \theta)|}{b_1} - \frac{|l\sin(\theta_f - \theta)|}{b_2}\right)\\
&\theta = \tan^{-1}\left(\frac{y_2 - y_1}{x_2 - x_1}\right) \quad \text{for} \quad x_2 - x_1 \geq 0\\
&\theta = \tan^{-1}\left(\frac{y_2 - y_1}{x_2 - x_1}\right) -\pi \quad \text{for} \quad x_2 - x_1 < 0 \quad \text{and} \quad y_2 - y_1 < 0\\
&\theta = \tan^{-1}\left(\frac{y_2 - y_1}{x_2 - x_1}\right) + \pi \quad \text{for} \quad x_2 - x_1 < 0 \quad \text{and} \quad y_2 - y_1 \geq 0
\end{aligned} \tag{5.13}$$

where $\mathbf{x}_1(x_1, y_1)$ and $\mathbf{x}_2(x_2, y_2)$ are coordinate vectors for two different points; $l = \sqrt{(x_1 - x_2)^2 + (y_1 - y_2)^2}$; θ_f is the angle of the fiber direction; b_1 and b_2 are the correlation lengths in fiber and transverse directions, respectively. For an example, an open-hole coupon ($[0/45/90/-45]_s$) with 152 (L) × 38(W) × 1.11(t) mm and diameter 6.35 mm was chosen. Correlation length 15 mm was assumed in the fiber direction and 2 mm in the transverse direction.

5.3 Mesh-Independent Discrete Damage Modeling in Laminate Composites

The matrix cracking was simulated by the regularized mesh independent cracking (MIC) framework [24–27] whereby the simulation begins without any initial matrix crack and the matrix cracks are inserted based on LaRC03 failure criterion [28]. The delamination between the plies was simulated by using a cohesive formulation at the ply interface. Fiber failure criterion, such as Critical Failure Volume (CFV) [10] is used to stop the computation if fiber macro-failure is detected. The stochastic simulation tool is integrated in a way that generated random realization data of ply level strength properties are read at the beginning of the simulation process at each integration point. This data is then used in the failure criterion to insert the cracking interface by using Rx-FEM. Damage propagation process is deterministic in this paper and requires additional effort for extension to stochastic analysis.

5.4 Probabilistic Discrete Damage Analysis of Laminate Composites

Effects of statistics of spatially varying strength properties on damage characteristics are investigated by the proposed random field modeling and mesh-independent DDM method. The example is damage simulation of $[45/90/-45/0]_s$ carbon/epoxy laminate composite (Hexply IM7/8552) with ply thickness of 0.5 mm, length of 120 mm and width of 32 mm under uniaxial

Table 5.1 Materials properties for simulations of IM7/8552

Elastic properties	Coefficients of thermal expansion	Strength (MPa)	Toughness (J/m^2)
E_{11} (GPa) = 161	α_{11} (°C^{-1}) = 0	$Y_t^a = 60$	$G_{IC} = 200$
E_{22}, E_{33} (GPa) = 11.38	α_{22}, α_{33} (°C^{-1}) = 3×10^{-5}	$Y_c = 260$	$G_{IIC} = 1000$
G_{23} (GPa) = 3.98	$T-T_0$ (°C) = −150	$S^a = 90$	
G_{12}, G_{13} (GPa) = 5.17		$X_t = 3116$	
$\nu_{23} = 0.436$		$X_c = 2068.3$	
ν_{12}, $\nu_{13} = 0.32$			

[a]Median of lognormal distribution

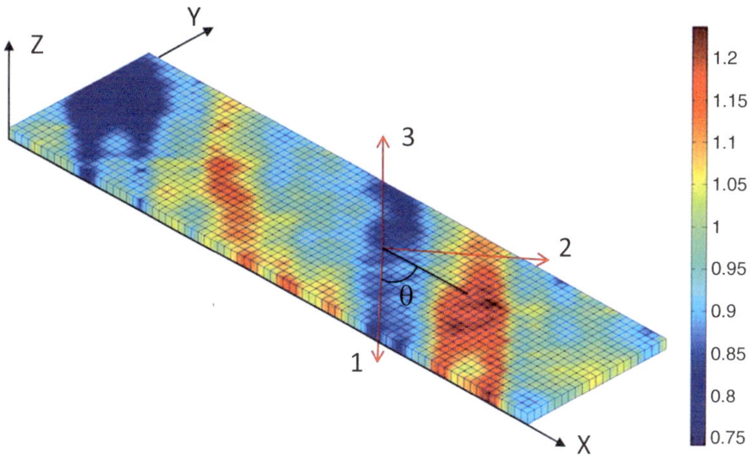

Fig. 5.1 Example realization of Yt in −45 lamina, b1 = 100 mm and b2 = 10 mm, c.o.v. of Yt = 12.2 %

Table 5.2 Statistical parameters for [45m/90m/−45m/0m]s m = 4 quasi-isotropic laminate composite

	Correlation length		Variance		c.c between Y_t and S	
Case 1	$d_1 = 10$ mm	c.o.v of	c.o.v of $Y_t = 2.2$ %	$d_1 = 100$ mm	0.0	$d_1 = 100$ mm
	$d_2 = 1$ mm	Y_t: 12.2 %;	c.o.v. of S = 1.2 %	$d_2 = 10$ mm		$d_2 = 10$ mm
Case 2	$d_1 = 100$ mm	S: 6 %;	c.o.v of $Y_t = 12.2$ %	c.c. between Y_t and S:	0.45	c.o.v of
	$d_2 = 10$ mm	c.c. between Y_t and S:	c.o.v. of S = 6 %	0.0		Y_t: 12.2 %
Case 3	$d_1 = 1000$ mm	0.0	c.o.v of $Y_t = 25$ %		0.95	S: 6 %
	$d_2 = 100$ mm		c.o.v. of S = 15 %			

tension [25, 29]. The transverse strength (Y_t) and in-plane shear strength (S) are considered as spatial random fields. Y_t and S values in Table 5.1 are assumed as median values of the lognormal distribution. Linear elasticity of each lamina is modeled with transversely isotropic material properties as summarized in Table 5.1. Boundary conditions in [25] were followed for simulations. Figure 5.1 depicts one realization of the transverse strength in −45 ply.

In order to investigate effects of statistical parameter changes on discrete damage patterns and failure behavior, three cases for each parameter (i.e. correlation lengths (C.L.), variances (c.o.v.) and correlation coefficient (c.c.) between Yt and S) are determined as summarized in Table 5.2. Therefore, a total of 108 simulations were conducted with 12 simulations per each case.

5.4.1 *Damage Patterns and Sequences*

Regardless of variations of statistical parameters, general damage events and sequences appeared to match well with experimental observations. A typical damage sequences is illustrated in Fig. 5.2. Generally, matrix cracking in 90° lamina starts first and it is followed by matrix cracking in lamina 45° and −45° almost at the same time. However, full development of matrix cracks of 45° lamina tends to be reached at earlier stage than −45° lamina since 45° lamina is used as top and bottom layer. Matrix cracks of 45° lamina fully developed before major delamination failure is progressed. Matrix cracks in −45° lamina continues even when delamination failure is under progression since it is interfaced with 0° lamina. Although matrix cracking in 90° begins at very earlier stage, like −45° lamina, it continues to develop even after major delamination

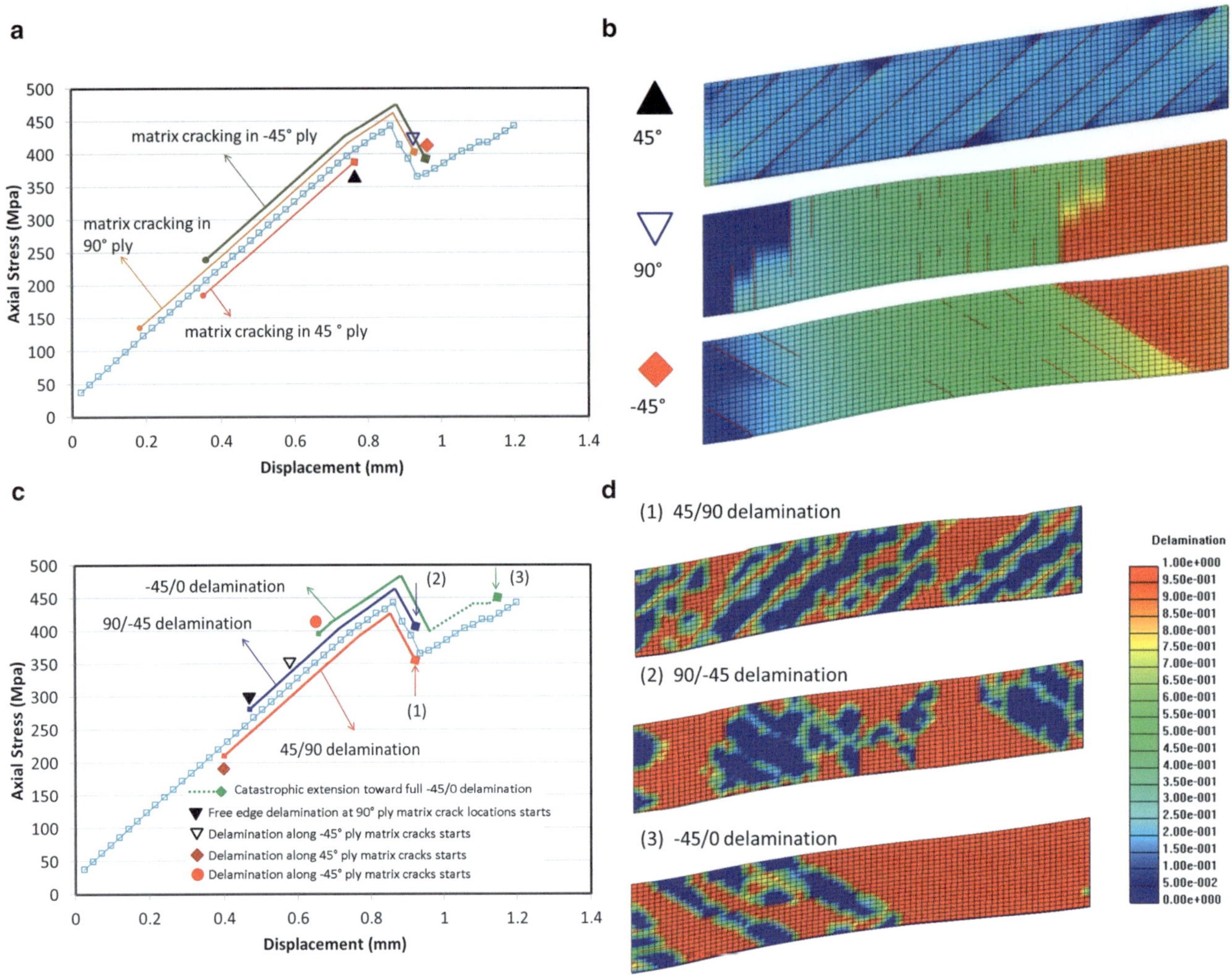

Fig. 5.2 (**a**) Matrix cracking events and sequences for quasi-isotropic laminate; (**b**) Cracks distributions in 45°, 90° and −45° lamina; (**c**) Delamination events and sequences for the quasi-isotropic laminate composite and (**d**) Contours for delamination damage from stochastic damage analysis with d1 = 100 mm, d2 = 10 mm, c.o.v. of Yt = 12.2 % and c.o.v. of S = 6 %

failure started. Delamination damage on the interface 45/90 occurred first at around 220 MPa and it gradually progresses along the matrix cracks in 45° lamina. The delaminated areas continued to grow until the major drop of axial stress. Growth of delamination damage on the 45/90 stops when catastrophic delamination on −45/0 interface begins. Delamination on the interface 90/−45 started at around 270 MPa along the free edge where 90° matrix cracks are located and as the loading increases, triangular-shape delaminated areas were formed along the free edge because of interactions with matrix cracks in −45° and 90° lamina and observed to gradually grow until catastrophic delamination on the interface 45/0 started. Delamination on the interface −45/0 started along the direction of matrix crack in −45° lamina at around 400 MPa and was followed by immediate and sudden catastrophic expansion. Eventually most of areas on the interface −45/0 were observed to be delaminated at the final loading stage.

5.4.2 Effect of Spatial Correlation Length in Strength Properties

Two thousand samples by non-parametric bootstrapping technique were obtained based on the simulation results of each case for its sufficient convergence to estimate realistic statistical parameters. Then the probability density functions (PDF) for three cases are plotted (Fig. 5.3b) showing increased dispersions and means of the ultimate failure stress with increasing correlation length. Axial stress versus displacement for Case 2 is plotted in Fig. 5.3a. It is also found that as the correlation

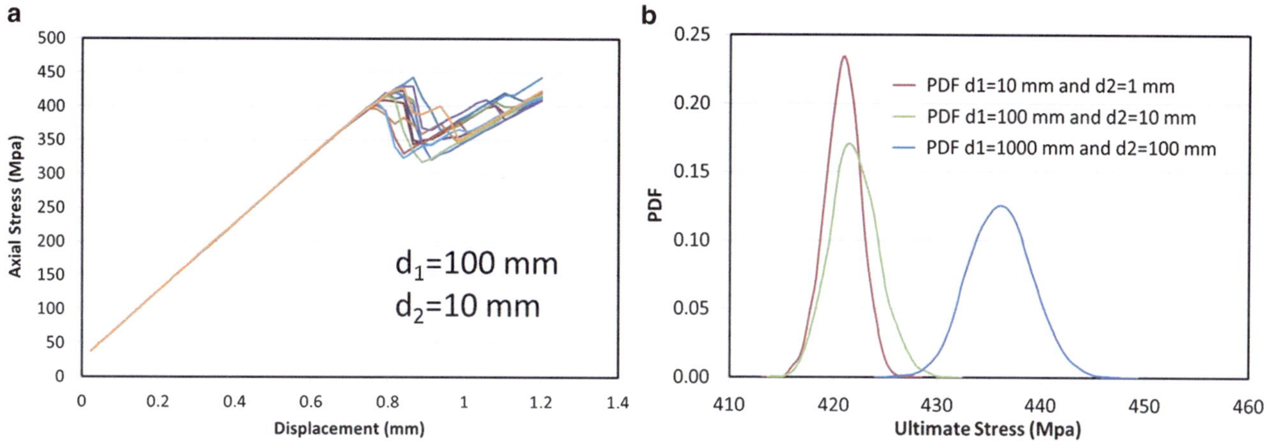

Fig. 5.3 (**a**) Axial stress versus displacement of Case 2; (**b**) PDF of ultimate stress for different correlation lengths of Yt and S

length increases, matrix cracking in 45°, 90° and −45° lamina started at later stages and matrix cracks tend to start at multiple steps rather than starting simultaneously. These effects of the correlation length on the matrix crack patterns result in increasing means and variances of the ultimate delamination failure stress as the correlation length increases. The best prediction (436.03 MPa) by current simulation shows less than 3.6 % underestimation error referring to experiment result (452 MPa [29]) whereas the probabilistic analysis with uncorrelated realization of elemental strength properties showed 9.5 % over prediction (495 MPa [25]) of the delamination failure stress.

5.4.3 *Effect of Variance of Matrix Strength Properties*

Similarly, effect of variance of matrix strength to failure behavior is studied. According to the results, the larger variances of the matrix strength properties are, the smaller mean of the ultimate failure stress is. However, the mean of the ultimate stress in Case 2 was slightly less than that of Case 3. It may be because of small difference of variances between Case 2 and 3 which are not well pronounced in the response variance. As expected, as the variances of the matrix strength increase, the variance of the ultimate stress also increases. Comparing with experimental test result (452 MPa [29]), the mean of the ultimate stress in Case 1 (432.16 MPa) showed 4.4 % underestimation error.

5.4.4 *Effect of Correlation Coefficients Between Cross-Correlated Strength Properties*

Strength in tension and shear is likely to be correlated, thus the effect of correlation coefficients is investigated. PDFs of the mean of major stress drop and mean of the ultimate stress determined from 2000 bootstrapped sample sets are generated based on simulation results to approximate true probabilistic models. Figure 5.4 shows plots of axial stresses versus displacements for three cases. The largest scattering of softening behavior was observed in Case 1 whereas the smallest scattering of softening behavior was observed in Case 3. It is found that as the correlation coefficient increases, the mean of ultimate strength also increases and the major drop of stress due to delamination failure increases.

Table 5.3 summarized statistical data analysis results from 2000 sets of bootstrapped samples. As the correlation coefficient increases from 0.0 to 0.95, the mean of the ultimate stress also increases from 418.02 to 430.93 MPa. Instead, the coefficient of variation (c.o.v.) of the ultimate stress decreased from 2.8272 to 1.9921. Comparing with experimental test result [29], the mean of the ultimate stress in Case 3 (430.93 MPa) showed 4.6 % underestimation error. It is worth noting that the correlated values of matrix strength components are physically more realistic than uncorrelated since most matrix defects will locally deplete normal and shear strength at the same time.

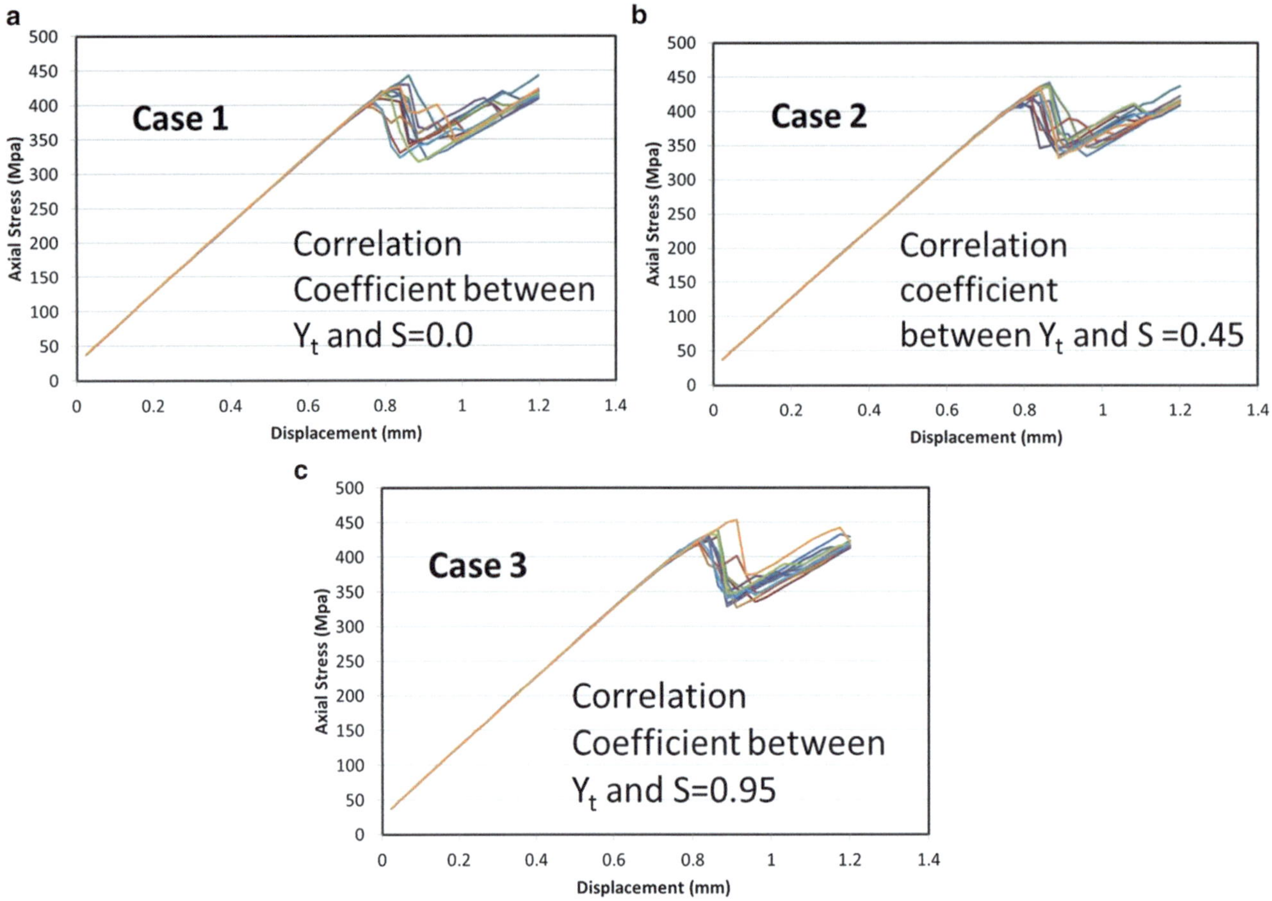

Fig. 5.4 Effects of the correlation between Yt and S on the failure behavior of [45/90/−45/0]s laminate composite (**a**) c.c.=0.0; (**b**) c.c.=0.45; (**c**) c.c.=0.95

Table 5.3 Statistical effects of correlation between Yt and S on failure stress of [45/90/−45/0]s

	Correlation coefficient between Y_t and S	Means of ultimate stress (MPa)	c.o.v. of ultimate stress (%)	Experimental test (MPa) [29]
Case 1	0.0	418.02	2.8272	452.00
Case 2	0.45	426.80	2.4879	
Case 3	0.95	430.93	1.9921	

5.5 Conclusions

In this paper, probabilistic failure response and damage pattern of PMC laminate composites has been investigated by considering spatially varying and cross-correlated strength properties. For this purpose, an efficient random field modeling framework for multiple cross-correlated random processes is proposed. Discrete damage modeling (DDM) based on Regularized eXtended-Finite Element Method (Rx-FEM) is utilized for failure modeling. Mean and variance of the ultimate stress increase as the correlation length increases. As the variance of the matrix strength properties increases, the mean of the ultimate stress tends to decrease. As the correlation coefficient increases, the mean of the ultimate stress shows increases and the variance of the ultimate stress shows decreases. Particularly, as the correlation coefficient increases, the mean of stress drop by major delamination shows increases and its variance showed decreases. Significant effects of the statistical parameters on the failure behavior and ultimate component strength was observed, manifesting importance of accurate definitions of the statistical properties for predicting probabilistic failure behavior and damage tolerance of laminate composites. The average strength values predicted by probabilistic analysis with spatially correlated strength values were closer to respective experimental data than the predictions with uncorrelated strength values.

References

1. Composite Materials Handbook: Volume 1. Polymer Matrix Composites Guidelines for Characterization of Structural Materials, D.o.D. Handbook. MIL-HDBK-17-1F, Department of Defense, Washington, DC (2002).
2. Zureick, A.H., Bennett, R.M., Ellingwood, B.R.: Statistical characterization of fiber-reinforced polymer composite material properties for structural design. J. Struct. Eng. ASCE. **132**(8), 1320–1327 (2006)
3. Mahadevan, S., Liu, X.: Probabilistic optimum design of composite laminates. J. Compos. Mater. **32**(1), 68–82 (1998)
4. Chen, N.Z., Soares, C.G.: Spectral stochastic finite element analysis for laminated composite plates. Comput. Methods Appl. Mech. Eng. **197**(51–52), 4830–4839 (2008)
5. Noh, H.C.: Stochastic finite element analysis of composite plates considering spatial randomness of material properties and their correlations. Steel Compos. Struct. **11**(2), 115–130 (2011)
6. Sasikumar, P., Suresh, R., Gupta, S.: Analysis of CFRP laminated plates with spatially varying non-Gaussian inhomogeneities using SFEM. Compos. Struct. **112**, 308–326 (2014)
7. Thieme, M., et al.: Probabilistic failure simulation of glass fibre reinforced weft-knitted thermoplastics. Compos. Sci. Technol. **90**, 25–31 (2014)
8. de Morais, A.B.: Prediction of the longitudinal tensile strength of polymer matrix composites. Compos. Sci. Technol. **66**(15), 2990–2996 (2006)
9. Gurvich, M.R., Pipes, R.B.: Probabilistic strength analysis of four-directional laminated composites. Compos. Sci. Technol. **56**(6), 649–656 (1996)
10. Iarve, E.V., Kim, R., Mollenhauer, D.: Three-dimensional stress analysis and Weibull statistics based strength prediction in open hole composites. Compos. A: Appl. Sci. Manuf. **38**(1), 174–185 (2007)
11. Gurvich, M.R., Pipes, R.B.: Reliability of composites in a random stress state. Compos. Sci. Technol. **58**(6), 871–881 (1998)
12. Sriramula, S., Chryssanthopoulos, M.K.: Quantification of uncertainty modelling in stochastic analysis of FRP composites. Compos. A: Appl. Sci. Manuf. **40**(11), 1673–1684 (2009)
13. Onkar, A.K., Upadhyay, C.S., Yadav, D.: Probabilistic failure of laminated composite plates using the stochastic finite element method. Compos. Struct. **77**(1), 79–91 (2007)
14. Carbillet, S., Richard, F., Boubakar, L.: Reliability indicator for layered composites with strongly non-linear behaviour. Compos. Sci. Technol. **69**(1), 81–87 (2009)
15. Wu, W.F., Cheng, H.C., Kang, C.K.: Random field formulation of composite laminates. Compos. Struct. **49**(1), 87–93 (2000)
16. Feng, X., et al.: Analysis on the variances of material and structural properties based on random field theory. Probab. Eng. Mech. **26**(2), 222–230 (2011)
17. Vanmarcke, E.: Random Fields: Analysis and Synthesis. MIT Press, Cambridge (1984)
18. Nader, J.W., et al.: Probabilistic finite element analysis of modified ASTM D3039 tension test for marine grade polymer matrix composites (vol 27, pg 6, 2008). J. Reinf. Plast. Compos. **30**(16), 1413–1413 (2011)
19. Nader, J.W., et al.: Probabilistic finite element analysis of ASTM D6641 compression test for marine grade polymer matrix composites. J. Reinf. Plast. Compos. **28**(8), 897–911 (2009)
20. Yun, G.J.: Uncertainty quantification of spatially varying lamina strength properties for probabilistic failure analyses of laminate composites. In: SFFP Final Report 2012, The University of Akron, Akron (2012)
21. Vorechovsky, M.: Simulation of simply cross correlated random fields by series expansion methods. Struct. Saf. **30**(4), 337–363 (2008)
22. Cho, H., Venturi, D., Karniadakis, G.E.: Karhunen-Loeve expansion for multi-correlated stochastic processes. Probab. Eng. Mech. **34**, 157–167 (2013)
23. Shang, S., Yun, G.J.: Stochastic finite element with material uncertainties: implementation in a general-purpose simulation program. Finite Elem. Anal. Des. **64**, 65–78 (2013)
24. Iarve, E.V.: Mesh independent modelling of cracks by using higher order shape functions. Int. J. Numer. Methods. Eng. **56**(6), 869–882 (2003)
25. Iarve, E.V., et al.: Mesh-independent matrix cracking and delamination modeling in laminated composites. Int. J. Numer. Methods. Eng. **88**, 749–773 (2011)
26. Mollenhauer, D., et al.: Simulation of discrete damage in composite overheight compact tension specimens. Compos. A: Appl. Sci. Manuf. **43**(10), 1667–1679 (2012)
27. Swindeman, M.J., et al.: Strength prediction in open hole composite laminates by using discrete damage modeling. AIA J. **51**(4), 936–945 (2013)
28. Davila, C.G., Camanho, P.P., Rose, C.A.: Failure criteria for FRP laminates. J. Compos. Mater. **39**(4), 323–345 (2005)
29. Hallett, S.R., et al.: Modelling the interaction between matrix cracks and delamination damage in scaled quasi-isotropic specimens. Compos. Sci. Technol. **68**(1), 80–89 (2008)

Chapter 6
Development of a Specimen for In-Situ Diffraction Planar Biaxial Experiments

G.M. Hommer and A.P. Stebner

Abstract In this paper, the design of cruciform shaped, planar biaxial loading specimens using finite element analysis, mechanical testing, and digital image correlation is discussed. The specimens were designed to be capable of arbitrary combinations of tension and compression loading. Digital image correlation results from uniaxial tension tests of first-generation specimen infer key design attributes of second-generation specimen. Finite element results are compared with a plane stress analytical formulation and differences between the two are attributed to stress concentration fields originating at the intersection of specimen arms. These results motivate a parametric finite element geometry optimization of second-generation specimen.

Keywords SEM • DIC • FEA • Planar biaxial • Compression

Nomenclature

E	Young's modulus
ε_{11}	Strain in the 11 direction
ε_{22}	Strain in the 22 direction
λ_a	Ratio of 11–22 direction applied surface tractions
λ_{ga}	Analytical formulation ratio of 11–22 direction gage stresses
λ_{gs}	FEA simulation ratio of 11–22 direction gage stresses
ν	Poisson's ratio
σ_{11}	Stress in the 11 direction
σ_{11a}	Applied surface traction in the 11 direction
σ_{11ga}	Analytical formulation gage stress in the 11 direction
σ_{11gs}	FEA simulation gage stress in the 11 direction
σ_{22}	Stress in the 22 direction
σ_{22a}	Applied surface traction in the 22 direction
σ_{22ga}	Analytical formulation gage stress in the 22 direction
σ_{22gs}	FEA simulation gage stress in the 22 direction

6.1 Introduction

6.1.1 Motivation

Advanced alloys, such as lightweight metals and shape memory alloys, are becoming increasingly important to the advancement of many industries. They often possess complex microstructures that result in anisotropic and asymmetric behaviors, often due to twinning and phase transformation of low symmetry crystal structures. Because of this, their three-dimensional mechanical properties and mechanisms of deformation cannot be fully understood through uniaxial loading. To elucidate these behaviors, a custom planar biaxial load frame capable of *in situ* X-ray and neutron diffraction

G.M. Hommer (✉) • A.P. Stebner
Mechanical Engineering Department, Colorado School of Mines, Brown Hall W350, 1610 Illinois Street, Golden, CO 80401, USA
e-mail: ghommer@mines.edu

A.M. Beese et al. (eds.), *Fracture, Fatigue, Failure and Damage Evolution, Volume 8*, Conference Proceedings of the Society for Experimental Mechanics Series, DOI 10.1007/978-3-319-21611-9_6

experimentation has been built. The instrument was also designed to study full plane stress yield and transformation loci. However, many previous planar biaxial experiments have primarily focused on tension-tension loading of sheet metals. Thus, the first challenge in executing these experiments was designing specimen geometries capable of planar biaxial compression-compression, tension-compression, and tension-tension.

6.1.2 Background

Cruciform-shaped geometry is the widely accepted standard for planar biaxial specimens [1–11]. The specimen gage, where material behavior is examined, is the center of geometry. A standard geometry capable of only tension-tension loading, due to arm buckling before gage yield in compression, has been established by ISO 16842:2014 "Metallic materials—Sheet and strip—Biaxial tensile testing method using cruciform specimen". In this study the general cruciform-shaped geometry was used and desired specimen behaviors were established. The attributes include: maximum stress in the gage, uniform elastic stress distribution in the gage, uniform and axial load transfer to the gage, and 1 mm gage thickness for diffraction capability.

6.2 Methods

The first-generation (G1) specimens were machined from 6061 aluminum alloy plate with the rolling direction along one of the loading axes. As a planar biaxial load frame was not available, uniaxial tensile tests to failure were conducted with a strain rate of 10^{-2} s^{-1} at room temperature on G1. Digital image correlation (DIC) data was collected during the tensile tests and used to infer design improvements for the second-generation (G2) specimen. Finite element analysis (FEA) using ABAQUS software was conducted on G2 using isotropic elastic-plastic model. FEA material properties were taken from experimental data for 2023-T3 aluminum alloy [12]. The FEA elastic behavior of G2 was compared with a plane-stress isotropic linearly elastic analytical solution.

6.3 Results and Discussion

6.3.1 First-Generation Specimen

The pertinent attributes of G1, shown in Fig. 6.1a are continuous elliptical arm intersection curvatures to minimize stress concentration, reduced gage thickness to maximize stress in gage and conform to 1 mm maximum gage thickness requirement for diffraction experimentation, and square gage to simplify macroscopic stress calculation. Tensile test DIC data shows a uniform elastic strain field in the gage in Fig. 6.1b. Development of principal stress planes in Fig. 6.1c show maximum stresses occurred in the gage. There are reduced stress zones next to the gage along the axis of applied load as a result of specimen geometry. Stress concentrations from elliptical-shaped arm width reduction, and load path disruption from 2 mm radius transition to reduced gage thickness, are responsible for the reduced stress zones. There are also stress concentrations present at the corner radii of the gage square. The DIC data was determined to have strain noise of ±0.02 %. The result of impedance on uniform and axial load transfer to the gage caused by aforementioned stress concentrations is shown Fig. 6.1d, G1 planar biaxial specimen failure from an applied load ratio of –0.5 on a 45° plane, which does not correspond to the applied load ratio under uniform and axial load transfer.

6.3.2 Second-Generation Specimen

The G1 design issues addressed in G2, shown in Fig. 6.2, were square gage section and 2 mm radius transition into the reduced gage thickness. Circular gage geometry was used to eliminate stress concentrations, and resulting load transfer disruptions. A 70 mm convex radius that tangentially transitions into a 7.5 mm concave radius leading into the gage prevents the reduced stress zone from occurring directly next to the gage, as in G1. Proportional tension-tension FEA results in Fig. 6.3 show reduced stress zone in G2 centered at minimum arm width, as expected from classical stress concentration and crack-tip

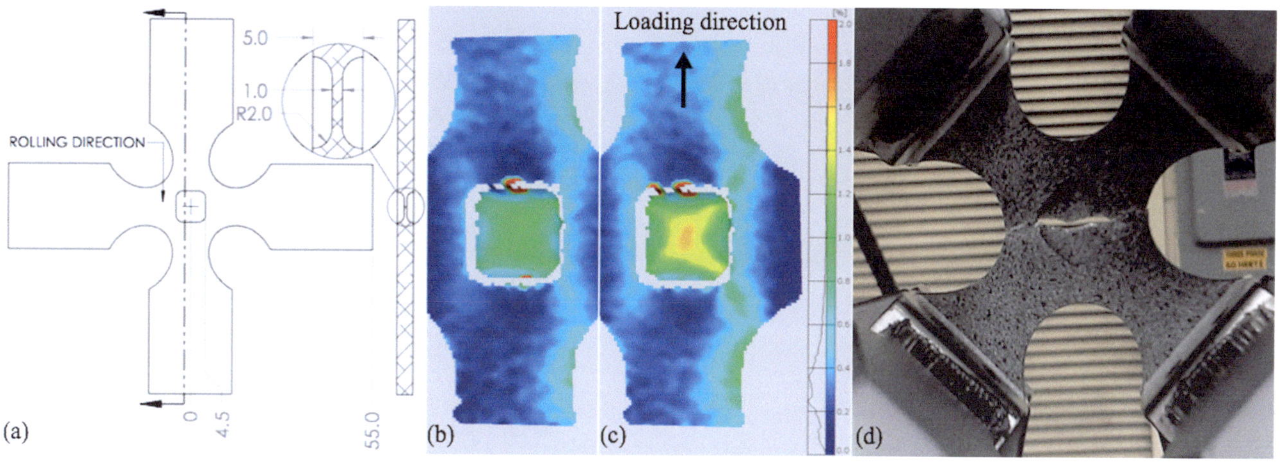

Fig. 6.1 First-generation specimen machine drawing with pertinent feature dimensions (**a**), DIC images from tensile test showing uniform elastic strain field in the gage (**b**) and development of principal strain planes indicating maximum stresses in the gage (**c**), and planar biaxial failure from an applied load ratio of −0.5 (**d**)

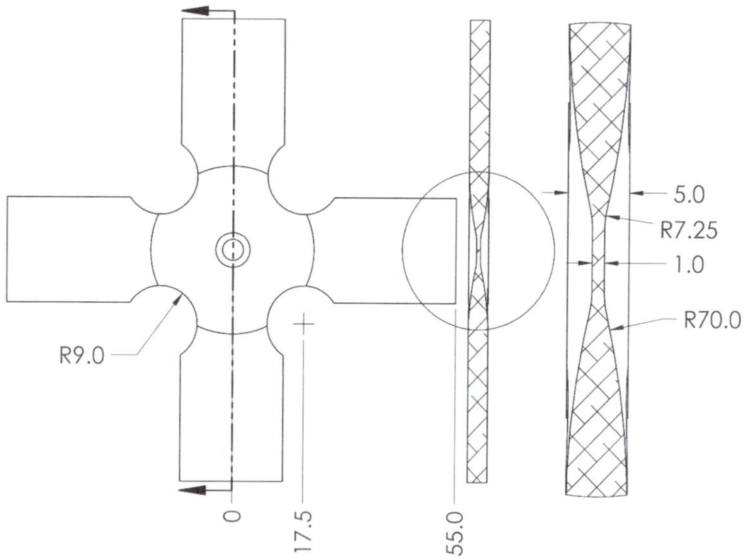

Fig. 6.2 Second-generation specimen machine drawing with pertinent feature dimensions

stress field formulations [13–15]. The load has sufficient distance after passing minimum arm thickness to redistribute and enter the gage uniformly.

FEA simulations were ran for 11 applied surface traction ratios:

$$\lambda_a = \frac{\sigma_{11a}}{\sigma_{22a}} \tag{6.1}$$

Resulting stress data from the center of the gage was taken to formulate the corresponding FEA gage stress ratios:

$$\lambda_{gs} = \frac{\sigma_{11gs}}{\sigma_{22gs}} \tag{6.2}$$

An analytical formulation for the gage stress ratio (6.7) as a function of the applied surface traction ratio was derived from isotropic linearly elastic plane-stress [16]:

$$\varepsilon_{11} = \frac{\sigma_{11}}{E} - \frac{\nu\sigma_{22}}{E} \tag{6.3}$$

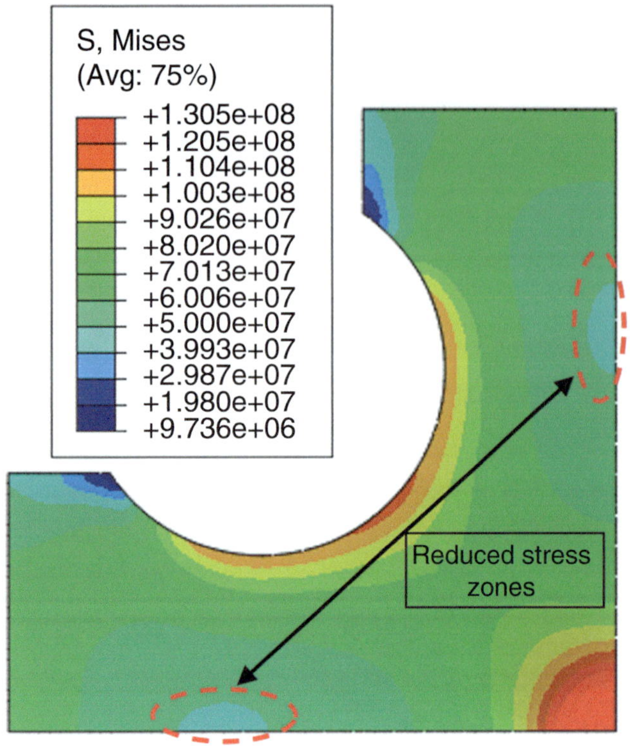

Fig. 6.3 FEA results for proportional tension-tension showing reduced stress zones at minimum arm width and uniform stress, units are Pa

$$\varepsilon_{22} = \frac{\sigma_{22}}{E} - \frac{\nu\sigma_{11}}{E} \tag{6.4}$$

Plane-stress assumption is supported by FEA results, as in-plane stresses in the gage were four orders of magnitude greater than out-of-plane stresses. Rearranging and adopting the current work's nomenclature, (6.3) and (6.4) become:

$$\sigma_{11ga} - \sigma_{11a} - \nu\sigma_{22a} \tag{6.5}$$

$$\sigma_{22ga} = \sigma_{22a} - \nu\sigma_{11a} \tag{6.6}$$

Through substitution of $\sigma_{11a} = \lambda_a\sigma_{22a}$ the analytical formulation for the gage stress ratio becomes:

$$\lambda_{ga} = \frac{\lambda_a - \nu}{1 - \nu\lambda_a} \tag{6.7}$$

The analytical and FEA gage stress ratios are plotted against the applied surface traction in Fig. 6.4. As the plane-stress assumption was previously justified, differences between gage stresses can be attributed to the stress concentration fields caused by the arm intersection radii. As λ_a decreases, the points of maximum stress concentration move along the radii towards the axis of larger applied surface traction, analogously rotating and translating the corresponding stress fields. The stress concentration fields have minimum interference with the gage for proportional loading. As loading deviates from proportional, the stress concentrations move towards the gage, increasing interference. This explains the gage stress difference trend between analytical and FEA. This stress concentration field behavior is illustrated in Fig. 6.5, a simplified cruciform geometry with uniform thickness to emphasize behavior of stress concentration fields, where the black lines are approximately tangent to point of maximum stress and rotate toward the vertical axis with decreasing λ_a. These results motivate a parametric optimization of G2 wherein the objective is to minimize the gage stress difference between the FEA results and analytical formulation. In doing so stress concentrations will be minimized since they are the source of load transfer disruptions and resulting discrepancy between actual and analytical specimen behavior.

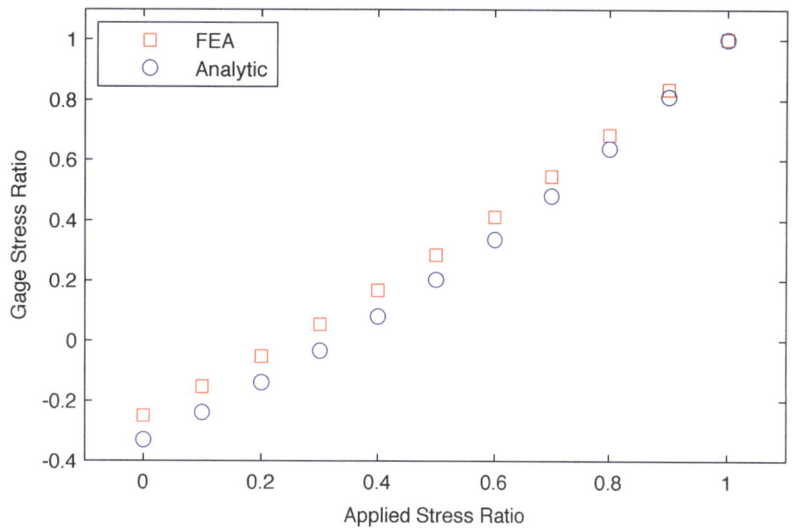

Fig. 6.4 Gage stress ratio vs. applied stress ratio for FEA and plane stress analytical formulation, results diverge as stress concentration fields interfere with specimen gage

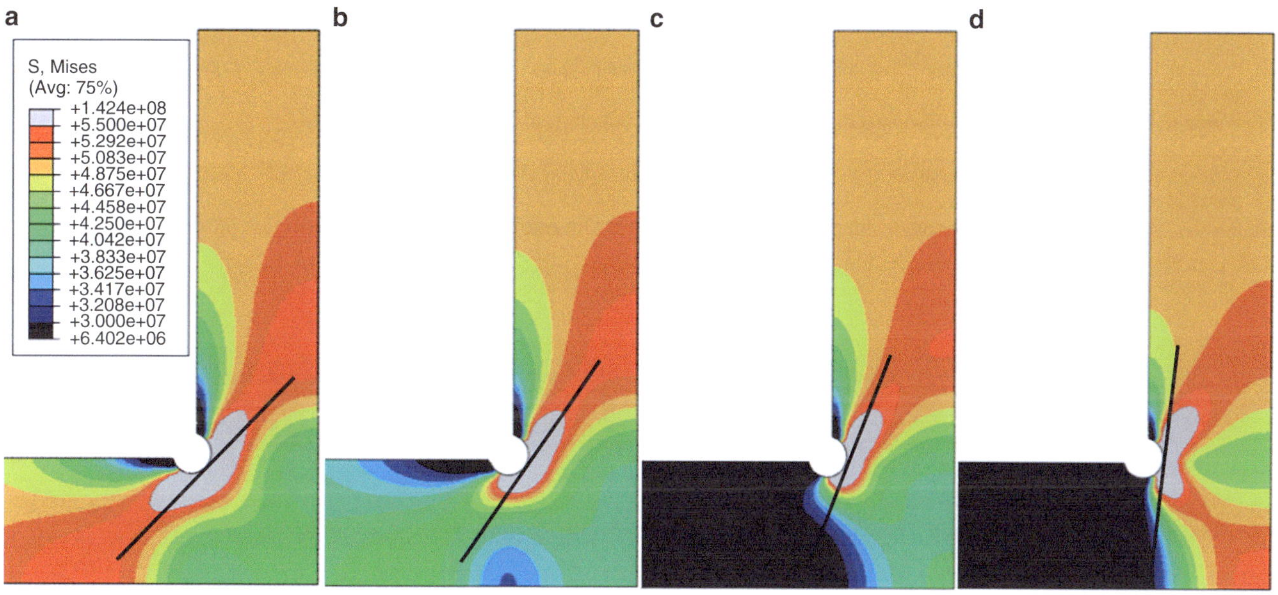

Fig. 6.5 Rotation and translation of stress concentration field with varying tension-tension applied surface tractions of 1.0 (**a**), 0.8 (**b**), 0.5 (**c**), and 0 (**d**), as applied surface traction ratio decreases, stress concentration field interference with the gage increases, the black lines are approximately tangent to point of maximum stress and rotate toward the vertical axis with decreasing λa, the geometry is a simplified cruciform specimen with uniform thickness, units are Pa

6.4 Conclusions

A cruciform-shaped planar biaxial specimen for *in-situ* diffraction experimentation capable of tension-tension, tension-compression, and compression-compression loading was designed using DIC and FEA techniques. It was found that circular gage with minimum rate of thickness reduction from full specimen thickness to gage thickness are specimen attributes that produce desirable performance under the criteria of maximum stress in, and uniform and axial load transfer to, the gage. The ratio of in-plane axial stresses in the specimen gage, σ_{11}/σ_{22} from FEA and plane-stress analytical solution was compared to the applied surface traction. Stress concentration fields originating from arm intersection radii caused differences between the two, as the plane-stress assumption was found to be consistent with FEA results. The trend in differences was explained by movement of stress concentration fields relative to the gage with decreasing applied surface traction ratio. The results motivate a parametric geometry optimization of G2 specimen.

References

1. Metallic materials—sheet and strip—biaxial tensile testing method using a cruciform test piece, ISO 16842:2014
2. Abu-Farha, F., Hector Jr., L.G., Khraisheh, M.: Cruciform-shaped specimens for elevated temperature biaxial testing of lightweight materials. JOM **61**(8), 48–56 (2009)
3. Demmerle, S., Boehler, J.P.: Optimal design of biaxial tensile cruciform specimens. J. Mech. Phys. Solids **41**(1), 143–181 (1993)
4. Hanabusa, Y., Takizawa, H., Kuwabara, T.: Numerical verification of a biaxial tensile test method using a cruciform specimen. J. Mater. Process. Technol. **213**(6), 961–970 (2013)
5. Hu, J.-J., Chen, G.-W., Liu, Y.-C., Hsu, S.-S.: Influence of specimen geometry on the estimation of the planar biaxial mechanical properties of cruciform specimens. Exp. Mech. **54**(4), 615–631 (2014)
6. Kuwabara, T., Kuroda, M., Tvergaard, V., Nomura, K.: Use of abrupt strain path change for determining subsequent yield surface: experimental study with metal sheets. Acta Mater. **48**(9), 2071–2079 (2000)
7. Makinde, A., Thibodeau, L., Neale, K.W.: Development of an apparatus for biaxial testing using cruciform specimens. Exp. Mech. **32**(2), 138–144 (1992)
8. Makris, A., Vandenbergh, T., Ramault, C., Van Hemelrijck, D., Lamkanfi, E., Van Paepegem, W.: Shape optimisation of a biaxially loaded cruciform specimen. Polym. Test. **29**(2), 216–223 (2010)
9. Shiratori, E., Ikegami, K.: Experimental study of the subsequent yield surface by using cross-shaped specimens. J. Mech. Phys. Solids **16**(6), 373–394 (1968)
10. Tiernan, P., Hannon, A.: Design optimisation of biaxial tensile test specimen using finite element analysis. Int. J. Mater. Form. **7**(1), 117–123 (2014)
11. Yu, Y., Wan, M., Wu, X.-D., Zhou, X.-B.: Design of a cruciform biaxial tensile specimen for limit strain analysis by FEM. J. Mater. Process. Technol. **123**(1), 67–70 (2002)
12. Problem FAC-2. http://www.afgrow.net [Online]. http://www.afgrow.net/applications/DTDHandbook/Examples/page1_1.aspx. Accessed 2 Mar 2015
13. Williams, M.L.: On the stress distribution at the base of a stationary crack. J. Appl. Mech. **24**, 111–114 (1957)
14. Irwin, G.R.: Analysis of stresses and strains near the end of a crack transversing a plate. J. Appl. Mech. **24**, 361–364 (1957)
15. Sanford, R.J.: A critical re-examination of the westergaard method for solving opening-mode crack problems. Mech. Res. Commun. **6**(5), 289–294 (1979)
16. Barber, J.R.: Plane strain and plane stress. In: Barber, J.R. (ed.) Elasticity, 3rd edn, pp. 40–41. Springer, New York (2010)

Chapter 7
V-Notched Rail Test for Shear-Dominated Deformation of Ti-6Al-4V

Sharlotte Kramer, John Laing, Thomas Bosiljevec, Jhana Gearhart, and Brad Boyce

Abstract Evermore sophisticated ductile plasticity and failure models demand experimental material characterization of shear behavior; yet, the mechanics community lacks a widely accepted, standard test method for shear-dominated deformation and failure of ductile metals. We investigated the use of the V-notched rail test, borrowed from the ASTM D7078 standard for shear testing of composites, for shear testing of Ti-6Al-4V titanium alloy sheet material, considering sheet rolling direction and quasi-static and transient load rates. In this paper, we discuss practical aspects of testing, modifications to the specimen geometry, and the experimental shear behavior of Ti-6Al-4V. Specimen installation, machine compliance, specimen-grip slip during testing, and specimen V-notched geometry all influenced the measured specimen behavior such that repeatable shear-dominated behavior was initially difficult to obtain. We will discuss the careful experimental procedure and set of measurements necessary to extract meaningful shear information for Ti-6Al-4V. We also evaluate the merits and deficiencies, including practicality of testing for engineering applications and quality of results, of the V-notched rail test for characterization of ductile shear behavior.

Keywords Shear • Failure • Metals • Mechanical properties • Model calibration

7.1 Introduction

Several ductile plasticity and failure models such as the shear-modified Gurson model [1, 2] require shear testing for calibration. The mechanics community has the ubiquitous uniaxial tensile test and compact tension fracture test for experimental characterization of tensile deformation and failure, but lacks a widely accepted method for shear-dominated behavior. Proposed experimental methods for this purpose include the Iosopescu geometry (ASTM D5379) [3, 4], the punch geometry (ASTM D732), the Butterfly geometry [5], the Arcan shear [6], tension-torsion fracture [7], and the V-notched rail shear (ASTM D7078) [8, 9]. These methods vary in complexity of setup, specimen design, measurements, boundary conditions, and data interpretation, all leading to a difficult choice of method depending on the material of interest, cost and time considerations, and the available experimental lab facilities.

The motivation for the shear testing in this paper was the second Sandia Fracture Challenge (SFC), which was a blind round robin prediction of ductile tearing by the computational mechanics community. Sandia National Laboratories hosted the first SFC in 2012, where computationalists were charged to predict ductile failure in a "crack-in-a-maze" geometry in 15-5PH stainless steel, based on standard material characterization experimental data from uniaxial tension and compact tension fracture tests [10]. The assessment of the first SFC showed that no computational team could properly predict all the features of the failure; one area of improvement for the predictions noted by the challenge participants was providing shear-dominated material characterization data. The second SFC, in 2014, was another "crack-in-a-maze" geometry, but in Ti-6Al-4V at two loading rates, one quasi-static and one at a modest transient rate, and the material characterization experimental data was from uniaxial tension tests and shear-dominated tests, as requested by the first SFC participants.

In this present paper, we discuss the use of the V-notched rail shear test to investigate the shear behavior of Ti-6Al-4V for quasi-static and transient loading rates for the second SFC. The V-notched rail shear test was first developed for composite materials, but we demonstrate that this method can be used for metals, provided careful procedures are followed. Local strain and displacement measurements, fixture compliance tests, careful gripping procedures, characterization of specimen slip in the grips, and specimen design are all factors for successful, repeatable tests, but interpretation of the data through the large deformations and failure regimes requires computational modeling to extract meaningful material characterization data.

S. Kramer (✉) • J. Laing • T. Bosiljevec • J. Gearhart • B. Boyce
Sandia National Laboratories, 1515 Eubank SE, Albuquerque, NM 87123, USA
e-mail: slkrame@sandia.gov

A.M. Beese et al. (eds.), *Fracture, Fatigue, Failure and Damage Evolution, Volume 8*, Conference Proceedings of the Society for Experimental Mechanics Series, DOI 10.1007/978-3-319-21611-9_7

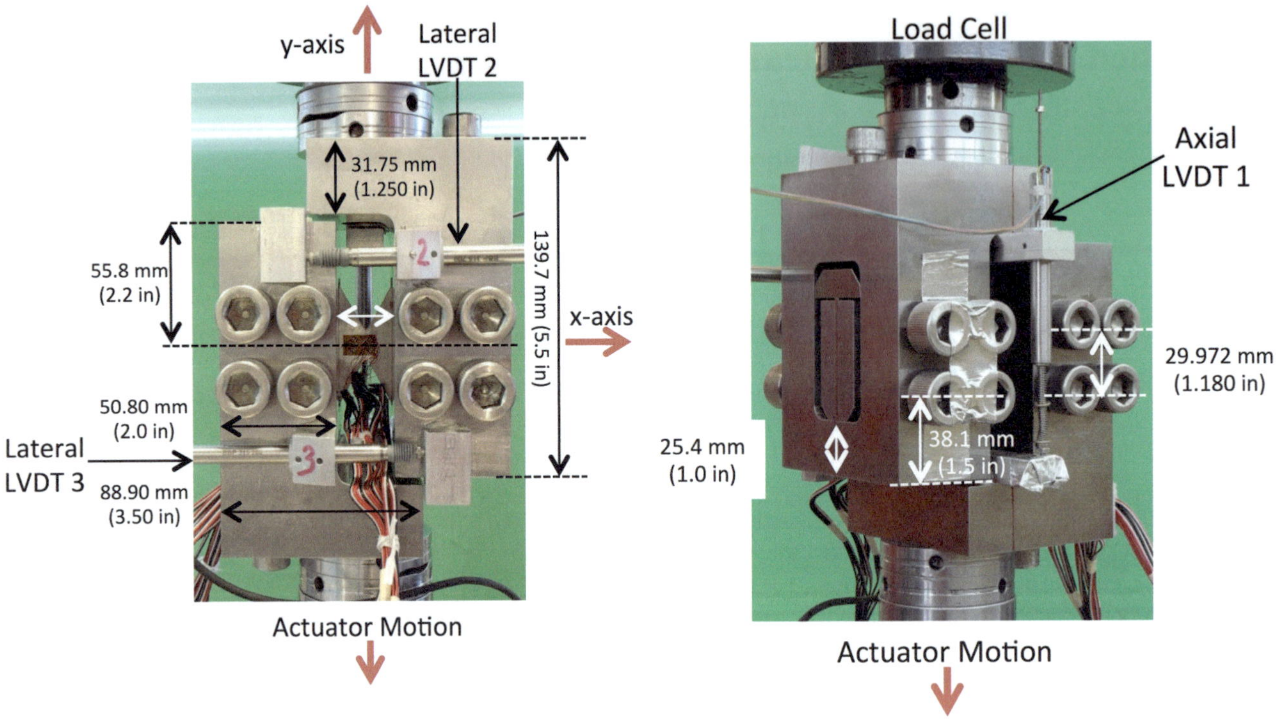

Fig. 7.1 Wyoming Test Fixtures adjustable combined loading shear fixture, as installed in a 100-kN MTS two-post, servo-hydraulic, single-actuator load frame, with a strain-gaged double V-notched specimen: (*left*) front view of the fixture with attached lateral LVDTs and reference blocks; (*right*) rear view of the fixture with attached axial LVDT and reference block

7.2 Experimental Setup and Testing Procedure

7.2.1 Fixture and Measurements

The V-notched rail test requires special grips, as prescribed by ASTM D7078; we obtained the adjustable combined loading shear (CLS) fixture from Wyoming Test Fixtures as shown in Fig. 7.1. The CLS fixture has approximately 50.8 mm × 50.8 mm of gripping area on each half of the fixture, adjusts to different specimen thicknesses, and provides a 25.4-mm central gap between the grip halves where the double V-notch feature of the specimens undergoes shear-dominated deformation. The 17-4 PH stainless steel fixture has sixteen 5/8-18 UNC stainless bolts (two tall, two wide on each face) to secure the specimens. To measure the local displacement of the two fixture halves relative to each other, we installed an axial linear variable differential transformer (LVDT) (Measurement Specialties MHR open core LVDT, ±12.7-mm range) across the vertical gap of the fixture halves in line with the actuator, as shown in Fig. 7.1. To measure relative rotation of the fixture about the z-axis (specimen thickness), we attached two lateral LVDTs (Macro Sensors BBP 315-100 LVDT, ±0.254-mm range) to the front face of the fixture, above and below the gripping bolts, perpendicular to the actuator. The fixture was installed in a 100-kN MTS, two-post, servo-hydraulic, single actuator load frame with a 100-kN MTS load cell.

7.2.2 Preliminary Testing and Test Procedure Modifications

When this fixture is used for composite materials, strains measured by strain gages, applied to the center of the specimen, and the load history are sufficient to characterize the shear stress-strain behavior. In the case of a ductile metal specimen, the strains are often larger than the range of conventional strain gages, requiring a computionalist to model the test to extract shear properties of the metal. Without local displacement/strain measurement on the surface of the specimens, the computationalist having only have global fixture measurements would need to compensate for fixture compliance and any other specimen behavior. Therefore, we needed to perform compliance testing and measure fixture motion during the tests. Prior to compliance testing, we performed preliminary tests to determine the load range and fixture behavior on a few specimens to develop an appropriate testing approach.

Preliminary testing of specimens with the standard 90° V-notches from the 3.124-mm thick Ti-6Al-4V plate material had several issues that had to resolved to get reasonable data. The large loads required to fail the specimens led to significant rotations about the z-axis of the fixture halves (through-thickness axis) as measured by the lateral LVDTs (on the order of 0.5 mm relative displacement), causing significant slip of the specimens in the grips and considerable galling of the specimen gripping surfaces. The large torque on the gripping bolts required to secure the specimen, approximately 67.8 N-m (50 ft lb), would sometimes cause the bottom half of the fixture to rotate with the actuator, bending the specimen about the actuator axis (y-axis). Alignment of the fixture halves and the specimen during gripping was challenging to maintain. These issues are presumably not present for composite materials, which would require much lower loads for failure and lower torque on the bolts for specimen installation.

Modification of the specimen geometry and development of careful specimen installation procedures resolved these issues with the setup. With the 25.4-mm V-notch width set by the 25.4-mm gap in the test fixture, we cut deeper V-notches at 60° into the specimens instead of the standard 90°, as shown in Fig. 7.2, in order to reduce the area between the notches, thus reducing the maximum load of the test. The lower maximum load in turn significantly reduced the fixture rotation about the z-axis; all tests with these modified tests led to negligible relative displacements of the lateral LVDTs. The specimen installation procedure utilized stacked rosette strain gages from Micro-Measurements (Model CA2-06-125WW-350), with gage length 3.18 mm (0.125 in.) and 3 % strain measurement range, attached to the front and back of each specimen. The strain gages were located in the center of the specimen and to the side of center closer to the left (bottom) fixture half. The installation procedure required that the strain not rise more than 100-microstrain on any gage as the gripping bolts were tightened, providing confidence that the specimen was not bent or twisted during installation. Also, the bolts on the left (bottom) fixture half were gradually tightened, such that the eight bolts were all tightened in 13.6 N-m (10 ft lb), increments up to 67.8 N-m (50 ft lb), as opposed to each bolt individually torqued to 67.8 N-m (50 ft lb). The specimen underwent precycling of the actuator to loads of ±4.448 kN (1000 lb) in order to assess how the strain gages and specimen are behaving in the elastic regime. If the strain gages on the two faces of the specimen indicated unwanted bending in the specimen, then the specimen was removed and then reinstalled. These changes to the experimental procedure and specimen allowed for reasonable test data, but the specimens did continue to slip slightly throughout the test. We characterized this specimen slip behavior so that the slip could be accounted for in the data, as will be described.

Eight V-notched Ti-6Al-4V specimens were tested: four each with the dominant shear loading direction aligned with the rolling direction (denoted VA) and perpendicular to the rolling direction (denoted VP). Specimens VA1, VA2, VP2, and VP6 were tested at an actuator rate of 0.0254 mm/s, and specimens VA3, VA4, VP3, and VP4 were tested at 25.4 mm/s. The testing protocol included strain gage installation on each specimen, the above specimen installation in the fixture, the precycles of ±4.448 kN (1000 lb), and then the monotonic pull to failure at the prescribed actuator displacement rate.

7.2.3 Compliance and Specimen Slip Characterization

The lateral stiffness of each fixture half was characterized. A 2.224-kN (500-lb) load cell was attached to a manual pull rod and a clevis. The clevis was attached to a rod end bearing gripped in one half of the fixture, as shown in Fig. 7.3. The manual pull rod allowed the operator to apply a lateral load to each half of the fixture to measure the lateral displacements of fixture. As seen in Fig. 7.3, the lateral stiffness of the upper fixture half (attached to the stationary part of the load frame) is greater than the lower fixture half (attached to the actuator), and overall the fixture is relatively soft laterally. The lower fixture half has more rotation for a given load than the upper fixture half as seen in the difference between the lateral LVDTs. This data shows how much lateral load is required to displace the fixture halves laterally; in turn, the lateral displacement measured during the V-notched specimen tests can be related to the lateral stiffness measured here. The V-notched specimen tests had negligible lateral displacements, implying that the specimens did not exhort significant lateral loads on the fixture.

The fixture had significant axial compliance that required characterization so that computationalists could utilize the load-axial displacement data for material characterization. The fixture compliance characterization tests were completed using two different plates: a generic alloy steel with dimensions 55.766-mm × 129.29-mm × 9.385-mm (2.1955-in × 5.090-in × 0.36975-in) and the Ti-6Al-4V with dimensions 55.88-mm × 127.36-mm × 3.061-mm (2.200-in × 5.014-in × 0.1205-in). Each plate had four sets of three strain gages similarly located as on the V-notched specimens. The plates were installed in the same manner as described for the V-notched specimens and then were subjected to a set of cyclic loading to various load levels at 0.1-Hz rate, to a maximum of 35.6 kN (8000 lb), which is larger than the maximum load of the V-notched

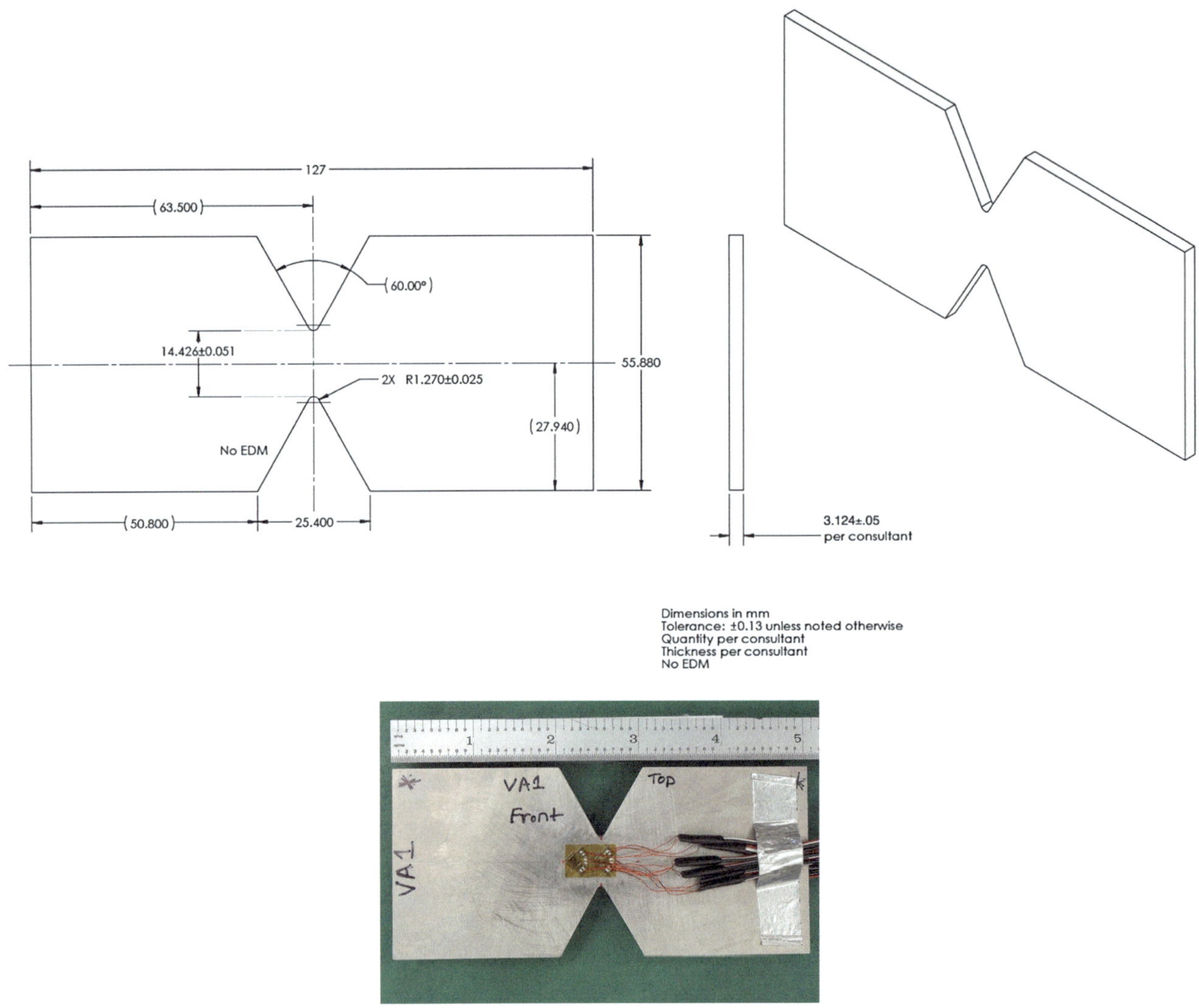

Fig. 7.2 Modified specimen geometry for the 3.124-mm thick Ti-6Al-4V plate with 60° deep V-notches: (*top*) specimen fabrication drawing; (*bottom*) specimen VA1 with attached strain gage rosettes in the center of the gage section, shown with a ruler in inches

Ti-6Al-4V specimens. Then the plates then underwent the same set of cyclic loads without removal of the plates. The load versus axial LVDT displacement data is given in Fig. 7.4. The axial compliance tests revealed two main effects: nonlinear compliance of the test fixture and specimen slip in the grips. It is possible that these two factors can be taken into account by applying a first-order linear correction to load vs. axial LVDT data so that the modulus matches the strain-gage data. We have also made an attempt to characterize these factors independently through the compliance tests, offering another more detailed possibility to account for specimen slip and fixture compliance through computational modeling of the experiment.

With regards to specimen slip, for the first set of cyclic loads in each plate, the amount of displacement for the first monotonic loading from zero to the upper bound of each load level is larger than each subsequent cycle segment for that load level; also the remaining cycle segments return to the same displacement. This implies that the specimen slip predominately occurred upon the initial positive loading through a load regime; in other words, the cyclic loading exhibits considerable slip on the first portion of increasing load with little accumulation of additional slip in subsequent cycles between the same two load end-points. When the Ti-6Al-4V plate underwent the a second set of cyclic loading without the plate being removed, the plate did not accumulate any more slip, as shown in Fig. 7.4 bottom right, so the behavior here is just due to fixture compliance. The slip behavior of the steel plate was less pronounced than that of the Ti-6Al-4V plate because the steel plate grip surfaces were visibly rough as compared to those of the smooth Ti-6Al-4V plate. The slip behavior of the Ti-6Al-4V compliance test plate is assumed to be the same as that of the V-notched specimens; thus, these Ti-6Al-4V plate compliance tests allow for empirical characterization of the slip behavior.

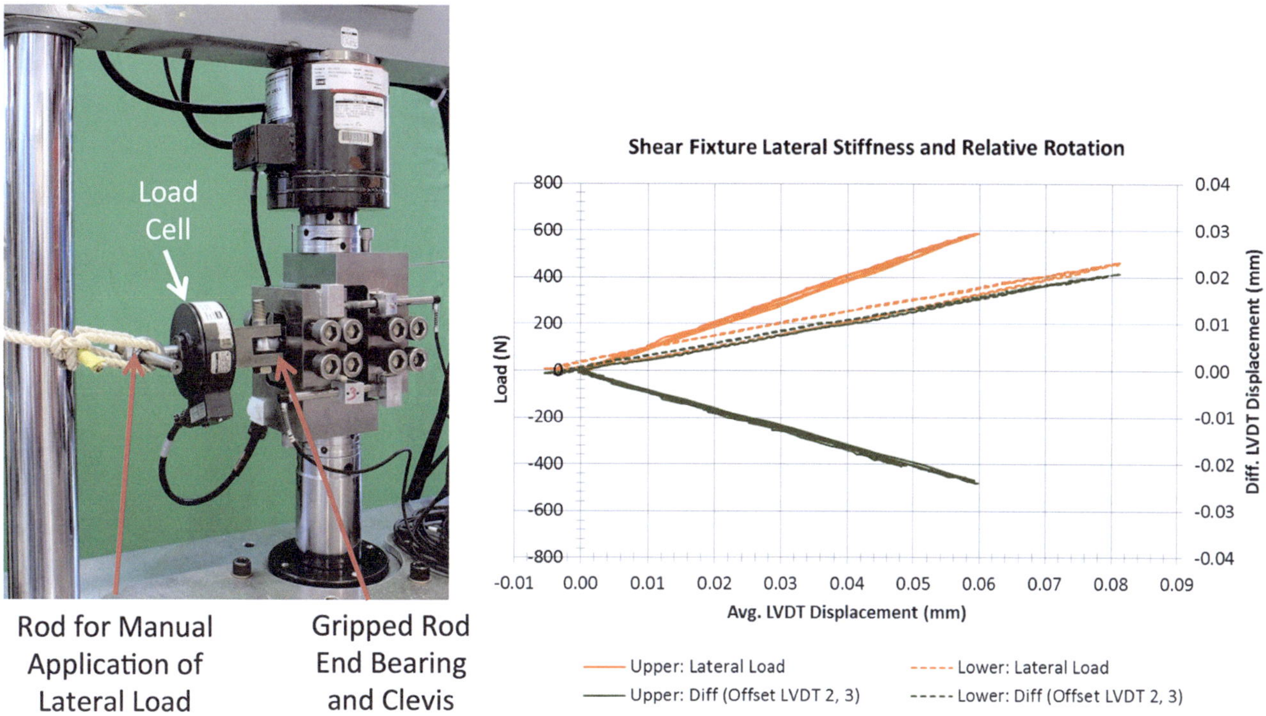

Fig. 7.3 Fixture lateral compliance test: (*left*) test setup; (*right*) load and relative displacement difference between the lateral LVDTs versus the average lateral LVDT displacement

This empirical characterization can be applied to the V-notched specimen tests to remove the effect of slip in the axial LVDT data, assuming the same slip behavior is present for the two loading rates of the V-notched specimen tests. As previously mentioned, the V-notched specimens underwent precycles of ±4.448 kN (1000 lb) to assess the strain gage and specimen behavior after installation prior to the monotonic pull to failure; the axial LVDT was zeroed after the precycles, so slip accumulated between 0 and 4.448 kN (1000 lb) was assumed to be removed from the monotonic pull data. Therefore, to remove the accumulated slip from the axial LVDT data for the monotonic pull to failure portion of the V-notched specimen data, an empirical formula based on the slip seen in the Ti-6Al-4V plate compliance test was applied to the V-notched monotonic pull data *above* the 4.448-kN (1000-lb) precycle level. The empirical formula for slip in the increasing load portion of the Ti-6Al-4V plate compliance test was derived in the following manner. The slip accumulated on the initial monotonic pull between two load levels (the maximum load of the prior precycle and the new maximum load of the current precycle), as seen by the axial LVDT, was assumed to be the difference in the axial LVDT values at zero load for the first complete cycle of the current precycle set. This was taken at zero load because there is no fixture compliance at zero load, and thus the difference in the axial LVDT measurement after one cycle was assumed to be due to specimen slip on the increasing load segment. For example, to calculate the slip between 4.448 and 8.896 kN (1000 and 2000 lb), one would subtract the axial LVDT values at the end and start of the first cycle of the 0–8.896-kN (0–1000-lb) precycle set. In other words, one assumes the slip between 0 and 4.448 kN (1000 lb) has already occurred in the prior 0 and 4.448-kN (1000 lb) precycle set, so comparing the end values of the Axial LVDT 1 of the first cycle of the 0–8.896-kN (0–2000 lb) precycle set should provide the slip between 4.448 and 8.896 kN (1000 and 2000 lb). This exercise was repeated for each precycle set in increments of 4.448 kN (1000 lb) up to 35.586 kN (8000 lb), and then a power-law curve was fit to the accumulated-slip-in-Axial-LVDT-1 vs. Load-above-the-precycle-set curve, shown in Fig. 7.5, for two separate Ti-6Al-4V plate compliance tests. [Note: The small amount of slip accumulated in subsequent cyclic loadings is ignored in this empirical relationship because the V-notched specimen tests have only a monotonic axial displacement, and thus extra slip in the compliance tests would not be present in the V-notched specimen tests. There is also an implicit assumption that no slip occurs for decreasing load in the monotonic pull segment in the V-notched specimen tests.] Since the first compliance test had unblemished surfaces, we used the curve fit from the first compliance test for slip removal:

$$(Slip) = 8.528 * 10^{-4} * (Load - 4.448)^{1.435} \text{ mm} \tag{7.1}$$

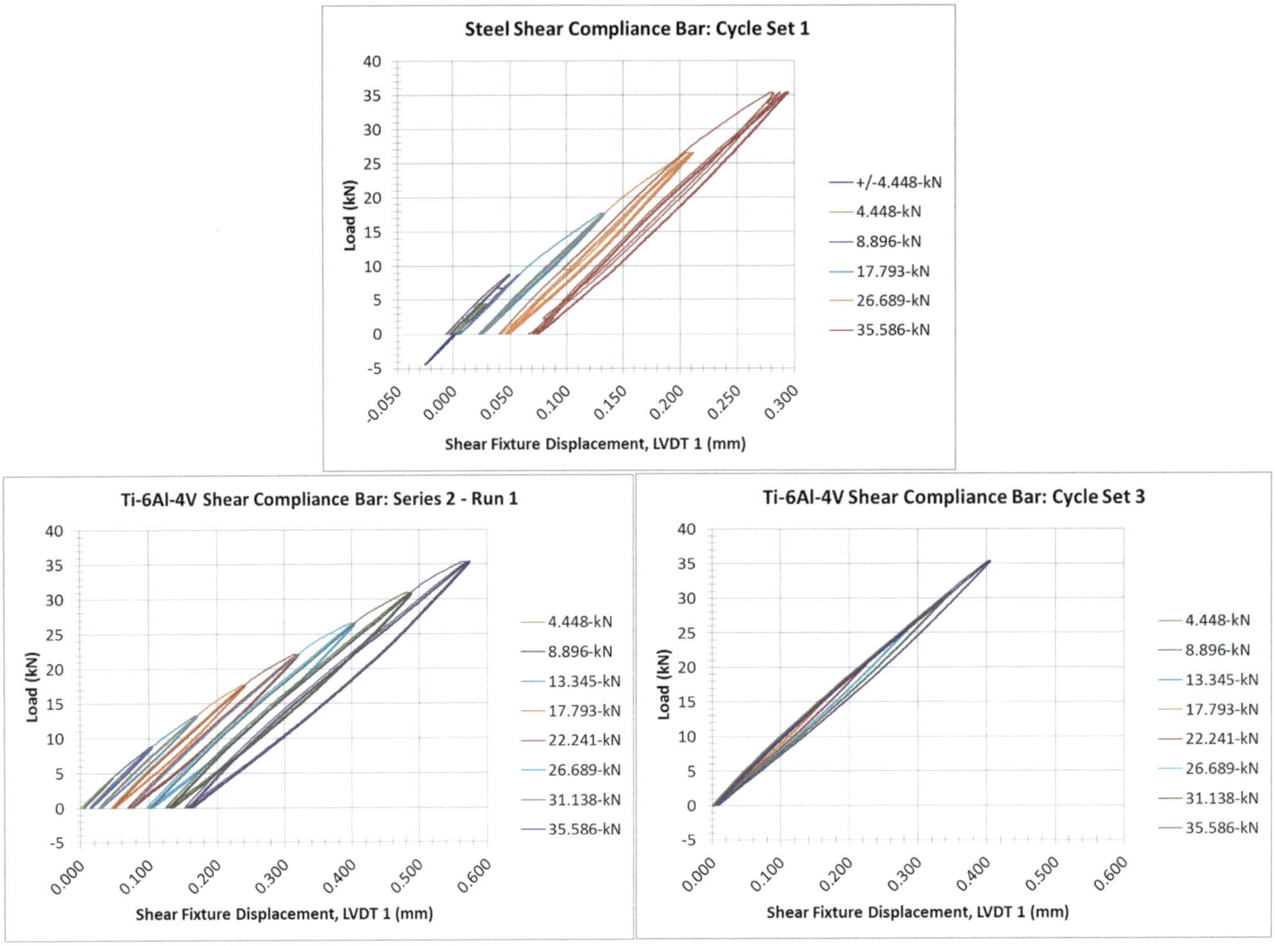

Fig. 7.4 Compliance behavior of the CLS fixture with steel and Ti-6Al-4V plates: (*top*) compliance behavior with a steel plate, including plate grip slip; (*bottom left*) compliance behavior with a Ti-6Al-4V plate for the first set of cyclic loads, with plate grip slip; (*bottom right*) compliance behavior with a Ti-6Al-4V plate for the third set of cyclic loads, without significant plate grip slip

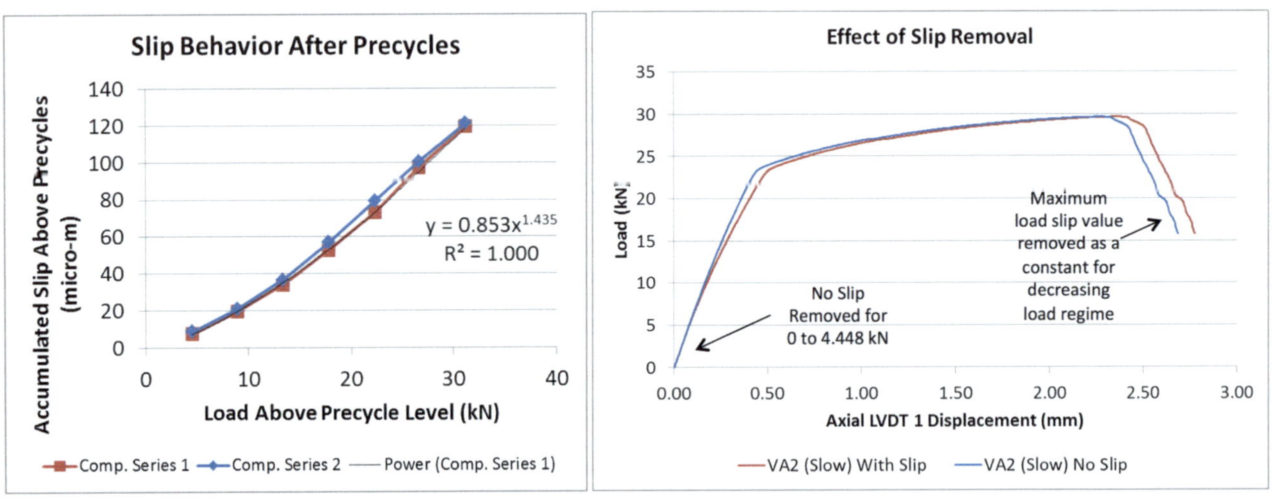

Fig. 7.5 Slip in the Ti-6Al-4V specimens: (*left*) empirical fit to characterize specimen slip; (*right*) load vs. axial LDVT displacement with and without slip removed

Once maximum load is reached, it is assumed that the slip does not decrease with the load drop, so the maximum slip value is subtracted from axial LVDT reading for decreasing loads after maximum loads. For example, in specimen VA2 shown in Fig. 7.5, the maximum load was 29.658 kN (6668 lb) with an associated slip of 0.0876 mm, and that amount of slip was removed from the axial LVDT for all displacements after maximum load had been achieved.

The compliance behavior of the fixture, uncorrupted by specimen slip, is apparent last cycle between 0 and 35.6 kN (8000 lb) and in the cycle sets after the first cycle set (see Fig. 7.4), as previously mentioned. The behavior in these cyclic loads included both the elastic deformation of the plates and fixture compliance. An estimate of these relative magnitudes of the fixture compliance and elastic deformation of the compliance plate is as follows. The fixture displacement is approximately 0.30 mm at 26.7 kN (6000 lb) for the Ti-6Al-4V plate. Using a simple calculation for simple shear in the elastic regime for the Ti-6Al-4V with a 44-GPa shear modulus, the relative vertical displacement, measured by the axial LVDT, of the 55.88-mm × 25.4-mm × 3.061-mm (2.200-in × 1.0-in × 0.1205-in) section in the grips, would be approximately 0.09 mm at 26.7 kN (6000 lb). This implies that the fixture compliance is about 0.21 mm at 26.7 kN (6000 lb), which is on the order of 10–20 % of the fixture displacement at maximum load of the VP V-notched specimens. A computationalist would be advised to account for this fixture compliance when modeling this experiment because the compliance is not negligible.

7.3 Results and Discussion

For all eight specimens, the calculated shear stress versus shear strain data for the portion of the test where the central strain gages were attached to the specimens are provided in Fig. 7.6. The elastic modulus from the shear stress-strain data is 44 GPa, which is consistent with literature values [11]; thus, this experimental approach does appropriately apply a shear-dominated deformation in the center of the specimens, allowing for characterization of shear-dominated behavior of Ti-6Al-4V using strain gages in the low-strain regime. The shear stress-strain data shows that Ti-6Al-4V is rate-sensitive, with higher a yield stress at an actuator rate of 25.4 mm/s versus that at quasi-static rates. The low-strain regime for the two rolling directions does not show significant anisotropic yield behavior in shear at both displacement rates.

The data from the strain gages gives confidence that the experimental technique is reasonable for metals, but is not sufficient to characterize the large shear deformation and failure behavior; the load-displacement data, despite the issues with specimen slip and fixture compliance, do provide the potential for this characterization in conjunction with computational modeling. The load-displacement data, still including the specimen slip and fixture compliance, are given in Fig. 7.7, showing anisotropic plastic behavior between the two rolling directions and strain-rate dependence of the plastic behavior, with different maximum loads, displacement at maximum load, and displacement to failure. For both rolling directions, the faster rate specimens reached maximum load earlier and failed earlier than the quasi-static rate specimens.

Figure 7.8 is a compilation of failed specimens viewed from the front of the specimen and along the through-thickness of the fracture surfaces. In all of the specimens, the crack did not propagate from the roots of the V-notches, but rather at the side of the roots at an angle to the opposite side. The crack paths of the VA specimens, with their rolling direction aligned with

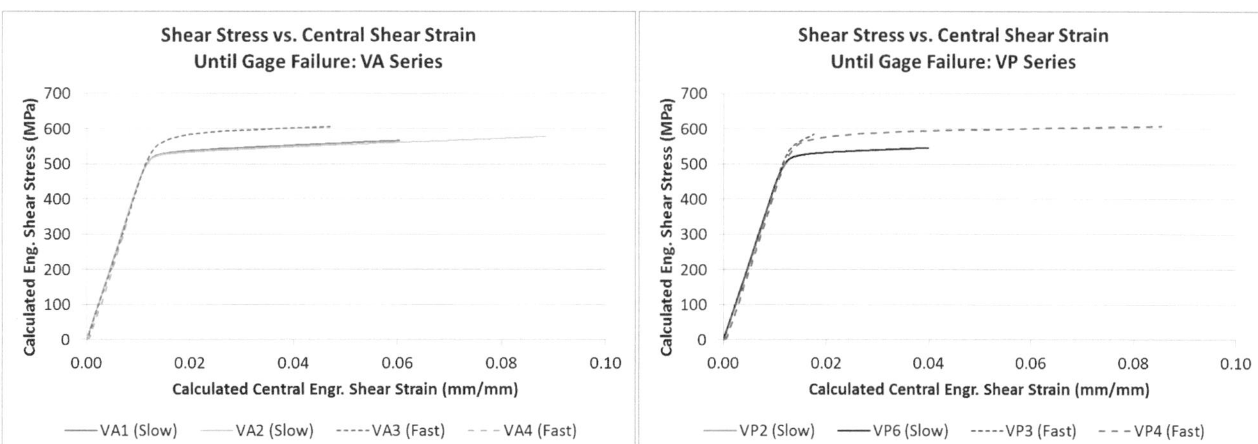

Fig. 7.6 Calculated shear stress vs. shear strain for the V-notched Ti-6Al-4V specimens: (*left*) VA series specimens; (*right*) VP series specimens [*Note*: the end of each line indicates when the strain gages delaminated from the specimen]

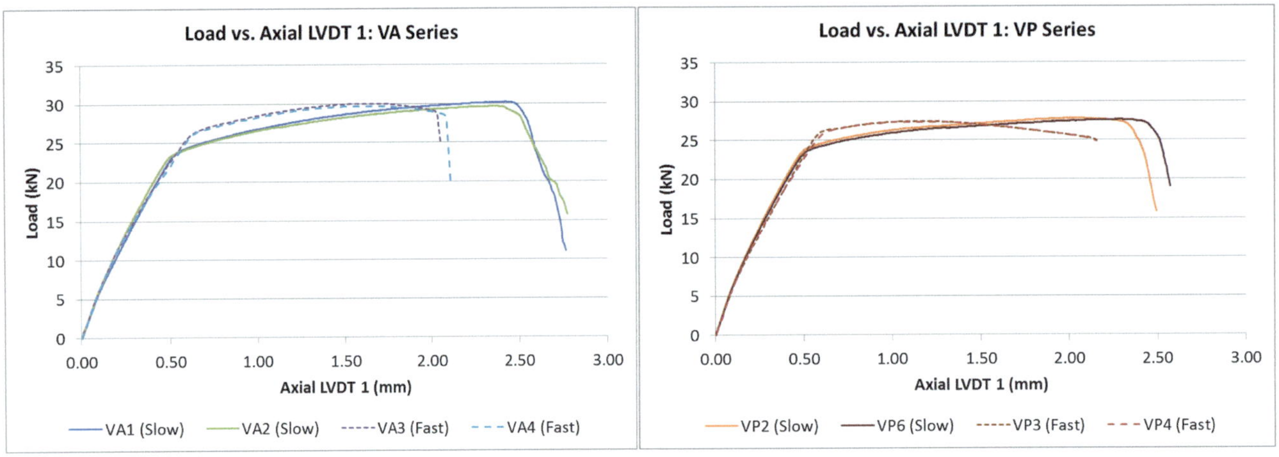

Fig. 7.7 Load vs. axial LVDT displacement for the V-notched Ti-6Al-4V specimens: (*left*) VA series specimens; (*right*) VP series specimens [*Note*: this data includes the fixture compliance and specimen slip, hence the elastic regime is not linear like the shear stress vs. shear strain data]

the shear directions, were rather jagged at the quasi-static loading rates; the other crack paths were straighter. We observed negligible lateral displacement of the fixture during the VA quasi-static tests, so this not due to rotation of the grips , but rather to some phenomenon that requires further investigation. The width of the shear-dominated deformation region in the center of the specimens was larger for the quasi-static rates for both rolling directions.

There are a few areas of improvement for this technique for shear-dominated loading. First, one potential method for extracting the calculated shear stress-stain behavior for the entire deformation is Digital Image Correlation (DIC). The relatively large visible area of interest of the V-notched specimens in this fixture would make DIC a straightforward method for local displacement/strain measurements, providing flexibility in selection of the virtual strain gage length and location. Second, the gripping mechanism could be greatly improved beyond the current bolt-fastened grips. Hydraulic gripping with a constant pressure would provide an even gripping pressure over the entire gripping area; the hydraulic pressure could be set to a greater value than that achievable with the bolt-fastening approach. Also, the grip faces could be surface-treated to have a rougher gripping surface to reduce slip. An improved gripping mechanism would increase testing throughput and enhance repeatability. Third, the specimen could be modified to have (1) a thinner thickness as to lower the overall load required, but this may lead to issues with specimen bending during installation, (2) other angles for the V-notches to alter the load required, and (3) a roughened grip surface to reduce specimen slip.

7.4 Conclusion

We evaluated the use of the V-notched rail test for characterization of shear-dominated deformation and failure of Ti-6Al-4V. Using a modified V notched specimen and careful procedures for specimen installation, we determined that the shear stress-strain data from strain gages gave credible elastic shear modulus measurements, providing confidence that the test was performing as expected, at least in the low-strain regime. Large fixture compliance and specimen slip obscured direct use of the load-displacement data for engineering calculations. Two approaches for compensation include (1) a first-order linear correction to the data such that the modulus from the corrected load-displacement data matches the strain-gage data, and (2) direct computational modeling of the experiment. The latter requires careful experimental characterization of the specimen slip and fixture compliance, which we completed through cyclic loading of stiff solid plates made of Ti-6Al-4V and alloy steel over the range of loads for the V-notched specimen tests. Modifications to the gripping mechanism and to the test specimen would help to reduce the specimen slip and reduce the testing loads in future tests. DIC could greatly increase the efficacy of the V-notched rail test: strain measurements over the gage section for the entire test could eliminate the need for fixture compliance compensation since we would have the shear stress-strain relationship for the entire test from the measured load, DIC data and test geometry. With these improvements, the V-notched rail test appears promising as a test methodology for characterization of shear-dominated deformation and failure of ductile metals. As is, the V-notched rail test requires considerable computational effort to interpret the test results at large deformations.

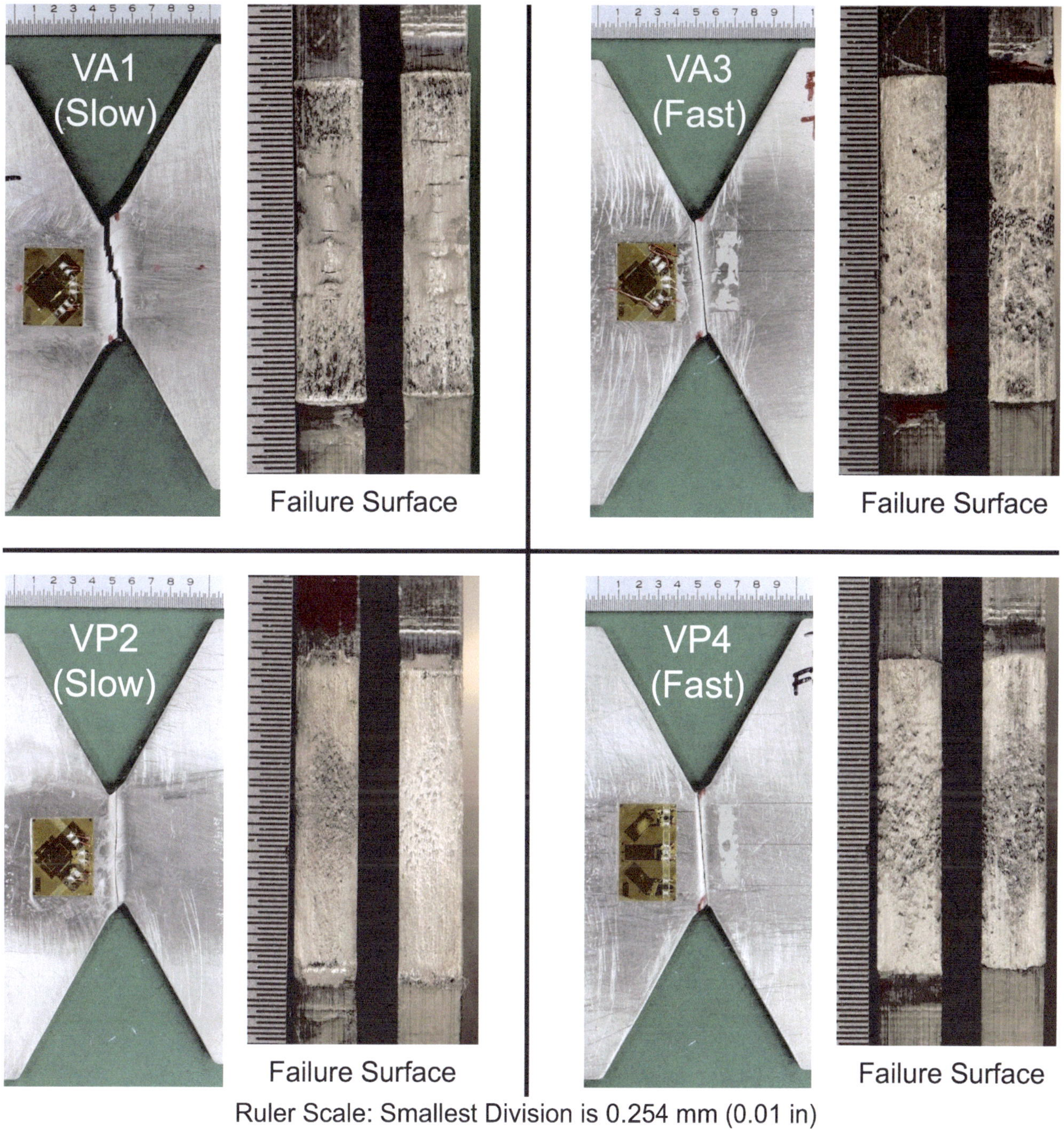

Fig. 7.8 Post-test images of four V-notched specimens, including the front face views and the failure surfaces

Acknowledgment Sandia is a multiprogram laboratory operated by Sandia Corporation, a Lockheed Martin Company, for the United States Department of Energy under contract DE-AC04-94AL85000.

References

1. Nahshon, K., Hutchinson, J.W.: Modification of the Gurson model for shear failure. Eur. J. Mech. A Solids **27**(1), 1–17 (2008)
2. Xue, Z., Pontin, M.G., Zok, F.W., Hutchinson, J.W.: Calibration procedures for a computational model of ductile fracture. Eng. Fract. Mech. **77**(3), 492–509 (2010)

3. Iosipescu, N.: New accurate procedure for single shear testing of metals. J. Mater. **2**(3), 537 (1967)
4. Adams, D.F., Walrath, D.E.: Further development of the Iosipescu shear test method. Eng. Fract. Mech. **27**(2), 113–119 (1987)
5. Dunand, M., Mohr, D.: Optimized butterfly specimen for the fracture testing of sheet materials under combined normal and shear loading. Eng. Fract. Mech. **78**(17), 2919–2934 (2011)
6. Ghahremaninezhad, A., Ravi-Chandar, K.: Ductile failure behavior of polycrystalline Al 6061-T6 under shear dominant loading. Int. J. Fract. **180**(1), 23–28 (2013)
7. Xue, Z., Faleskog, J., Hutchinson, J.W.: Tension–torsion fracture experiments—part II: simulations with the extended Gurson model and a ductile fracture criterion based on plastic strain. Int. J. Solids Struct. **50**(25–26), 4258–4269 (2013)
8. Adams, D.O., Moriarty, J.M., Gallegos, A.M., Adams, D.F.: Development and evaluation of thed rail shear test for composite laminates. Federal Aviation Administration Report DOT/FAA/AR-03/63, FAA Office of Aviation Research, Washington, DC (2003)
9. Adams, D.O., Moriarty, J.M., Gallegos, A.M., Adams, D.F.: The V-notched rail shear test. J. Compos. Mater. **41**(3), 281–297 (2007)
10. Boyce, B.L., et al.: The sandia fracture challenge: blind round robin predictions of ductile tearing. Int. J. Fract. **186**(1–2), 5–68 (2014)
11. Fukuhara, M., Sanpei, A.: Elastic-moduli and internal frictions of Inconel 718 and Ti-6Al-4V as a function of temperature. J. Mater. Sci. Lett. **12**(14), 1122–1124 (1993)

Chapter 8
A Statistical/Computational/Experimental Approach to Study the Microstructural Morphology of Damage

J.P.M. Hoefnagels, C. Du, T.W.J. de Geus, R.H.J. Peerlings, and M.G.D. Geers

Abstract The fractural behavior of multi-phase materials is not well understood. Therefore, a statistic study of micro-failures is conducted to deepen our insights on the failure mechanisms. We systematically studied the influence of the morphology of dual phase (DP) steel on the fracture behavior at the onset in two ways: (i) in a numerical setting by statistically averaging over the micro-structural arrangements around the damage sites in no less than 400 randomly-generated idealized microstructural models loaded in pure shear; and (ii) in an experimental setting by statistically averaging, similar to the numerical simulations, over the damage sites found in a large collection of large field-of-view SEM images of DP steel deformed in uniaxial tension, where deliberately-overexposed backscattered electron images sharply mark the damage location, while simultaneously-recorded secondary electron images are used to identify the material phases. The numerical and experimental analyses were validated and tested for accuracy. Application of both techniques to DP showed a similar single topological feature to be most sensitive to damage: a small region of soft matrix material with hard inclusion particles on opposing sides. These results are representative for and yield insight in damage evolution in a wide variety of multi-phase materials.

Keywords Ductile damage • Dual phase steel • Damage statistics • Damage characterization • Damage modelling

8.1 Introduction

Dual phase (DP) steel is applied widely in the automotive industry because of its mechanical properties: high strength-to-weight ratio, high strength and ductility at the same time, low yield-to-ultimate strength ratio and good formability. However, the fractural behavior of DP steel is not understood completely. Debate over critical failure mechanism still remains. Therefore, a statistic study of micro-failures is conducted to deepen our insights on the failure mechanisms. We systematically study the influence of the morphology of DP steel on the fracture behavior at the onset in two ways: (i) in a numerical setting by averaging over the damage sites in a large collection of randomly-generated idealized microstructural models loaded in pure shear [1] and (ii) in an experimental setting by averaging over the damage sites found in a large collection of large field-of-view SEM images of DP steel deformed in uniaxial tension.

8.2 Numerical Damage Characterization

The numerical model is highly idealized to determine the influence of geometrical arrangement of the hard phase on the fractural behavior independently from other influences. This influence has been studied in a structural fashion, which is not easily achieved experimentally, or with realistic microstructures: 400 microstructures of a two-phase material are randomly generated and numerically deformed under pure shear assuming, for both phases an isotropic elasto-plastic constitutive model but with different initial yield stresses and hardening laws and using periodic boundary conditions, see Fig. 8.1. The onset

J.P.M. Hoefnagels (✉) • C. Du • T.W.J. de Geus • R.H.J. Peerlings • M.G.D. Geers
Department of Mechanical Engineering, Eindhoven University of Technology, P.O. Box 513, Eindhoven 5600, The Netherlands
e-mail: j.p.m.hoefnagels@tue.nl

A.M. Beese et al. (eds.), *Fracture, Fatigue, Failure and Damage Evolution, Volume 8*, Conference Proceedings of the Society for Experimental Mechanics Series, DOI 10.1007/978-3-319-21611-9_8

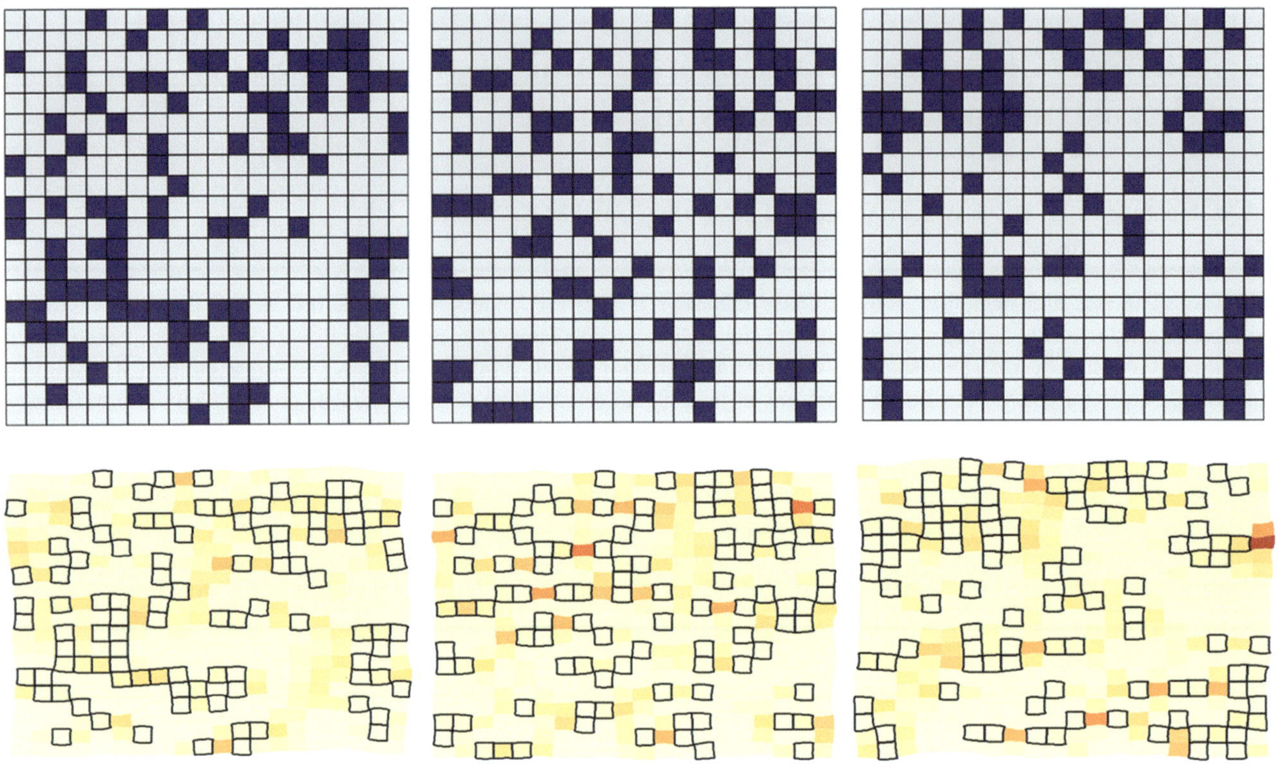

Fig. 8.1 (*Top*) Three examples of the 400 randomly-generated micro-structural two-phase arrangements, which have been numerically deformed under pure shear. (*Bottom*) Maps of the damage indicator after deformation; the damage indicator depends linearly on the effective plastic strain and the non-negative part of the hydrostatic stress

Fig. 8.2 The average indicator function weighted by the damage in each element (of all 400 random cells). The origin is arbitrarily chosen in the center. Notice that the neutral color is chosen to coincide with the volume fraction of hard phase, which is 0.25 for martensite in dual phase steel; *red* and *blue* indicate, respectively, an elevated likelihood of hard and soft phase

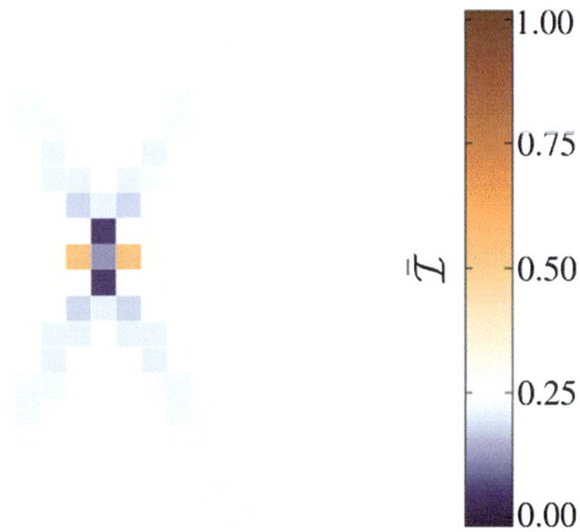

of damage is signaled using a simple indicator, which depends linearly on the effective plastic strain and the non-negative part of the hydrostatic stress. The 400 micro-structural arrangements around the damage sites are statistically averaged, as explained in Fig. 8.2.

8.3 Experimental Damage Characterization

In the experimental work, deformed DP steel DP600 specimens are observed using an intelligent SEM imaging strategy, in which secondary electron (SE) and backscattered electron (BSE) images are taken simultaneously in each scan such that the pixels in SE and BSE images at the same position match each other exactly. While the SE images are used to

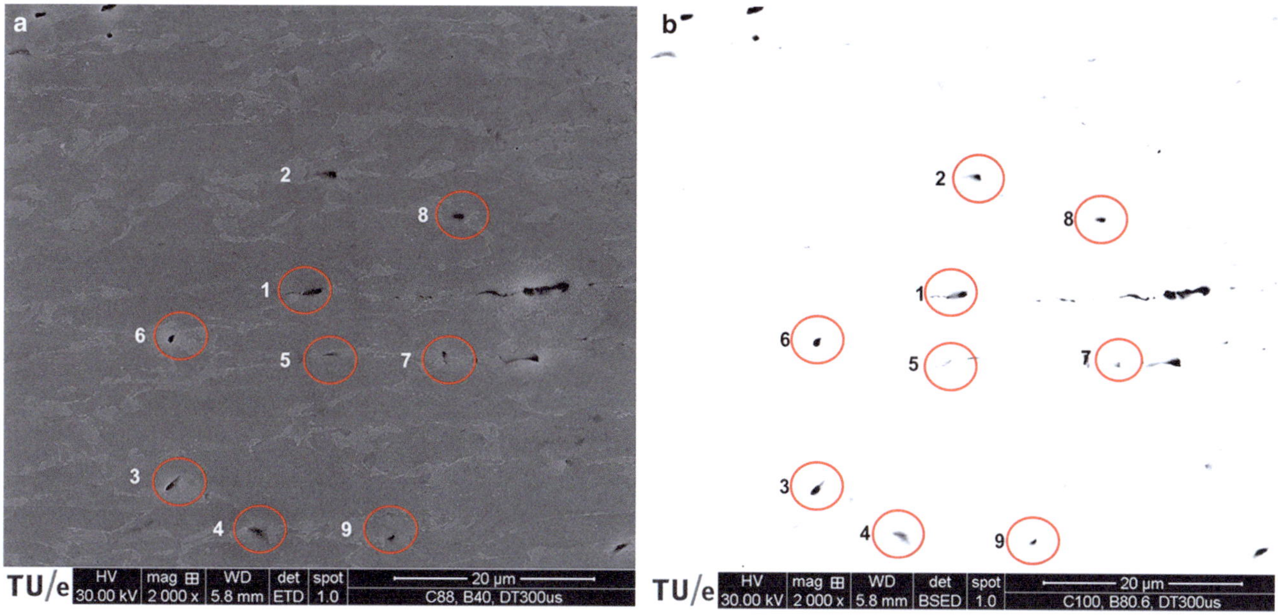

Fig. 8.3 (**a**, **b**) Simultaneously recorded SEM images of the same large field of view, measured in (**a**) secondary electron (SE) mode which shows the ferrite and martensite locations and (**b**) back-scatter electron (BSE) mode which 'only' shows the damage sites

identify the material phases, see Fig. 8.3a, the BSE images are deliberately overexposed in such a way that the topography contrast between ferrite and martensite disappears such that the locations of micro-failures are easily identified, see Fig. 8.3b. A Matlab program automatically processes both BSE images and SE images to identify the failure mechanisms and phases. Particularly, the white edges around the damage sites, which would disrupt the statistical averaging, are automatically identified and turned into the neutral grey value, see Fig. 8.4. Finally, the SE images are used to automatically average the micro-structural arrangements around the damage sites, similar to the numerical simulations, see Fig. 8.5.

8.4 Discussion

A single topological feature was found which is most sensitive to damage, see Figs. 8.2 and 8.5. In this feature, a small region of soft matrix material has hard inclusion particles on opposing sides. A similar observation was made by [2] for dual-phase steel, while for the special case where additional hard phase particles are located on both sides, this feature resembles the interrupted bands frequently encountered in industrial DP steel [3]. The imaging strategy applied in the experimental work shows promising result. Particularly, optimum settings for yielding simultaneously over-exposed BSE images and good-contrast SE images were found (Figs. 8.3 and 8.4), and the identification of damage sites from the SE images was carefully validated based on cross section made by focused ion beam milling (not shown). Moreover, the automatic Matlab program to identify all damage sites from the large field-of-view SEM images and statistically average over the surrounding micro-structural phases was found to operate successfully. A large amount of SEM images with proper field of view have been measured and are currently being analyzed automatically by the program. The resulting experimental image of the statistics of micro-failure (Fig. 8.5) can directly be compared to the numerical results (Fig. 8.2), a study which is ongoing. The similarities and differences between the numerical and experimental statistic damage averaging results will be discussed at the conference.

Finally, it is noted that the experiments have been focused on DP steel so far. However, these results are representative for and yield insight in damage evolution in a wide variety of multi-phase materials.

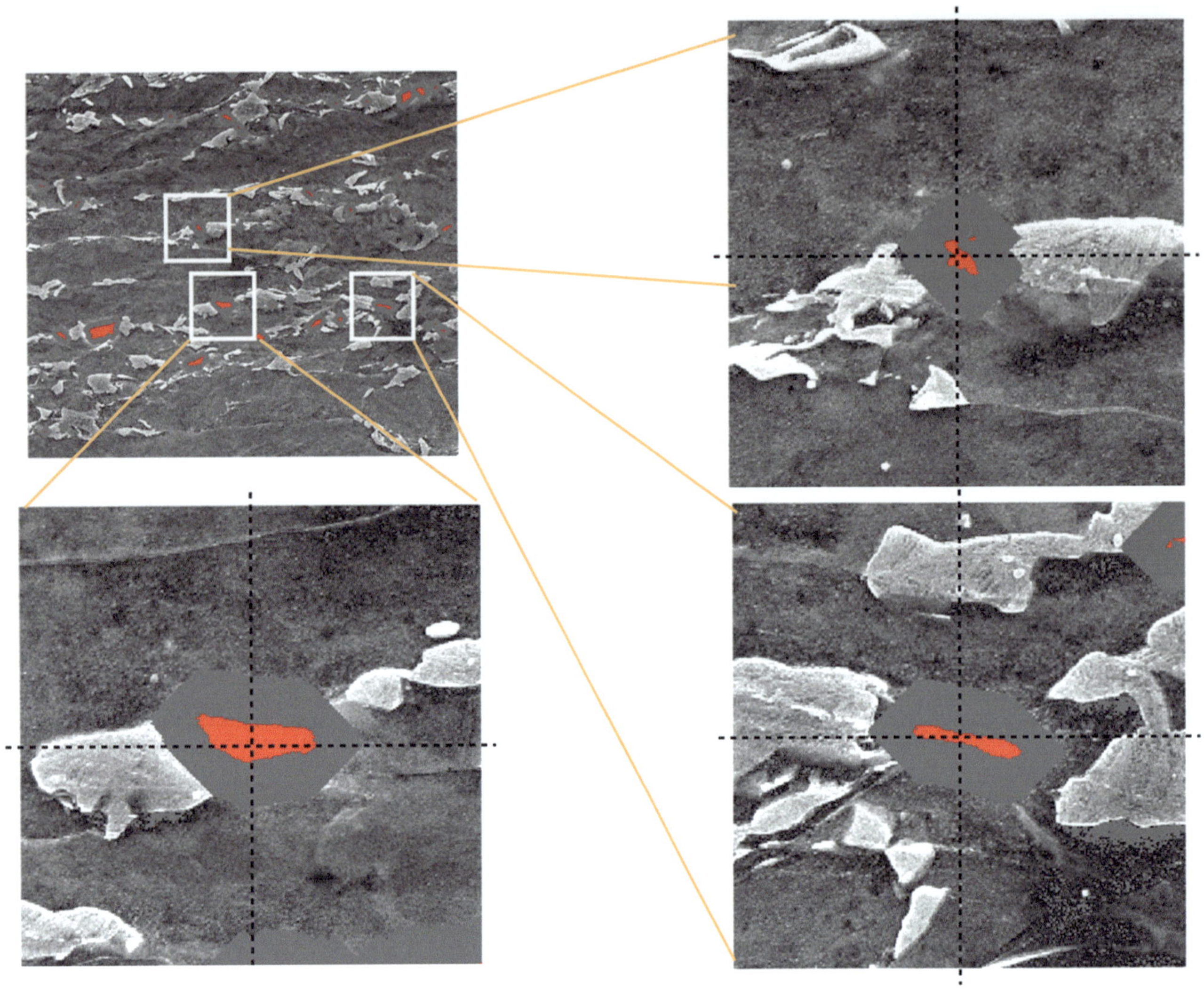

Fig. 8.4 A Matlab program automatically identifies the damage sites, here shown as *red* areas in the SE images. The *white* edges around the damage sites in the SE images, which would disrupt the statistical averaging, are automatically identified and turned into the neutral grey value

8.5 Conclusions

We systematically studied the influence of the morphology of dual phase (DP) steel on the fracture behavior at the onset in two ways: (i) in a numerical setting by statistically averaging over the micro-structural arrangements around the damage sites in 400 randomly-generated idealized microstructural models loaded in pure shear; and (ii) in an experimental setting by statistically averaging over the damage sites found in a large collection of large field-of-view SEM images of DP steel deformed in uniaxial tension. The numerical and experimental procedures and analyses were validated and tested for accuracy. Application of both techniques to DP showed a similar single topological feature to be most sensitive to damage: a small region of soft matrix material with hard inclusion particles on opposing sides. These results are representative for and yield insight in damage evolution in a wide variety of multi-phase materials.

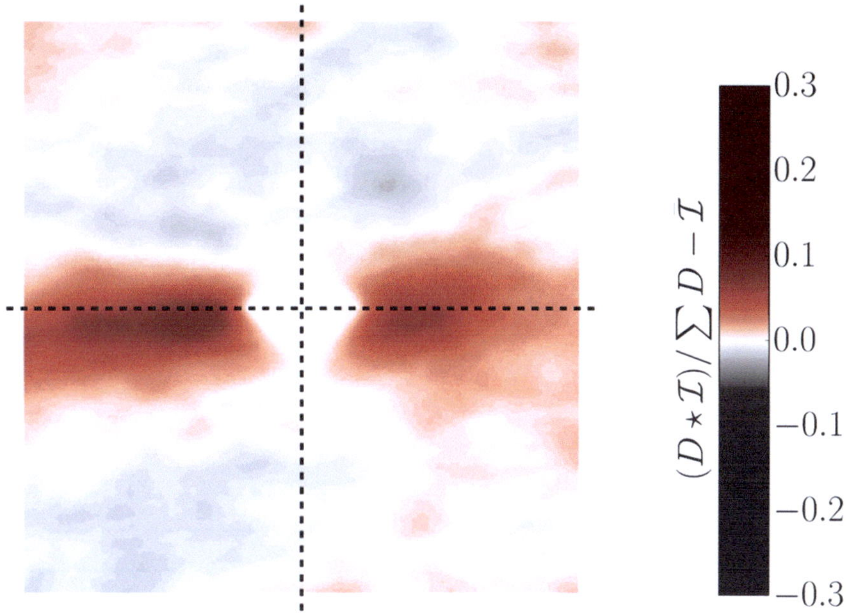

Fig. 8.5 The SE images, corrected for the *white* edges around the damage sites (see Fig. 8.4) are directly averaged, by superposing the grey value distribution around each damage sites. Notice that the neutral color is chosen to coincide with the volume fraction of hard phase, which is 0.25 for martensite in dual phase steel; *red* and *black* indicate, respectively, an elevated likelihood of martensite and ferrite phase

References

1. de Geus, T.W.J., Peerlings, R.H.J., Geers, M.G.D.: Microstructural topology effects on the onset of ductile failure in multi-phase materials: A systematic computational approach. Int. J. Solids Struct. **67–68**, 326–339 (2015)
2. Kadkhodapour, J., et al.: Mechanisms of void formation during tensile testing in dual-phase steel. Acta Mater. **59**, 2575 (2011)
3. Tasan, C.C., Hoefnagels, J.P.M., Geers, M.G.D.: Microstructural banding effects clarified through micrographic digital image correlation. Scr. Mater. **62**, 835–838 (2010)

Chapter 9
Prediction of Ductile Fracture Through Small-Size Notched Tensile Specimens

L. Farbaniec, H. Couque, and G. Dirras

Abstract In this study, a combined experimental and numerical procedure that based on extended analysis of stresses, strains, and damage of the specimen notch region, was proposed to investigate high-strength steel and pure nickel metals. Miniaturized notched tensile specimens with different notch radii were used to generate various levels of triaxial stress, and to evaluate stress-dependent failure. It was shown that the triaxiality plays a major role in the damage evolution demonstrated by decreasing ductility. The experimental investigation was supplemented by scanning electron microscopy observations of fractured surfaces. The deformation mechanisms leading to the failure ware linked with the extensions of the Gurson model for porous ductile metals. The evolution of damage in both materials was compared and discussed.

Keywords Ductility • Notched tensile specimen • Stress triaxiality • Gurson model

9.1 Introduction

Current technology development leads to further miniaturization of engineering components. In many cases, the available quantity of these materials is insufficient to predict mechanical properties, such as fracture toughness, by conventional testing procedures, and in accordance with ASTM standards. This is particular true for ductile materials, and attracts the interest in small-scale mechanical testing techniques.

At the root of the ductile failure mechanisms lay nucleation, growth and coalescence of voids. The nucleation usually initiates through the debonding of particle-matrix interfaces or cracking of particles. Once the process has started, voids grow and coalesce readily with increasing plastic deformation. The first attempt to describe the growth of cylindrical and spherical voids in ductile materials dates back to the 1960s [1, 2] To this day, a significant consideration in a continuum-mechanics framework has attracted the model developed by Gurson [3]. Several improvements and extensions were made to the model and discussed elsewhere [4–11].

This study presents a program of tensile tests of two ductile materials involving small-size notched tensile specimens. The evolution of ductile damage under a wide range of triaxial stress states was investigated. Finite element simulations using the extended Gurson model were performed to support the experimental evaluation of the controlling mechanisms of fracture.

L. Farbaniec (✉)
Université Paris 13, Sorbonne Paris Cité, LSPM-CNRS, 99 avenue Jean Baptiste Clément, Villetaneuse 93430, France

Hopkins Extreme Materials Institute, Johns Hopkins University, Baltimore, MD 21218, USA
e-mail: lfarban1@jhu.edu

H. Couque
Nexter Munitions, route de Guery, Bourges 18000, France

G. Dirras
Université Paris 13, Sorbonne Paris Cité, LSPM-CNRS, 99 avenue Jean Baptiste Clément, Villetaneuse 93430, France

A.M. Beese et al. (eds.), *Fracture, Fatigue, Failure and Damage Evolution, Volume 8*, Conference Proceedings of the Society for Experimental Mechanics Series, DOI 10.1007/978-3-319-21611-9_9

Table 9.1 Chemical composition of the materials

	Elements [wt % max.]									
Nickel	Co	Cu	Zn	Fe	P	C	S	Pb		
	0.0003	0.0010	0.0005	0.003	0.0002	0.005	0.0005	0.0001		
Steel	C	Si	Mn	Ni	Cr	Mo	Cu	S	P	Al
	0.41	0.25	0.94	0.26	0.46	0.18	0.23	0.03	0.02	0.04

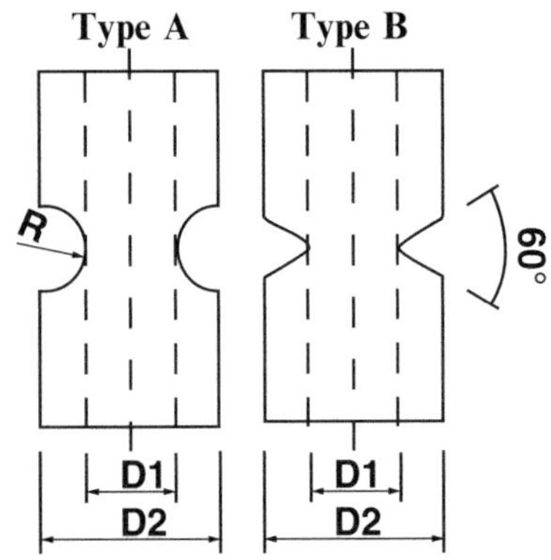

ID	Type	D1	D2	R	Config.I	Config.II
Smooth	Smooth	4	-	-	×	×
D4R0.3	B	4	8	0.3	×	
D4R1.5*	A	4	8	1.5*	×	×
D4R4.0	A	4	8	4	×	
D5R1.5*	A	5	8	1.5*		×
D6R1.5*	A	6	8	1.5*		×

Fig. 9.1 Nomenclature and dimensions of the samples (∗ specimens made of steel have a radius of 1.25 mm)

9.2 Experimental Setup

The study was focused on two ductile materials with different characteristics: nickel and steel. The nominal composition of these materials is listed in Table 9.1. Small tensile specimens were cut from stock materials along the same direction to minimize the effect of the loading direction on the mechanical behavior. Figure 9.1 shows the nomenclature and nominal dimensions of the specimens. The geometry of these specimens was selected in such a way to obtain a wide range of stress triaxiality states. In the first case, all specimens had the same diameter *D1* of minimum cross section and the same outer diameter *D2*; the radius of curvature *R* of the notch root varied from 0.3 to 4 mm. The description '*Configuration I*' was adopted for this case. In the second experimental procedure, referred as '*Configuration II*', the radius *R* and the outer diameter *D2* were constant and the diameter *D1* varied from 4 to 6 mm. The experimental study was supplemented by testing plain specimens with gage diameter of 4 mm. A total of three tests for each type of geometry were conducted. The tests were carried out on a Wolpert-Werke Testatron machine with a capacity of 50 kN in a speed control mode. The longitudinal deformation of the specimen was measured using a strain gage extensometer. The tests were executed at room temperature and strain rate regime of 1.5×10^{-3} s^{-1}. Afterwards, fracture surfaces were investigated using a scanning electron microscope (SEM). The numerical calibration procedure and parameters characterizing the constitutive model are described in [12, 13].

9.3 Results and Discussion

The true stress–strain curves of both investigated materials and their specimen configurations are presented in Fig. 9.2. The mechanical response of the specimen was sensitive to the presence of the notch. In general, the decrease in the notch radius (Configuration I) caused increase in the maximum stress and decrease in the failure strain. In the case of the Configuration II, the maximum stress remained constant, and the increasing diameter of minimum cross-section postponed the failure.

Fracture surfaces of both materials examined by the SEM technique are shown in Fig. 9.3. The characteristic 'cup and cone' surface was typical for all specimens. The inner circles of the specimens showed characteristic Mode I failure with the fracture surface normal to the loading direction. The outer circle appeared as a shear failure, in which the surface was separated at an angle of ±45° with respect to the loading direction. From a micro-mechanic perspective, the crack extension and subsequent failure was governed by initiation, growth, and coalescence of voids. This characteristic fracture surface populated by dimples was found in 'cup' regions of the specimens. In both materials, the voids nucleated from the second-phase particles either due to particle cracking (subfig. (1) in Fig. 9.3b) or particle-matrix debonding (subfig. (2) in Fig. 9.3a).

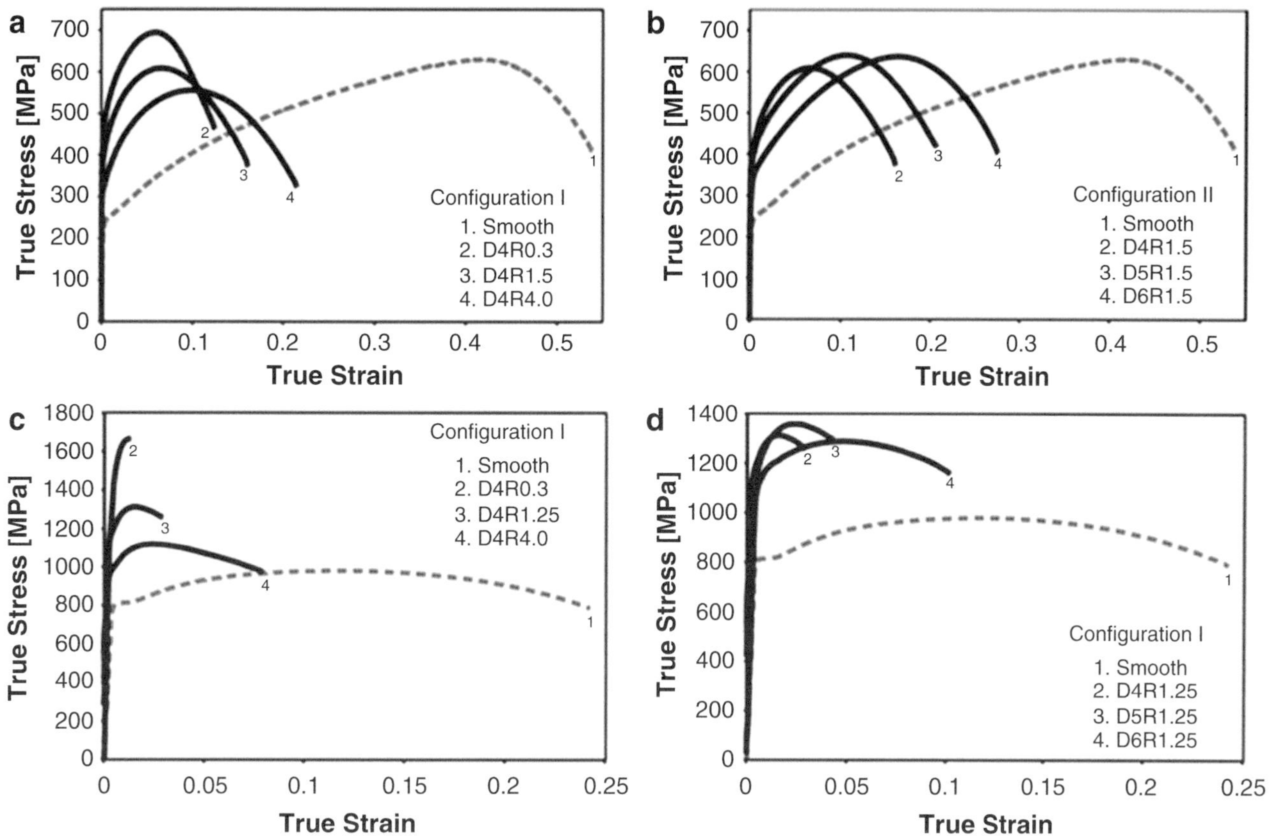

Fig. 9.2 True stress—strain curves for: (**a**) nickel—configuration I; (**b**) nickel—configuration II; (**c**) steel—configuration I; (**d**) steel—configuration II

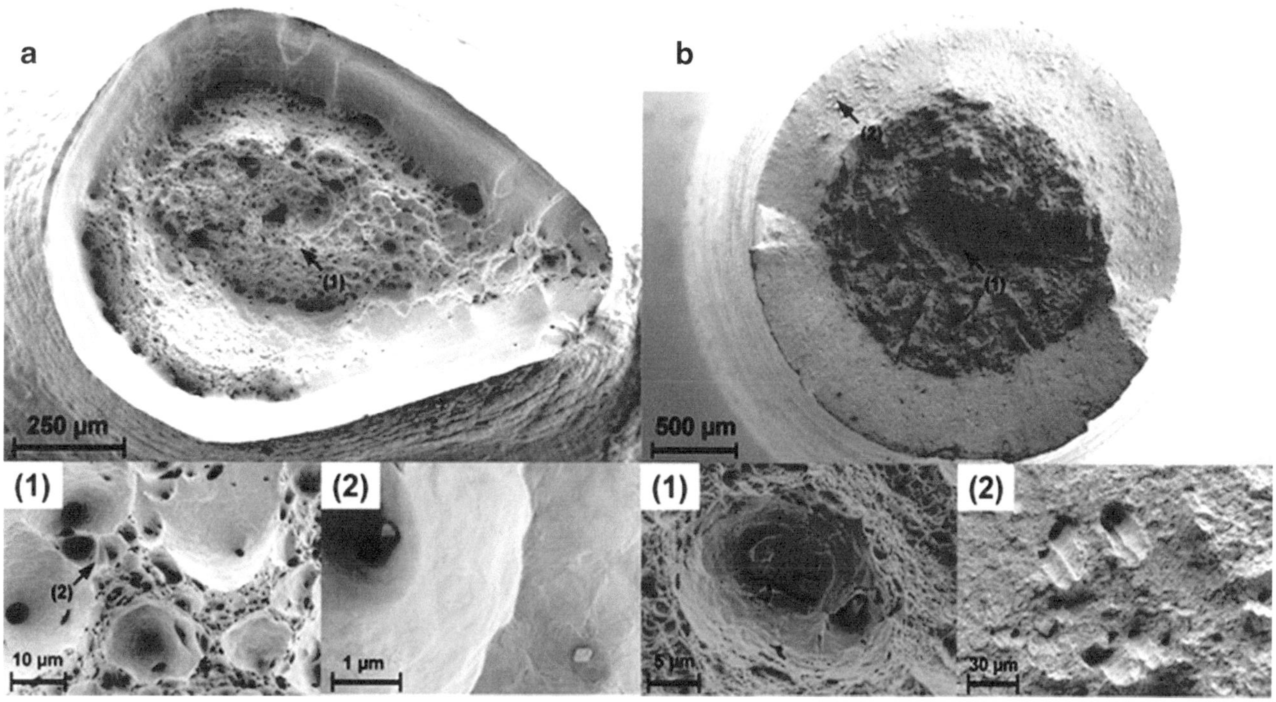

Fig. 9.3 (**a**) Fracture surface of the nickel sample with a characteristic 'cup and cone' failure mode; (1) dimples in the center of the sample; (2) nanoparticles inside the dimples; (**b**) Fracture surface of the steel sample with a characteristic 'cup and cone' failure mode; (1) nucleation of voids by particles fracture and debonding; (2) shear mode fracture surface

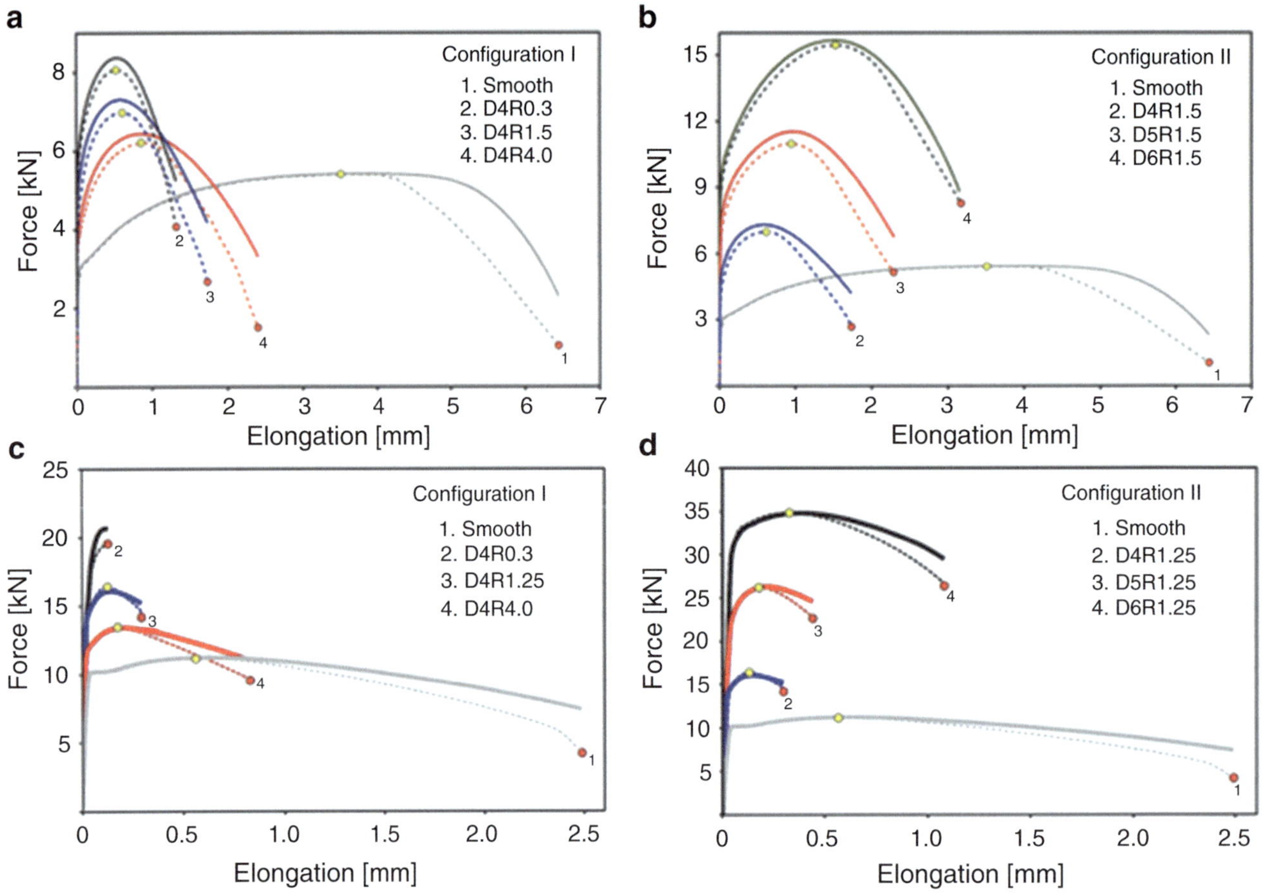

Fig. 9.4 Comparison of experimental (*solid lines*) and numerical (*dashed lines*) results: (**a**) nickel—configuration I; (**b**) nickel—configuration II; (**c**) steel—configuration I; (**b**) steel—configuration II. Critical points at the onset of coalescence and failure for void volume fracture are marked by *yellow* and *red dots*, respectively

The numerical force–displacement curves (and the corresponding experimental results) are plotted in Fig. 9.4. A good agreement between the experimental and numerical results was found for both materials and their specimen configurations. However, the numerical results were slightly underestimated. This is due to the limitation of the Gurson model that is related to the von Mises plasticity theory. In other words, the results should be irrespective of the hydrostatic component of the stress tensor. This is not the case here because the notched specimens generate higher hydrostatic pressure in the center of the sample. In addition, the last stage of the deformation should be interpreted as an approximation only. This is because the Gurson model cannot predict the shear failure mechanism.

The experimental failure strain and the corresponding initial stress triaxiality levels generated by the notch of the sample are presented in Fig. 9.5. It appears that the higher stress triaxiality generated by the notch, the lower value of the failure strain. The data sets of fracture initiation associated with the stress triaxiality fit power trend lines in the investigated range. This trend lines differ between both investigated materials.

The evolution of void volume fraction is different for the investigated materials and each specimen geometry. Note that with the increasing plastic strain the volume occupied by voids is larger. The values of these parameters form trend lines that gives a good approximation for both f_C and f_F against equivalent plastic strain (Fig. 9.6). Therefore, it can be concluded that the evolution of the void volume fraction also depends on the notch geometry. Since each notch geometry generates a particular stress triaxiality gradient in the specimen, the relationship between these parameters can be related. However, a more in-depth study is needed to better understand the ductile behavior of both materials at higher triaxial stress states. This can provide better estimates for the Gurson model parameters that can be eventually used to simulate the crack tip situation.

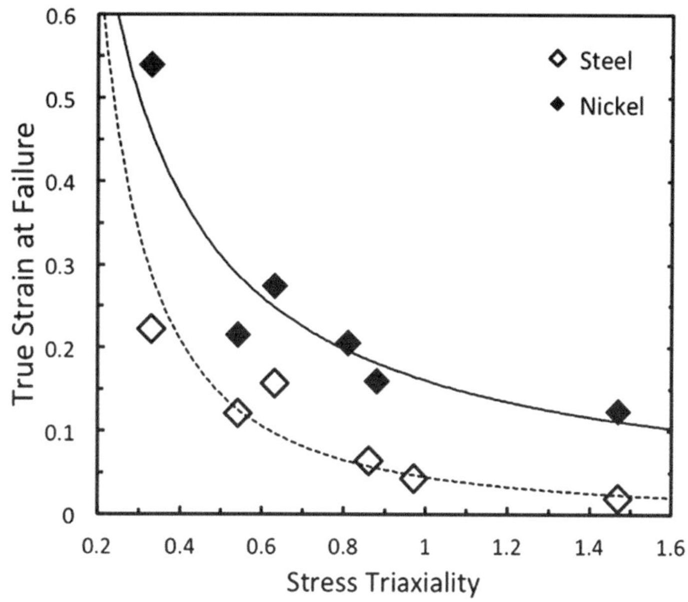

Fig. 9.5 True strain at failure strain and the corresponding stress triaxiality ratios generated by the specimen

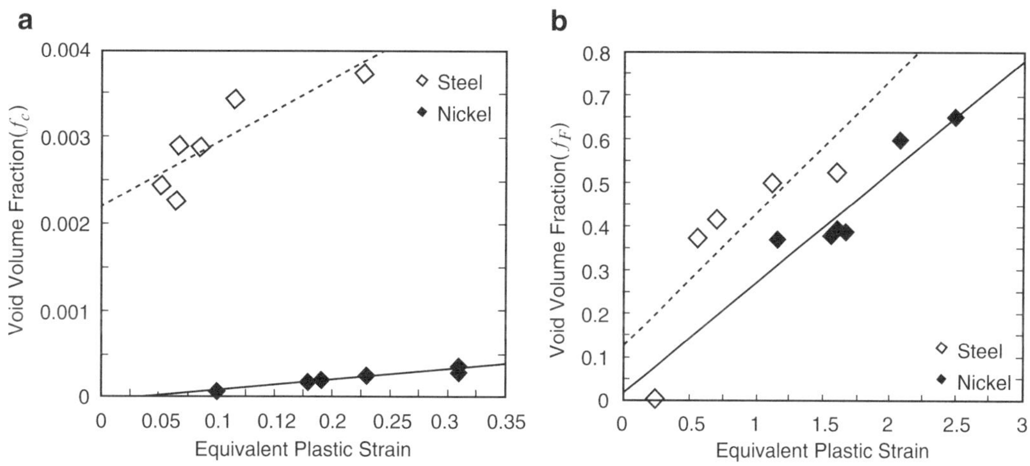

Fig. 9.6 Void volume fraction (**a**) at the onset of coalescence f_C and (**b**) at failure, f_F

9.4 Conclusions

The experimental results have shown that the stress triaxiality plays an important role in the ductile failure process. The evolution of damage and ductile fracture patterns were found to be comparable in both materials. The numerical results have confirmed that the chosen Gurson model parameters were able to reproduce the behavior of the materials. The estimated macroscopic load–displacement curves have been affected only slightly by the porosity parameters. Generally, the numerical calibration reaches a very good agreement with the experimental results. We found that a critical porosity (f_C) and the porosity at failure (f_F) show trends over the equivalent plastic strain, and these can be related to the stress triaxiality. The applied Gurson model parameters have shown a good prediction of the crack growth.

References

1. McClintock, F.A.: A criterion for ductile fracture by growth of holes. J. Appl. Mech. **35**, 363–371 (1968)
2. Rice, J.R., Tracey, D.M.: On the ductile enlargement of voids in triaxial stress fields. J. Mech. Phys. Solids. **17**, 201–217 (1969)
3. Gurson, A.L.: Continuum theory of ductile rupture by void nucleation and growth. Part I: yield criteria and flow rules for porous ductile media. J. Eng. Mater. Technol. **99**(1), 2–15 (1977)

4. Tvergaard, V.: Influence of voids on shear band instabilities under plane strain conditions. Int. J. Fract. **17**, 389–407 (1981)
5. Tvergaard, V.: On localization in ductile materials containing spherical voids. Int. J. Fract. **18**, 237–252 (1982)
6. Koplik, J., Needleman, A.: Void growth and coalescence in porous plastic solids. Int. J. Solids. Struct. **24**, 835–853 (1988)
7. Brocks, W., Sun, D.Z., Honig, A.: Verification of the transferability of micromechanical parameters by cell model calculation with visco-plastic materials. Int. J. Plast. **11**(8), 971–989 (1995)
8. Faleskog, J., Gao, X., Shih, C.F.: Cell model for nonlinear fracture analysis–I. Micromechanics calibration. Int. J. Fract. **89**(4), 355–373 (1998)
9. Gao, X., Faleskog, J., Shih, C.F.: Cell model for nonlinear fracture analysis–II. Fracture-process calibration and verification. Int. J. Fract. **89**, 375–398 (1998)
10. Kim, J., Gao, X., Srivatsan, S.: Modeling of void growth in ductile solids: effects of stress triaxiality and initial porosity. Eng. Fract. Mech. **71**, 379–400 (2004)
11. Zhang, Z.L.: A sensitivity analysis of material parameters for the Gurson constitutive model. Fatigue. Fract. Eng. Mater. Struct. **19**(5), 561–570 (1996)
12. Farbaniec, L.: Deformation mechanisms and fracture strength of polycrystalline ultrafine-grained materials: experimental and numerical investigations [PhD Thesis], University of Paris 13; (2012)
13. Farbaniec, L., Couque, H., Dirras, G.: Influence of triaxial stress state on ductile fracture strength of polycrystalline nickel. Int J Fract **182**, 267–274 (2013)

Chapter 10
Development of a Generalized Entropic Framework for Damage Assessment

Anahita Imanian and Mohammad Modarres

Abstract Data-driven stochastic and probabilistic methods that underlie reliability prediction and structural integrity assessment remain unchanged for decades. This paper develops an alternative approach to reliability assessment in terms of fundamental concepts of science within the irreversible thermodynamic framework. The common definition of damage, which is widely used to measure the reduction of reliability over time, is based on observable markers of damage at different geometric scales. Observable markers are typically based on evidences of any change in the physical or spatial properties or the materials, and exclude unobservable and highly localized damages. Thermodynamically, all forms of damage share a common characteristic: "energy dissipation". Energy dissipation is a fundamental measure of irreversibility that within the context of non-equilibrium thermodynamics is quantified by "entropy generation". The definition of damage in the context of thermodynamics allows for incorporation of all underlying dissipative processes including unobservable markers of damage. Using a theorem relating entropy generation to energy dissipation associated with damage producing failure mechanisms, this paper presents an approach that formally describes and measures the resulting damage.

Having developed the proposed damage model over time, one could determine the time that damage accumulates to a level where the component or structure can no longer endure the damage and fails. Existence of any uncertainties about the parameters and independent variables in this thermodynamic-based damage model leads to a time-to-failure distribution. Accordingly, such a distribution can be derived from the thermodynamic laws rather than estimated from the observed failure histories.

Keywords Irreversible thermodynamic • Energy dissipation • Damage • Entropy • Structural integrity

10.1 Introduction

Manufacturing transforms nature's raw materials into highly organized useful components. Aging or degradation, on the other hand, tends to return these components back to their natural states and reduces the integrity of material properties to the point that the components are no longer functional. An immediate question that arises is how to characterize the reduction of integrity? Alternatively, what is an appropriate measure of degradation? The term "damage" is widely used for this purpose in various disciplines. The definition of damage due to the physical mechanisms varies at different geometric and scales. For example, the definition of fatigue damage can vary from nano-scale through the macro-scale. At the atomic level the grain boundary is a likely location where atoms are more loosely packed. At the micro-scale damage is the accumulation of micro-stresses in the neighborhood of cracks. At the meso-scale level, damage might be defined as growth and coalescence of micro-cracks to meso-cracks. However, measuring damage is subject to the physically measurable variables (i.e., observable marker) when dealing with specific failure mechanisms. For example, in the fatigue mechanism material density, change of hardness, module of elasticity, accumulated number of cycles-to-failure, and crack length may be used as "observable markers" that measure the damage. Therefore, defining a consistent and broad definition of damage is necessary and plausible. To reach this goal, we elaborate on the concept of material damage within the thermodynamic framework.

Thermodynamically, all forms of damage share a common characteristic, which is the dissipation of energy. In thermodynamics, dissipation of energy is the basic measure of irreversibility, which is the main feature of the degradation processes in materials [1]. Chemical reactions, release of heat, diffusion of materials, plastic deformation, and other means of energy production involve dissipative processes. In turn, dissipation of energy can be quantified by the *entropy generation* within the context of irreversible thermodynamics for example in processes such as plasticity, dislocations [2, 3], erosion-corrosion

A. Imanian (✉) • M. Modarres
Department of Mechanical Engineering, Center for Risk and Reliability, University of Maryland, College Park, MD 20742, USA
e-mail: aimanian@umd.edu; modarres@umd.edu

A.M. Beese et al. (eds.), *Fracture, Fatigue, Failure and Damage Evolution, Volume 8*, Conference Proceedings of the Society for Experimental Mechanics Series, DOI 10.1007/978-3-319-21611-9_10

[4], wear [5], fracture [6], fatigue [7], fretting-corrosion [8], thermal degradation [9], and associated failure of tribological components [10, 11]. Therefore, dissipation (or equivalently entropy generation) can be considered as a substitute for characterization of damage. We consider this characterization of damage highly general, consistent and scalable.

The common practice in damage analysis and prediction of structural life and integrity is based on the traditional generic handbook-based reliability prediction methods, data driven prognostics approaches and Physics-of-Failure (PoF) methods. The traditional generic handbook-based reliability prediction methods such as those advocated in MIL-HDBK-217F [12], Telcordia SR-332 [13], and FIDES [14] rely on the analysis of field data (with incoherent operating and environmental conditions), with the assumption that the failure rates are constant. Numerous studies have shown that these methods cause misleading and inaccurate results and can lead to poor design and incorrect reliability prediction and operating decisions [15, 16]. The Physics of Failure (PoF) models [17–21] are more rigorous in terms of employing the specific knowledge of products, such as failure mechanism, material properties, loading profile and geometry. However, such empirical methods are limited to simple failure mechanisms and are hard to model when multiple competing and common cause failure mechanisms are involved. Finally, the data driven methods such as neural networks [22], decision tree classifiers [23] and Bayesian techniques [24] do not capture the difference between failure modes and mechanisms, although they can obtain the complex relationship and degradation trend in the data without the need for the particular product characteristics such as degradation mechanism or material properties.

In this paper, we introduce an entropy-based approach to assess the structural integrity of components and structures. This approach is based on the second law of thermodynamics and defines entropy as a more consistent measure of damage. The benefit of employing thermodynamic entropy to characterize materials damage is that the entropy generation can provide a unified measure of damage which as compared to other existing PoF or fusion prognostics methods [25, 26], can capture the effect of multiple failure mechanisms, more effectively. For example in the case of corrosion-fatigue, consider the physically measurable quantities such as stress and electrochemical affinity of the oxidation-reduction electrode reaction ($Me \Leftrightarrow Me^{z+} + ze$) of a metal. The entropy as a state function is independent of the path of the failure (which commonly depends on factors such as geometry, load, frequency of load, etc.) from the initial state to the final failed state of the material, considering a known failure threshold (endurance limit) [7]. Furthermore, it provides a powerful technique for studying the synergistic effects arising from interaction of multiple processes [27, 28].

Having developed the entropic damage model over time, one could determine the entropic damage accumulates to a level where the component or structure can no longer endure the damage and fails. The application of entropy enables the identification of all the absorptive and dissipative parts of energies that from the intuitive sense can be managed to yield the maximum life. The remainder of this paper is organized as follows. Section 10.2 describes our construction of the entropy model. Section 10.3 describes an entropic based framework for reliability assessment. Section 10.4 provides a case study that explores the application of the proposed framework, and Sect. 10.5 offers concluding remarks.

10.2 Total Entropy Produced in a System

Consistent with the second law of thermodynamics, entropy does not obey a conservation law. Therefore, it is essential to relate the entropy not only to the entropy crossing the boundary between the system and its surroundings, but also to the entropy produced by the processes taking place inside the system. Processes occurring inside the system may be reversible or irreversible. Reversible processes inside a system may lead to the transfer of the entropy from one part of the system to other parts of the interior, but do not generate entropy. Irreversible processes inside a system, however, result in generation of the entropy, and hence in computing the entropy they must be taken into account.

Using the second law of thermodynamic, it is possible to express the variation of total entropy flow per unit volume, dS, in the form of

$$dS = d^r S + d^d S \tag{10.1}$$

where, the total entropy S is defined for a domain g by means of specific entropy, s, per unit mass as $S = \int_g \rho s dV$, and the super scribes r and d represent the reversible and irreversible part of the entropy, respectively. The term $d^r S$ is the entropy supplied to the system by its surroundings through transfer of mass and heat (e.g., in an open system where wear and corrosion mechanisms occur). The rate of exchanged entropy is obtained as

$$\frac{d^r S}{dt} = -\int^{\Omega} \boldsymbol{J}_s . \boldsymbol{n}_s dA \tag{10.2}$$

where, $\boldsymbol{J}_s$ is a vector of the total entropy flow per unit area, crossing the boundary between the system and its surroundings, and $\boldsymbol{n}_s$ is a normal vector. Similarly, d^dS is the entropy produced inside of the system, which can be obtained from (10.3)

$$\frac{d^d S}{dt} = \int^{V} \sigma dV \tag{10.3}$$

where, σ is the entropy generation per unit volume per unit time. The second law of thermodynamics states that d^dS must be zero for reversible transformations and positive ($d^d\mathrm{S} > 0$) for irreversible transformations of the system.

The balance equation for entropy shown in (10.4) can be derived using the conservation of energy and balance equation for the mass.

$$\frac{ds}{dt} + \nabla \boldsymbol{J}_s = \sigma \tag{10.4}$$

This gives us an explicit expression for total entropy in terms of reversible and irreversible processes as [27, 28]

$$\frac{ds}{dt} = -\nabla.\left(\frac{\boldsymbol{J}_q - \sum_{k=1}^{n}(c_m\psi + \mu_k)\boldsymbol{J}_k}{T}\right) + \frac{1}{T^2}\boldsymbol{J}_q.\nabla T - \Sigma_{k=1}^{n}\boldsymbol{J}_k\left(\nabla\frac{\mu_k}{T}\right) + \frac{1}{T}\boldsymbol{\tau}:\dot{\epsilon_p} + \frac{1}{T}\Sigma_{j=1}^{r}v_j\,A_j + \frac{1}{T}\Sigma_{m=1}^{h}c_m\boldsymbol{J}_m(-\nabla\psi) \tag{10.1}$$

where, T is the temperature, μ_k the chemical potential, $\boldsymbol{J}_q$ the heat flux, $\boldsymbol{J}_k$ the diffusion flow, $\boldsymbol{J}_m$ any fluxes resulting from external fields (magnetic and electrical) such as electrical current, v_i the chemical reaction rate, $\boldsymbol{\tau}$ the stress tensor, $\dot{\epsilon_p}$ the plastic strain rate tensor, $A_j = -\Sigma_{i=1}^{u}\mu_i v_{ji}$ the chemical affinity or chemical reaction potential difference, ψ the potential of the external field such as electrical potential difference, and c_m the coupling constant. External forces may be resulted from different factors including electrical field, magnetic field, gravity field, etc., where the corresponding fluxes are electrical current, magnetic current and velocity. For example, in the case of an electric field, $\boldsymbol{E} = -\nabla\psi$ is the electric potential, $\boldsymbol{I} = \Sigma_{m=1}^{h}c_m\boldsymbol{J}_m$, the current density and $c_m = Fz_m$, where F is the Faraday constant and z_m is the number of ions. Each term in (10.5) is derived from the various mechanisms involved, which define the macroscopic state of the complete system.

By comparing (10.5) with (10.4) we can conclude

$$\boldsymbol{J}_s = \frac{\boldsymbol{J}_q - \sum_{k=1}^{n}(c_m\psi + \mu_k)\boldsymbol{J}_k}{T} \tag{10.6}$$

$$\sigma = \frac{1}{T^2}\boldsymbol{J}_q.\nabla T - \Sigma_{k=1}^{n}\boldsymbol{J}_k\left(\nabla\frac{\mu_k}{T}\right) + \frac{1}{T}\boldsymbol{\tau}:\dot{\epsilon_p} + \frac{1}{T}\Sigma_{j=1}^{r}v_jA_j + \frac{1}{T}\Sigma_{m=1}^{h}c_m\boldsymbol{J}_m(-\nabla\psi) \tag{10.7}$$

where, (10.6) shows the entropy flux resulted from heat and material exchange. Equation (10.7) represents the total energy dissipation terms from the system that from left to the right include heat conduction energy, diffusion energy, mechanical energy, chemical energy, and external force energy. Equation (10.7) is fundamental to non-equilibrium thermodynamics, and represents the entropy generation σ as the bilinear form of forces ($\boldsymbol{X}_i$) and force-dependent fluxes $\boldsymbol{J}_i(\boldsymbol{X}_j)$ as

$$\sigma = \Sigma_{i,j}\boldsymbol{X}_i\boldsymbol{J}_i\left(\boldsymbol{X}_j\right); \quad (i,\ j = 1, \ldots, n) \tag{10.8}$$

It is through this form that the contribution from the applicable thermodynamic forces and fluxes are expressed. When multiple failure mechanisms are involved in a degradation process such as corrosion fatigue, summing the contributions of the mechanical and electrochemical processes, one can write the total entropy generation for combined effect of plastic deformation and anodic and catholic dissolution as:

$$T\sigma = \boldsymbol{\tau}:\dot{\epsilon_p} + \tilde{A}i_{corr} \tag{10.9}$$

where $\tilde{A}$ is the electrochemical potential losses (over-potential). Additionally, using forces and fluxes enables one to take into account complex loading scenarios and operating conditions in computing entropy produced in degradation processes.

10.3 Reliability Assessment Using Entropy as an Index of Damage

It was stated earlier that damage caused through a degradation process could be viewed as the consequence of dissipation of energies that can be measured and expressed by entropy such that:

$$\text{Damage} \equiv \text{Entropy}$$

In the earlier discussion in this paper it was shown (Equation (10.5)) that one could express the total entropy per unit time per unit volume for individual dissipation processes resulting from the corresponding failure mechanisms. Therefore, the evolution trend of the damage, D, is obtained from

$$D|t \sim \int_0^t [\sigma|\boldsymbol{X}_i(u),\ \boldsymbol{J}_i(u)]\,du \tag{10.10}$$

where, $D|t$ is the monotonically increasing cumulative damage starting at time t from a theoretically zero value or practically some initial damage value. In this study, the evaluation of damage is performed relative to the initial damage value. The initial damage can be calculated using the correlation between the rate of damage and damage at different stage of degradation [29].

When D reaches a predefined (often subjective) level of damage endurance, D_f, it may be assumed that beyond that point the component or structure will fail. Because entropy as a parameter of degradation includes all observable damage markers (cracks, wear debris and pit densities) and unobservable damages such as subsurface dislocations, slip and micro-cavities, definition of a single failure threshold might not be possible due to long stretch of damage measurement from nano-scale to macroscopic scale. In this case, the cumulative damage and alternatively entropy endurance level can be estimated through the measurement of certain observable damage markers. The correlation between the observable damage markers and entropy, justified by several studies [7, 30], enables the definition of failure threshold on the basis of observable markers. In the other word, the damages grow, coalesce and eventually the weakest link among all coalesces damages manifests itself as an observable damage which causes failure. Additionally, records of the entropy data from historical data can be used to obtain the entropy to failure values. Entropy, as a thermodynamic state function is independent of the path to failure (loading values, frequency and geometry) and provides an overall constant failure criterion [28, 30].

The rate of entropy or damage can vary according to the type of degradation. However, damage in the system mounts up over time. For example, in the case of fatigue crack closure, while the crack as an observable marker of damage disappears, causing damage rate decrease, the damage accumulation keeps rising as unobservable markers of damage such as loading asymmetry, hardening properties, residual stresses and loading ratio increase [31].

Material, environmental, operational and other types of variability in degradation forces impose uncertainties on the cumulative damage, D. Existence of any uncertainties about the parameters and independent variables in this thermodynamic-based damage model leads to a time-to-failure distribution as depicted in Fig. 10.1. This figure shows how the interfaces of the accumulated damage and endurance level result in the probability density function (PDF) of the time-to-failures, which is the variable of interest in most reliability analysis. It is self-evident that the probability that the random variable, $D|t$, (i.e., the cumulative damage at time, t) exceeds the constant endurance level, D_f (thus causing the failure), must be equal to the probability that the random variable time-to-failure is less than t. Accordingly, Fig. 10.1 depicts the cumulative probability of damage as the yellow shaded area, and the cumulative time to failure probability as the pink shaded area.

Accordingly, assuming the constant endurance limit, D_f, one can derive the time-to-failure distribution, $g(t)$, from the thermodynamically based damage relation expressed by,

$$\int_0^{T_c} g(x)dx = \int_{D_f}^{\infty} f\,(D|t)\,dD \tag{10.11}$$

where, $f\,(D|t)$ is the PDF of the damage at time t. The corresponding time to failure PDF, $g(t)$, would be

$$g(t) = \frac{d}{dt}\int_{D_f}^{\infty} f\,(D|t)\,dD \tag{10.12}$$

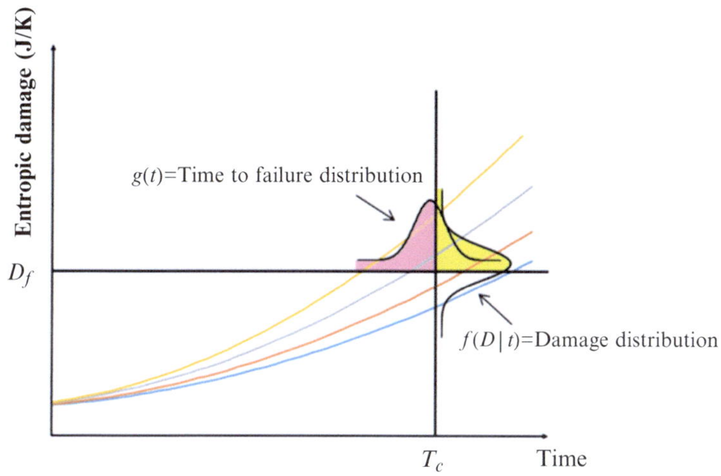

Fig. 10.1 Entropic damage-endurance modeling

Obviously, the reliability function can be expressed as

$$R(t) = \int_{T_c}^{\infty} g(t)dt = 1 - \int_{D_f}^{\infty} f\left(D|t\right) dD \tag{10.13}$$

Similar to the uncertainties about the amount of damage, the endurance limit, D_f, may also be uncertain because composition of materials and thus their strengths varies from sample to sample. In this case, a failure occurs when the damage in a component exceeds its endurance level. The probability that no failure occurs is equal to the probability that the random variable, D, is less than the random variable, D_f, describing the component endurance level.

$$R(t) = Pr\left(D|t < D_f|t\right) \tag{10.14}$$

where, $R(t)$ is the reliability of component at time t. Knowing the PDFs of the random variables D and D_f expressed by $f(D)$ and $h\left(D_f|t\right)$, the PDF of the time-to-failure distribution can be obtained by

$$R(t) = 1 - \int_{0}^{\infty} h\left(D_f|t\right) dD_f \int_{D_f}^{\infty} f\left(D|t\right) dD, \tag{10.15}$$

and

$$g(t) = -\frac{d}{dt} R(t) \tag{10.16}$$

It is possible that the PDF of the endurance limit is independent of time $\left(\text{i.e., } D_f|t \approx D_f\right)$. If multiple dissipative forces are at work then using the weakest link principle

$$R(t) = \prod_{i=0}^{n} \left\{ 1 - \int_{0}^{\infty} h\left(D_{f_i}|t\right) dD_{f_i} \int_{D_{f_i}}^{\infty} f\left(D_i|t\right) dD_i \right\} \tag{10.17}$$

Again (10.17) serves as the basis to derive the PDF of the time-to-failure.

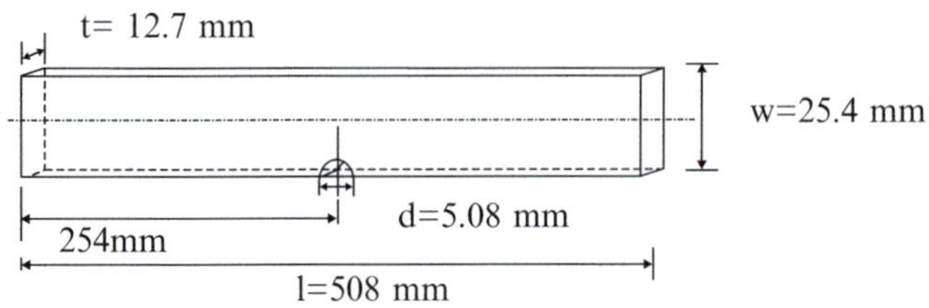

Fig. 10.2 Al7075-651 edge notch specimen

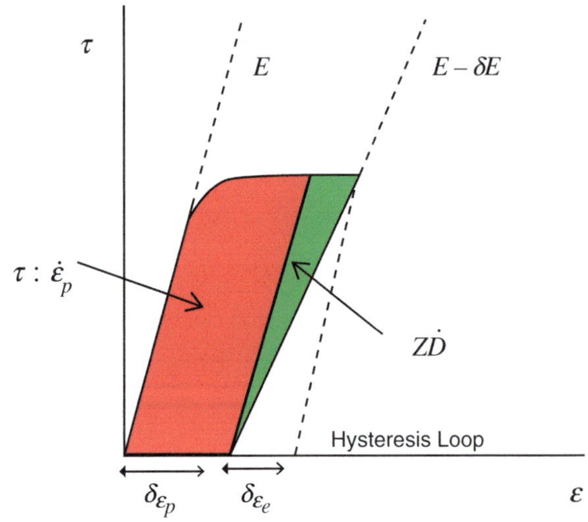

Fig. 10.3 Hysteresis energy (reproduced from [32])

10.4 Case Study

The entropy based approach was employed to obtain the reliability of the AL7075-T651 coupons subjected to fatigue loading, using an MTS servo-hydraulic uni-axial load frame, from Ontiveros et al. experimental results [32]. Geometries of the coupons used are shown in Fig. 10.2. All tests were performed at peak stress of 245 MPa with load ratio of 0.1 and frequency of 2 Hz. Since the focus of Ontiveros et al. study was crack initiation, so most of experiments were stopped when, a crack was detected at the notch by visual inspection.

The formulation for entropy generation using (10.7) can be derived as

$$\sigma = \frac{\tau : \dot{\varepsilon}_p}{T} + \frac{1}{T} Z \dot{D} + \frac{1}{T^2} J_q \cdot \nabla T \tag{10.18}$$

where, Z is the elastic energy release rate and $\dot{D}$ is the elastic damage rate variable.

In (10.18), the first two terms can be captured directly from the hysteresis loop as depicted in Fig. 10.3. In this plot, the largest area represents the energy dissipated due to plastic deformation. The remaining portion represents the energy dissipation as a result of elastic damage which can be observed as degradation of the Young's modulus [33].

Results of Ontiveros et al. [32] analysis showed that when compared to the plastic and elastic energy dissipations the fraction of the entropy generation due to heat conduction is considered to be negligible. Therefore, the third term does not take into account in the entropy calculation.

Figure 10.4 depicts the linear evolution of the entropic damage in fatigue aging mechanisms for seven samples. Considering linear relationship between entropic damage and time random variables, relations $a_i x = D_f$ and $\frac{D}{a_i} = T_c$ hold at damage and time thresholds, respectively, where a_i's are corresponding coefficients of entropy-time lines. Rewriting the cumulative time to failure distribution in terms of damage threshold and line coefficients as

$$\Pr(T < t) = \int_0^t g\left(\frac{D_f}{a\ i}\right) d\left(\frac{D_f}{a\ i}\right) \tag{10.19}$$

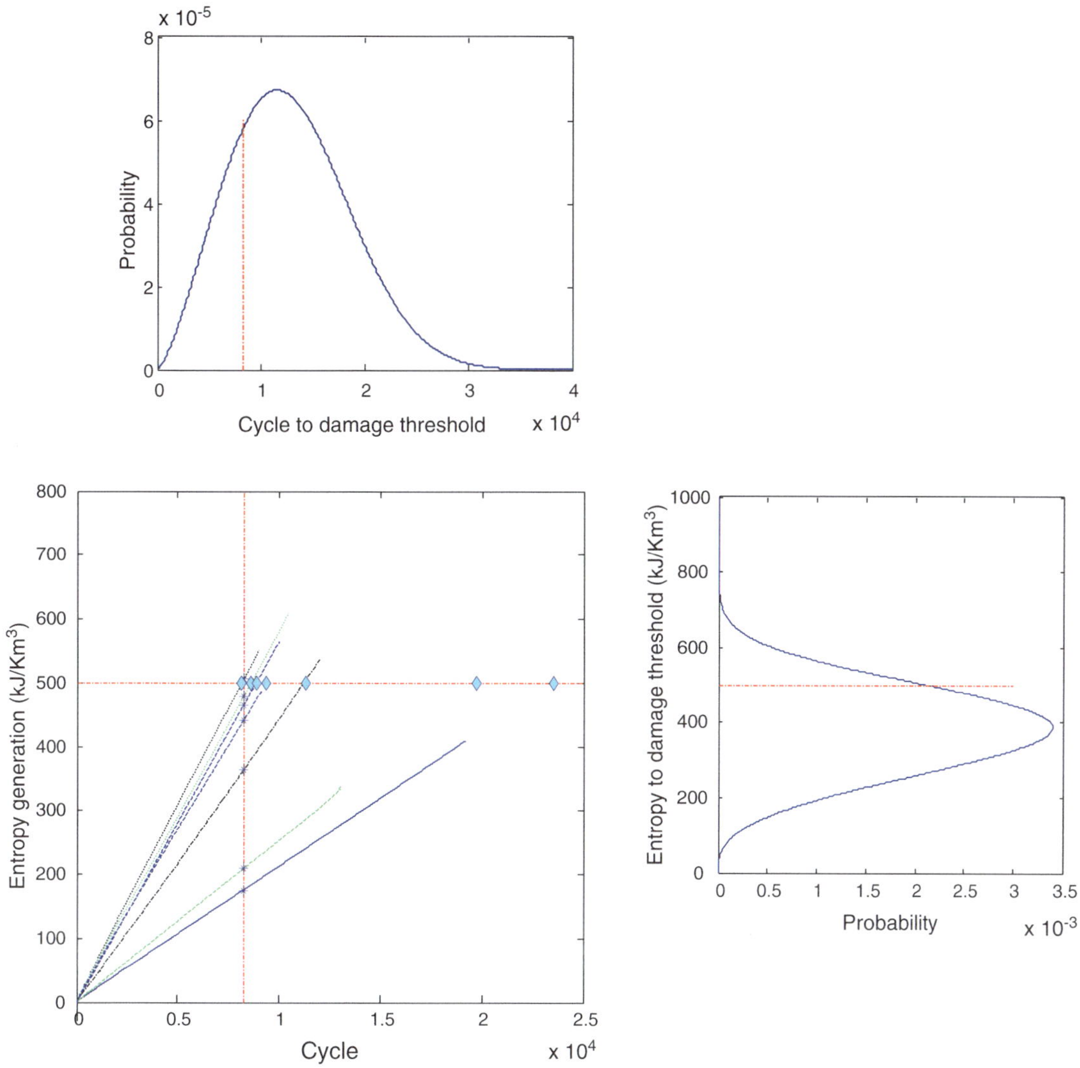

Fig. 10.4 Entropy evolution of Al 7075-T651 specimens

Damage distribution can be obtained through the distribution of the reciprocal, $a_i = 1/\left(\frac{1}{a_i}\right)$, by the inverse distribution relation. According to this relation, if the distribution of an arbitrary random variable Z is continuous with density function $k(z)$ and cumulative distribution function $K(z)$, then the cumulative distribution function, $W(u)$, of the reciprocal $U = 1/Z$, is found by noting that

$$W(u) = Pr(U \leq u) = \Pr\left(Z \geq \frac{1}{u}\right) = 1 - \Pr\left(Z \leq \frac{1}{u}\right) = 1 - K\left(\frac{1}{u}\right) \tag{10.20}$$

Failure data of Fig. 10.4 is employed to represent (10.13) and specifically (10.20). Fitting a Weibull distribution to the entropy to damages data at $T_c = 8872$ with the coefficient of determination equal to 0.96, the reliability of specimens calculated at damage threshold of $D_f = 500 \ \left(\mathrm{KJ/Km^3}\right)$ is

$$R(D) = 1 - \int_{D_f}^{\infty} \frac{\beta_D}{\alpha_D}\left(\frac{t}{\alpha_t}\right)^{\beta_D - 1} \exp\left(-\left(\frac{t}{\alpha_D}\right)^{\beta_D}\right) dD = 0.74 \tag{10.21}$$

where, $\alpha_D = 434.6$ and $\beta_D = 2.5$. Accordingly, fitting a Weibull distribution to the cycles to entropic damage threshold with $\alpha_t = 14789$ and $\beta_t = 2.2$, leads to the same amount of reliability as was shown in (10.13).

$$R(t) = \int_{T_c}^{\infty} \frac{\beta_t}{\alpha_D}\left(\frac{t}{\alpha_t}\right)^{\beta_t - 1} \exp\left(-\left(\frac{t}{\alpha_t}\right)^{\beta_t}\right) dt = 0.74 \tag{10.22}$$

It should be noted that, the values of cycle and damage thresholds are selected, arbitrarily in this case.

10.5 Conclusions

This paper presents a thermodynamic framework for the degradation level assessment using entropy generation as a measure of damage. It suggests that a unified measure of damage can be defined based on the entropy generation concept. The general entropy generation function is derived in terms of energy losses due to heat conduction, diffusion losses, mechanical dissipations, chemical losses and external force field effects (e.g. magnetic, electrical and gravity fields). It is shown that entropy generation and damage are related. However, the entropy generation function is subject to various stochastic variations of forces that cause damage.

A fundamental foundation for a science-based explanation of damage as a source of material failure and thus materials reliability is proposed. As such we offer an alternative approach for reliability assessment based on the second law of thermodynamics. The reliability model is built off of the relationship between damage PDF, endurance limit distribution and time-to-failure PDF. A case study analysis represents the application of thermodynamic entropy in reliability analysis. However, further work is required to construct a unified entropic damage paradigm that adequately elucidates single damage mechanism as well as mechanisms with multiple degradation processes. Towards this goal, we are now studying the entropy growth rate as a degradation parameter for the corrosion-fatigue mechanism in materials.

Acknowledgment This work is part of an ongoing research through grant number N000141410005 from the Office of Naval Research (ONR).

References

1. Tang, H., Basaran, C.: A damage mechanics based fatigue life prediction model for solder joints. J. Electron. Packag. **125**, 120–125 (2003)
2. Weertman, J., Weertman, J.: Elementary Dislocation Theory. Oxford University Press, New York (1992)
3. Gutman, E.: Mechanochemistry of Materials. Cambridge International Science, Cambridge (1998)
4. Lu, B., Luo, J.: Synergism of electrochemical and mechanical factors in erosion-corrosion. J. Phys. Chem. **110**, 4217–4231 (2006)
5. Amiri, M., Khonsari, M.: On the thermodynamics of friction and wear—a review. J. Entropy. **12**(5), 1021–1049 (2010)
6. Maugis, D.: Contact, Adhesion, and Rupture of Elastic Solids. Springer, Berlin (2000)
7. Naderi, M., Amiri, M., Khonsari, M.: On the thermodynamic entropy of fatigue fracture. Proc. R. Soc. Lond. A **466**, 423–438 (2010)
8. Uhlig, H., Revie, R.: Corrosion and Corrosion Control, 3rd edn. Wiley, New York (1985)
9. Li, S., Basaran, C.A.: A computational damage mechanics model for thermomigration. J. Mech. Mater. **41**, 271–278 (2009)
10. Nosonovsky, M., Bhushan, B.: Thermodynamics of surface degradation, self-organization, and self-healing for biomimetic surfaces. Philos. Trans. R. Soc. A **367**, 1607–1627 (2009)
11. Abdel-Aal, H.: On the role of intrinsic material response in failure of tribo systems. Wear **259**, 1372–1381 (2005)
12. U. S. Department of Defence: Military Handbook for Reliability Prediction of Electronic Equipment, Version A (1965)
13. Telcordia Technologies: Reliability Prediction Procedure for Electronic Equipment, Telcordia Customer Service, Piscataway (2001)
14. FIDES Guidance Issue: Reliability Methodology for Electronic Systems, FIDES Group (2004)
15. IEEE Standard 1413: IEEE Standard Methodology for Reliability Prediction and Assessment for Electronic Systems and Equipment (1998)
16. IEEE Standard 1413.1: IEEE Guide for Selecting and Using Reliability Predictions Based on IEEE 1413 (2002)
17. Manson, S.S.: Thermal Stress and Low Cycle Fatigue. McGraw-Hill, New York (1996)
18. Norris, K., Landzberg, A.: Reliability of controlled collapse interconnections. IBM J. Res. Dev. **13**, 266–277 (1969)
19. Bayerer, R., Hermann, T., Licht, T., Lutz, J., Feller, M.: Model for power cycling lifetime of IGBT modules—various factors influencing lifetime. In: Integrated Power Electronics Systems Conference, Nuremberg (2008)
20. Shi, R., Mahadevan, S.: Damage tolerance approach for probabilistic pitting corrosion fatigue life prediction. J. Eng. Fract. Mech. **68**, 1493–1507 (2001)
21. Harlow, D.G., Wei, R.P.: A probability model for the growth of corrosion pits in aluminum alloys induced by constituent particles. J. Eng. Fract. Mech. **59**, 305–325 (1998)
22. Byington, C., Watson, M., Edwards, D.: Data-driven neural network methodology to remaining life predictions for aircraft actuator components. In: Proceedings of the IEEE Aerospace Conference, New York (2004)

23. Schwabacher, M., Goebel, G.: A survey of artificial intelligence for prognostics. In: AAAI Fall Symposium: AI for Prognastics (2007)
24. Bhangu, B.S., Bentley, P., Stone, D.A., Bingham, C.M.: Nonlinear observers for predicting state-of-charge and state-of-health of lead-acid batteries for hybrid-electric vehicles. IEEE Trans. Veh. Technol. **54**(3), 783–798 (2005)
25. Held, M., Jacob, P., Nicoletti, G., Scacco, P., Poech, M.H.: Fast power cycling test of IGBT modules in traction application. Int. J. Electron. **86**(10), 1193–1204 (1999)
26. Cheng, C., Pecht, M.: A fusion prognostics method for remaining useful life prediction of electronic products. In: 5th Annual IEEE Conference on Automation Science and Engineering, Bangalore (2009)
27. De Groot, S.R., Mazor, P.: Non-Equilibrium Thermodynamics. Wiley, New York (1962)
28. Kondepudi, D., Prigogine, I.: Modern Thermodynamics: From Heat Engines to Dissipative Structures. Wiley, Chichester (1998)
29. Liakat, M., Khonsari, M.: An experimental approach to estimate damage and remaining life of metals under uniaxial fatigue loading. J. Mater. Des. **57**, 289–297 (2014)
30. Bryant, M.D., Khonsari, M.M., Ling, F.F.: On the thermodynamics of degradation. Proc. R. Soc. A Math. Phys. Eng. Sci. **464**, 2001–2014 (2008)
31. Romaniv, O.R., Nikiforchin, G.N., Andrusiv, N.N.: The effect of cracks closure and the estimation of cyclic crack resistance of structural alloys. J. Phys. Chem. Mech. Mater. **3**, 47–61 (1983)
32. Ontiveros, V.L.: Strain energy and thermodynamic entropy as prognastic measures of crack initiation in aluminum. Doctoral dissertation, University of Maryland (2013)
33. Lemaitre, J., Chaboche, J.L.: Mechanics of Solid Materials. Cambridge University Press, Cambridge (1990)

Chapter 11
Modelling of Experimental Observations of Electrical Response of CNT Composites

K. Shkolnik and V.B. Chalivendra

Abstract A physics based analytical models on prediction of electro-mechanical behavior of carbon nanotubes (CNTs) embedded epoxy composite are developed to investigate the change in resistance under quasi-static tensile and compression loading conditions. Two different types of contacts namely in-line and lateral contacts between CNTs were considered in predicting electrical tunneling of current. It was identified from experiments that the extent of these contacts vary during deformation and lateral contacts predominates after composite reaches maximum stress during deformation, resulting decrease in resistance. The non-linear constitutive response of the composite obtained from experimental results is incorporated into model to accommodate decrease/increase in distance between above contacts. Under tensile loading conditions, the model made decent predictions against experimental results for composites of three different weight fractions (0.1, 0.3 and 0.5 %) of CNTs. Later, this model is extended to predict electro-mechanical response for intermediate weight fractions. Our efforts are currently focused on developing model to predict electrical response under compression loading and the results of this model along with under tensile loading will be presented at the conference.

Keywords Carbon nanotubes • Electro-mechanical response • Damage detection • Epoxy • Physics based model

11.1 Introduction

Due to their high conductivity and greater aspect ratio, Carbon nanotubes (CNTs) have been used for in-situ sensing of damage in various kinds of polymer composites under different mechanical loading situations [1–6]. Vadlamani et al. [1] investigated the effect of CNTs weight fraction on electro-mechanical response under tensile loading conditions. In their study the authors reported a decreasing resistance trend up to 45 % due to matrix's molecular chains declustering and straightening at lower weight fractions of 0.1 and 0.3 % weight fractions. Later the same authors [2] studied the electro-mechanical response of rubber toughened and rigid particles epoxy systems under quasi-static tensile loadings. They observed threshold strain value of around 17 % for damage propagation in 20 phr rubber toughened epoxy system from change in electrical resistance measurements. This threshold strain value decreased to 13 % for 30 phr rubber toughened epoxy system. Cardoso et al. [3], reported damage detection within the process zone in a single edge notch tension specimen of CNTs embedded, rubber toughened epoxy composites under mode-I fracture conditions. At a high concentration of 30 phr rubber in epoxy, the electrical response at the fracture process zone resulted in unique variations of electrical response due to a large amount of stress whitening ahead of the crack-tip. Heeder et al. [4, 5], examined the electrical response of CNTs embedded epoxy materials under uniaxial compression at quasi-static, medium strain rate and high strain rate loading conditions. Although a monotonic decrease in electrical resistance up to 45 % was observed under quasi-static and high strain rate conditions, a decreasing (up to 60 %) and sudden increasing in resistance (up to 1 %) was noticed under medium strain rates. Chaurasia et al. [7] conducted computational micromechanics modeling to Study the effective macroscale piezoresistivity of carbon nanotube-polymer nanocomposites for strain and damage sensing. They identified the formation and disruption of the electron hopping pathways are to be one of the dominant mechanisms affecting macroscale effective piezoresistive response. Kuronuma et al. [8] performed an analytical and experimental study on strain sensing behavior of CNT/polycarbonate composites under tensile loading. Their model deals with inter-nanotube matrix deformation at the micro/nano scale due to the macroscale deformation of the composite. Their model is restricted to elastic deformation of the composite. In the current paper, we aim to extend their model to non-linear deformation and compare the model predictions with experimental observation.

K. Shkolnik • V.B. Chalivendra (✉)
Department or Mechanical Engineering, University of Massachusetts, North Dartmouth, MA 02747, USA
e-mail: vchalivendra@umassd.edu

A.M. Beese et al. (eds.), *Fracture, Fatigue, Failure and Damage Evolution, Volume 8*, Conference Proceedings of the Society for Experimental Mechanics Series, DOI 10.1007/978-3-319-21611-9_11

11.2 Experimental Details

CNT embedded epoxy materials are fabricated with a simple procedure discussed in our previous paper [1]. The modified four circumferential ring probes (FCRP) technique was implemented to capture the electrical resistance change during tensile loading [1]. A constant current was passed through the outer probes and voltage drop across two inner probes was measured to determine the electrical resistance of the middle section under tensile loading. Since the current flow, i, was constant and the voltage change, V, between the two inner probes was measured, the instantaneous resistance between the two inner probes, R', was calculated using Ohm's law. Percent change in resistance was then taken as,

$$\frac{R' - R0}{R0} \times 100 \tag{11.1}$$

where R_0 is the initial resistance between the two inner probes. Due to the complexity and variability of the networks present within each specimen, the initial R_0 value served as a baseline resistance for each experiment.

11.3 Experimental Results

The electro-mechanical response of the 0.1 % weight fraction of CNTs embedded in epoxy under quasi-static tensile load is shown in Fig. 11.1. In stage-I (until 1.5 % strain), as the applied strain increases, there is a steady rate of increase in resistance up to 1.25 % strain. Later between 1.25 and 1.5 % strain, the percentage resistance increase is very slow and reaches a maximum value of 1.8 % at an axial strain of 1.5 %. The maximum increase in resistance occurs well before the specimen reaches the peak stress shown in the stress-strain diagram. This increase in resistance is attributed to the dimensional change of the test specimen that increases distance between CNTs along the loading direction. As reported in the literature, CNTs have a maximum tunneling distance of 1.8 nm in polymer matrices. A slight increase in distance beyond the tunneling distance between neighboring CNTs causes a decrease in electron hopping density in the epoxy matrix. In stage-II (between 1.5 and 3.35 % strain), the resistance decreases at a moderate rate. During this stage, the specimen approaches maximum stress and induces a considerable amount of elongation of the clusters of entangled epoxy. This elongation causes a decrease in the distance between dispersed CNTs (lower than maximum tunneling distance of 1.8 nm) and hence increases the electron hopping density. The change in CNT networks with increased electron flow facilitates a significant increase in the conductivity of the test specimen. In stage-III (between 3.35 and 5.7 % strain), the resistance decreased further at a faster rate compared to the second stage. The rate at which electron hopping occurs is higher in this stage compared to stage-II due to further elongation of the clusters of entangled epoxy. In stage-IV (between 5.7 % and break), the percentage change in resistance decreased at a slow rate to a minimum value of −44 %. During this stage there is a competition between the straightening and slipping of the uncoiled epoxy matrix and formation of micro-cracks originating from the agglomerated CNTs. Agglomerates act like a defect under far-field tensile loading. The formation of micro-cracks breaks the local CNT conductive paths, thus decreases electrical conductivity of the composite. This can be noticed as a decrease in the rate of change of resistance before the specimen breaks.

11.4 Theoretical Modeling

Having a polymeric composite that contains an evenly dispersed CNT's, electrically connected networks of CNTs can be formed to create a continuous conductive pathway network for the transmission of electrical current. The electrical current within a CNT matrix can be broken up into two different categories. The first category focuses on the inherent electrical resistance of each individual CNT fiber. The second category of which is the main focus of this model is the electron tunneling effect between two adjacent CNT fibers. In order for this electron tunneling effect to correctly react the thickness of the inter-nanotube matrix must be very small [9]. When a nanocomposite is deformed under applied loading, the change of nanotube resistance is expected to be negligible because of the extremely small elastic deformation in nanotubes resulting from their high modulus while the large deformation in the inter nanotube matrix due to its low modulus contributes to the electrical resistance change of the composite [10]. For this model, two different CNT junction or contact types are explored. The contact junctions that are implicated are the overlapping and inplane types which are referred as (Type I) and (Type II) respectively. Figure 11.2 demonstratestheideal differences between these two different contact junctions. The effective

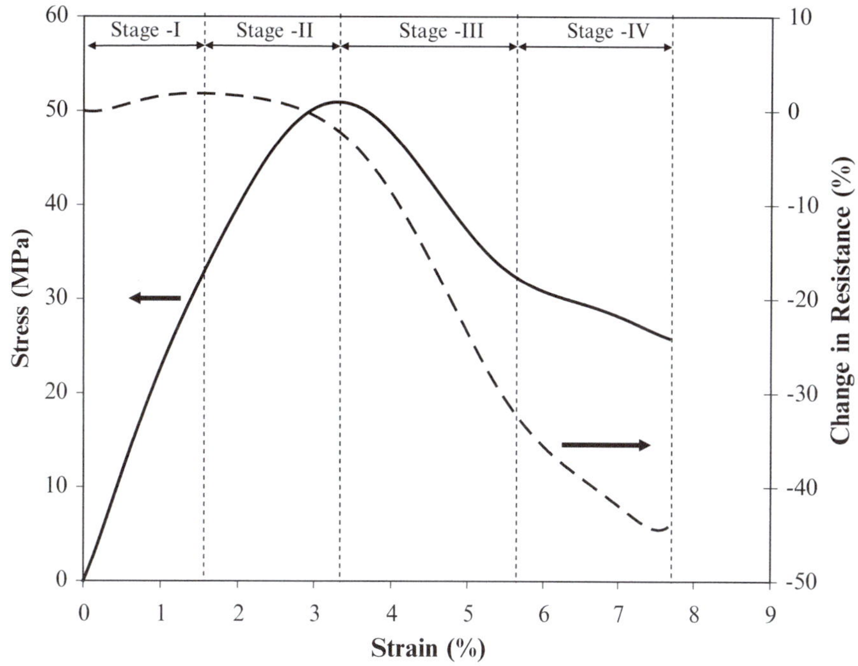

Fig. 11.1 Electro-mechanical response of 0.1 % weight fraction of CNTs embedded in epoxy

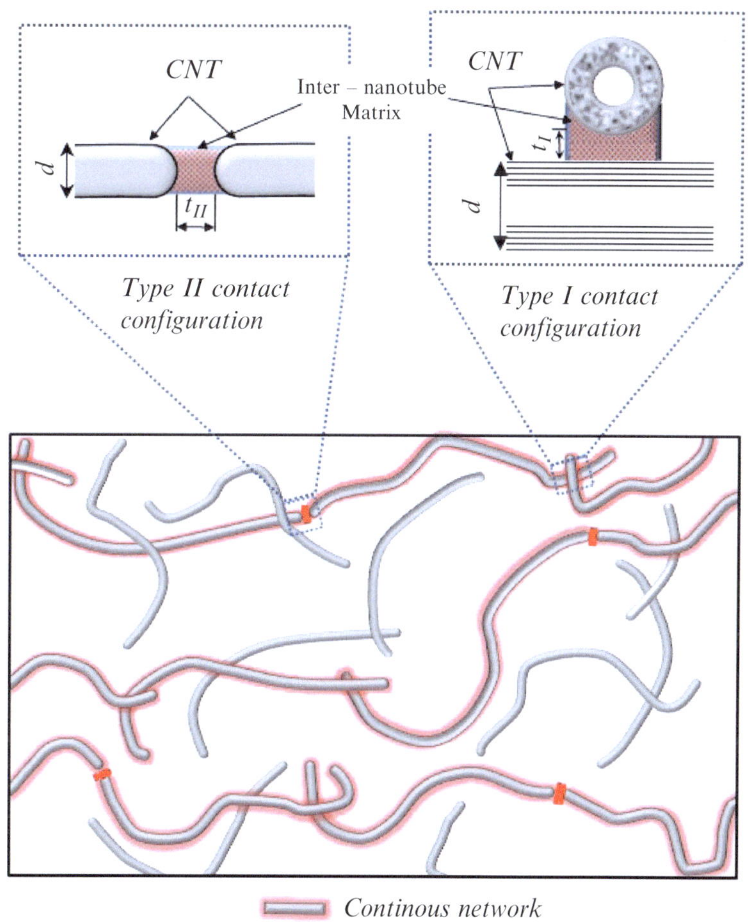

Fig. 11.2 Types of contacts on CNTs embedded conductive network

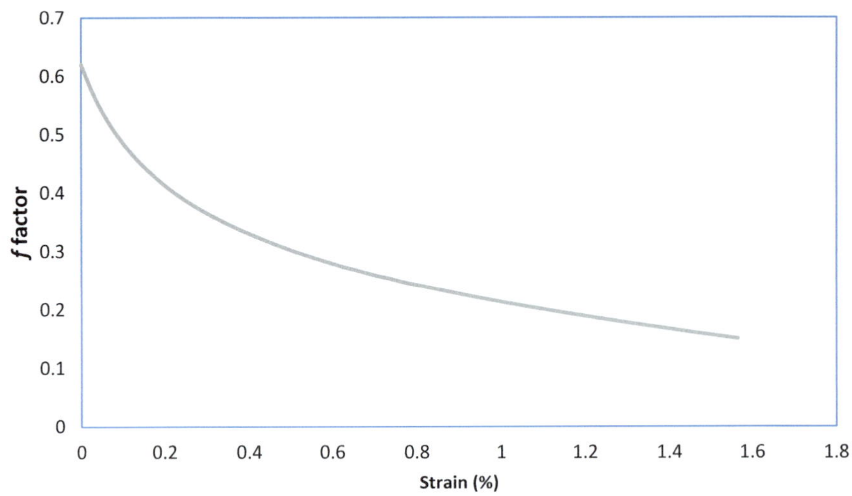

Fig. 11.3 Variation of f factor as a function of strain for stage-1 of the electro-mechanical response for 0.1 % weight fraction of CNTs embedded epoxy under tensile loading

CNT resistance which is the assumed successful conductive pathway for each contact type is donated R_i^{EN} where (I = I, II). This resistance represents the sum of nanotube resistance and the tunneling-type contact resistance due to the inter-nanotube matrix region R_i^{IM} [11]. The superscript EN indicates effective CNT and IM represents inter-nanotube matrix. It is important to comment that the tunneling type contact resistance plays a dominant role in the electrical transport properties of nanocomposites [9].

Within the CNT matrix of the polymer, it was assumed that the CNT networked had formed successful networks of conductive pathways above the *percolation threshold* consisting of both Type I and Type II contacts. To better assess the multiple possibilities of Type II pathways, the continuous conductive pathways were modeled as parallel conductive wires connected in an electrical series. Any effective pathway could have any number of Type I and Type II contacts. It was also assumed that these pathways have the same fraction of the number of Type II effective CNTs to the number of total effective CNTs in one effective conductive pathway [8]. A parameter f was formed to represent the fraction of Type II contacts. A modified expression [8], expressed the composites total resistance R^c of effective CNTs where C denotes a carbon nanotube composite.

$$R^C = (1 - f)\, R_{II}^{EN} + f R_I^{EN} \tag{11.2}$$

From this study, we found that f factor is not constant as opposed to previous work [8] and varies with the strain values. The varying f-factor is obtained by fitting the theoretical equation with experimental data shown in Fig. 11.1. Figure 11.3 shows a typical situation of variation of f factor as a function of strain for stage-1 of the electro-mechanical response for 0.1 % weight fraction of CNTs embedded epoxy under tensile loading conditions. The varying f factor is used to predict the electrical response for different weight fraction. The details will be discussed and complete results will be presented at the conference.

Acknowledgements Authors acknowledge the discussion made with summer intern Mr. E. John during the summer 2014.

References

1. Vadlamani, V.K., Chalivendra, V.B., Shukla, A., Yang, S.: Sensing of damage in carbon nanotubes and carbon black reinforced epoxy composites under tensile loading. Polym. Compos. **33**, 1809–1815 (2012)
2. Vadlamani, V.K., Chalivendra, V.B., Shukla, A., Yang, S.: In-situ sensing of non-linear deformation and damage in epoxy particulate composites. Smart Mater. Struct. **21**, 075011 (2012)
3. Cardoso, S., Chalivendra, V.B., Shukla, A., Yang, S.: Damage detection in the fracture process zone of rubber toughened epoxy using carbon nanotube sensory network. Eng. Fract. Mech. **96**, 380–391 (2012)
4. Heeder, N., Shukla, A., Chalivendra, V.B., Yang, S.Z., Park, K.: Electrical response of carbon nanotube reinforced nanocomposites under static and dynamic loading. Exp. Mech. **52**, 315–322 (2012)
5. Heeder, N., Shukla, A., Chalivendra, V.B., Yang, S.Z.: Sensitivity and dynamic electrical response of CNT-reinforced nanocomposites. J. Mater. Sci. **47**, 3808–3816 (2012)

6. Cardoso, S., O'Connell, C.D., Pivonka, R., Mooney, C., Chalivendra, V.B., Shukla, A. Yang, S.: Effect of external loads on damage detection of rubber-toughened nanocomposites using carbon nanotubes sensory network. Polym. Compos. (2014). doi: 10.1002/pc.23188
7. Chaurasia, A.K., Seidel, G.D., Ren, X.: Computational micromechanics model to study the effective macroscale piezoresistivity of carbon nanotube-polymer nanocomposites for strain and damage sensing. In: ASME Proceedings of Mechanics and Behavior of Active Materials, p. V002T02A018 (2013)
8. Kuronuma, Y., Takeda, T., Shindo, Y., Narita, F., Wei, Z.: Electrical resistance-based strain sensing in carbon nanotube/polymer composites under tension: analytical modeling and experiments. Compos. Sci. Technol. **72**, 1678–1682 (2012)
9. Li, C., Thostenson, E.T., Chou, T.W.: Dominant role of tunneling resistance in the electrical conductivity of carbon nanotube-based composites. Appl. Phys. Lett. **91**, 223114 (2007)
10. Li, C., Chou, T.-W.: Modeling of damage sensing in fiber composites using carbon nanotube networks. Compos. Sci. Technol. **68**, 3373–3379 (2008)
11. Takeda, T., Shindo, Y., Kuronuma, Y., Narita, F.: Modeling and characterization of the electrical conductivity of carbon nanotube-based polymer composites. Polymer **52**, 3852–3856 (2011)

Chapter 12
Effect of Micro-Cracks on the Thermal Conductivity of Particulate Nanocomposite

Addis Tessema, Dan Zhao, Addis Kidane, and Sanat K. Kumar

Abstract The effect of micro-cracks on the thermal conductivity of particle-reinforced nanocomposites is investigated. Two different particles (Carbon nanotube and Silicon dioxide) with different geometries are considered to account for the effect of particle aspect ratio. Three batches of specimens, two with and one without nano-fillers are fabricated. First, the thermal conductivity of the as-fabricated samples were measured using steady state linear heat transfer unit. Afterwards, the samples were subjected to cyclic loading and at the end of every 5000 cycles the samples were taken out and the thermal conductivity was measured. At the same time, the Modulus of Elasticity of the specimens were determined using uniaxial compression test. Based on these results, the effect of micro-cracks on the thermal conductivity of the nanocomposites is presented. In addition, the relation between micro-cracks, stiffness, and thermal conductivity are presented.

Keywords Micro-cracks • Nanocomposite • Thermal conductivity • Fatigue loading • Stiffness

12.1 Introduction

It has been well recognized that different types and amounts of damage could occur in composites during their service time, which negatively affects the physical properties of the composite materials. Thus it is critical to investigate the damage characteristics and its effect on the mechanical, thermal and electrical properties of the bulk materials. The presence of damage, such as micro to macro cracks, and discontinuity inside the material will hinder flow of load (mechanical, thermal and electrical) through the material network. Therefore, the extent and amount of damage has direct implication on the response of the materials. In fact, many studies [1, 2] have shown that it is possible to detect the damage size and sate based on the change in property of the material relative to the undamaged state.

Most common types of damage in fiber reinforced composite are, Fiber breaking, interface failure and matrix micro-cracking [3]. However, for particle filled composites, interface delamination and matrix micro-cracking are the main dominant damages under quasi-static loading [4]. Under fatigue loading the quantity and size of these micro-cracks grows in progress as the number of cycles grows [1, 3, 5]. In parallel as the loading cycle marches, due to the increase in the quantity and size of the cracks developed, the stiffness of the material degrades proportionally [6].

Another material property which could be affected by the presence of damage is thermal conductivity. In general, heat is transported in two ways within the material, one through vibration of lattice (phonon) and other is through electrons wave (similar to electrical conductivity in metals) [6]. In polymers and their composites, phonon is the dominant way to conduct heat [7]; furthermore since polymers are amorphous or semi-crystalline in the solid state, vibration of molecular chains is the primary heat carrier instead of lattice vibration in crystalline materials. The presence of any damage such as, interface between the matrix and filler particles, voids and cracks results in the discontinuity of the polymer chain which induces phonon scattering. Indeed, micromechanics models have shown that presence of micro-cracks has direct influence on the thermal conductivity [2], and also the amount and shape of cracks affects the effective conductivity of the materials.

Studying the effect of micro-cracks/damages within the material on the thermal conductivity, and interrelate the damage extent with variation in effective thermal conductivity using experimental techniques is the motive of this study. Former studies [1, 8] showed that, under fatigue loading the damage has been encountered and these damages has reduced

A. Tessema (✉) • A. Kidane
Department of Mechanical Engineering, University of South Carolina, 300 Main Street, Columbia, SC 29208, USA
e-mail: atessema@email.sc.edu

D. Zhao • S.K. Kumar
Department of Chemical Engineering, Columbia University, 500 West 120th Street, New York, NY 10027, USA

A.M. Beese et al. (eds.), *Fracture, Fatigue, Failure and Damage Evolution, Volume 8*, Conference Proceedings of the Society for Experimental Mechanics Series, DOI 10.1007/978-3-319-21611-9_12

the stiffness and electrical conductivity, "how about the thermal conductivity?" is the question we are going to resolve. Furthermore interaction between amount of damage, stiffness reduction and thermal conductivity variation will be developed.

12.2 Material and Sample Preparation

12.2.1 *Materials Used*

Two types of nanofillers have used to prepare nanocomposites, one is Carbon Nanotube (CNT) and the other is Silica Nanopowder. The nanocomposites are made with weight ratio of 0.5 % for CNT and 1 % for Silica nanofillers. The matrix material used is Bisphenol-A epoxy (Buehler Epothin 2 Resin, Part-A) and its hardener which is recommended by the manufacturer (Buehler Epothin 2 Hardener, Part-B). The Silica fillers have spherical shape with particle size of 10–20 nm (BET) and 99.5 trace metal basis and it is purchased from Sigma Aldrich. On the other hand, CNT fillers are tubular in shape and 95 % pure Carbon Nanotubes (CNT) with average outer diameter of 30 ± 15 nm and length of 5–20 μm, it is purchased from NanoLab Inc. The fabrication processes are briefly discussed in the next section.

12.2.2 *Sample Preparation*

Samples preparation procedures are as followed, first as-received epoxy resin (Part-A) and CNT are weighted based on the predetermined weight ratio (0.5 %). Following the weighted constituents added in to a glass beaker and mixed mechanically until the mixture turns to uniform black color. Furthermore the mixture is brought to ultrasonic mixing for uniform distribution of the filler CNT in to the matrix as shown Fig. 12.2. The ultrasonic mixing is made under pulse mode of 2 s ON and 1 s OFF for 30 min at maximum amplitude of 40 %. Since ultrasonic waves generate heat within the mixture, the beaker was kept under ice bath during ultrasonication. Next, the air trapped (bubbles) within the mixture during ultrasonication and mechanical mixing is removed by keeping the mixture in a vacuum chamber for about 20–30 min. Following, hardener is added in to the mixture (i.e. 1:2 ratio to the epoxy) and mixed thoroughly again. Then, the mixture is brought back in to the vacuum chamber and kept for few minutes again to remove any further air trapped. At last the mixture is poured into silicon rubber molds which are coated with release agent, and the mold is placed in controlled hood to cure for 24 h. As finishing, the top and bottom surfaces of the samples are polished and two holes are drilled along radial direction to place thermocouples. The silica nanocomposites samples are also prepared following similar procedures except that the ultrasonic mixer is not used in this case and the weight ratio is 1 %.

12.3 Experimental Method

12.3.1 *Thermal Conductivity*

A linear unidirectional heat transfer apparatus (Armfield HT11) is used to perform the thermal conductivity experiment based on ASTM C177 standard. As it shown in Fig. 12.1a, there is an electrical heat source (Heater) at the top, followed by 25.4 mm diameter polycarbonate (PC) rod, which is used to transfer the heat to the specimen below. The specimen is sandwiched between the two 25.4 mm diameter polycarbonate (PC) rods above and below. The bottom PC rod is in contact with a 25.4 mm diameter copper rod immersed in constantly running water at one end, which is serve as a heat sink. The apparatus incorporates row of thermocouples to monitor the temperature flow from the source to the sink through the specimen. The sectioned schematic diagram and actual image of the experimental set up are shown in Fig. 12.1b, where T1-T8 represents the thermocouple positions.

The amount of heat is controlled by adjusting the input voltage and in this study all experiments are conducted at 5 V electric supply. The temperature of the specimen during the test was below the glass transition temperature ($Tg = 43$ °C) in order to reduce the errors related to thermal effects. The temperatures are measured using the thermocouples attached directly to the center of the brass rods and specimen. The thermal conductivity of each specimen can be calculated from the temperature records and geometries, using the equation (12.1).

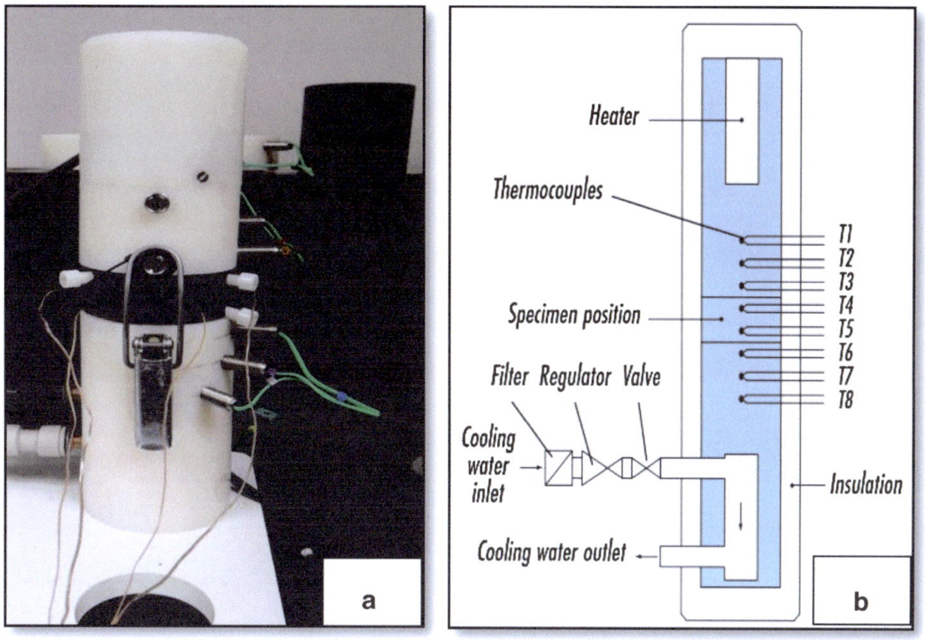

Fig. 12.1 (**a**) Actual and (**b**) schematic diagram of unidirectional/linear heat transfer device

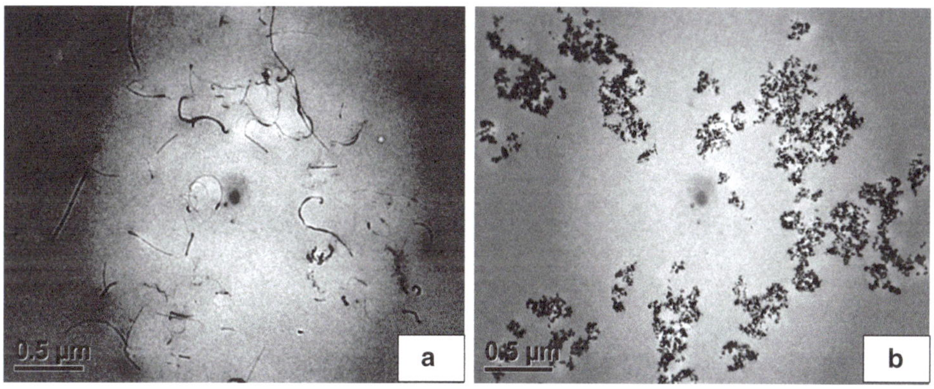

Fig. 12.2 TEM image of ultrasonicated (**a**) CNT-epoxy and (**b**) silica-epoxy composite

$$K = K_{Brass}\left[\frac{(\Delta X_{Brass}) * ((T_2 - T_3) + (T_6 - T_7))}{2 * (\Delta X_{Specimen}) * (T_4 - T_5)}\right] \quad (12.1)$$

Where K_{Brass} is thermal conductivity of the brass conductors, $\triangle X_{Speci} = 13$ mm is the distance between the thermocouple T_4 and T_5; $\triangle X_{Brass} = 15$ mm is the distance between the thermocouple T_2 and T_3 and also between T_6 and T_7.

12.3.2 Fatigue Loading

Damage (Micro-cracks) were generated within the specimen by subjected it to fatigue loading (compression-compression). MTS 810 machine was used for this purpose, as it can be seen from Fig. 12.3a the sample was placed between two flat plates and these flat plates are attached to the corresponding upper and lower grips of the machine. The fatigue test was performed with load ratio of 0.1 (i.e. P_{min}/P_{max}) at 2 Hz frequency in a sine wave form and the load was fluctuating between 890 and 8900 N. The specimen was set to thermal conductivity and unidirectional compressive test to obtain the variation in conductivity and stiffness of the material after every 5000 cycles for the first 20,000 cycles and every 10,000 cycles afterwards. First the as fabricated material is tested and then the thermal conductivity and the compressive stiffness of the specimen as a function of cycle loading is obtained.

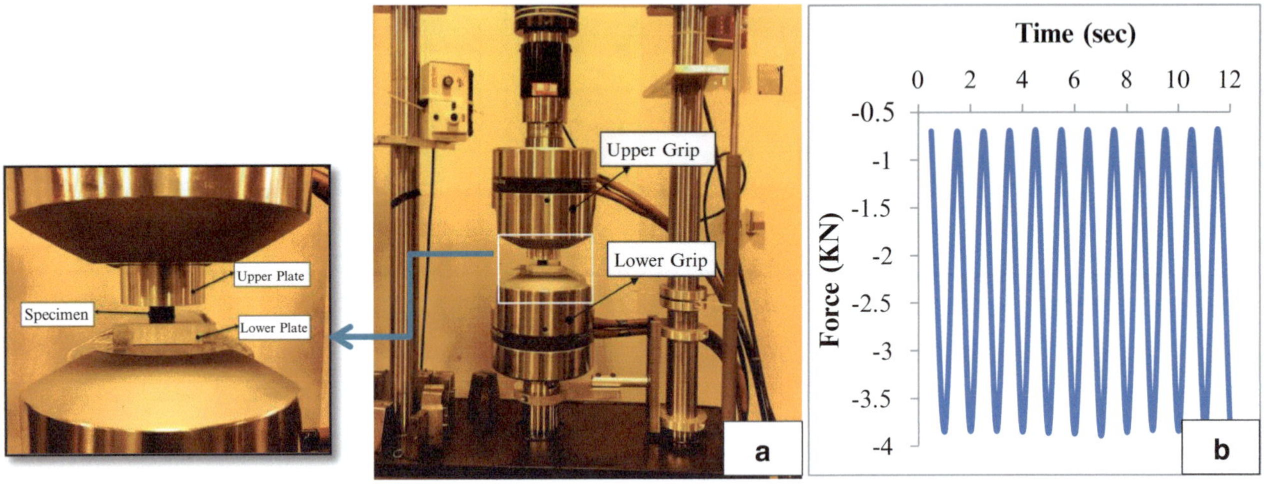

Fig. 12.3 (**a**) Fatigue loading (MTS 810) experimental set-up and (**b**) loading profile

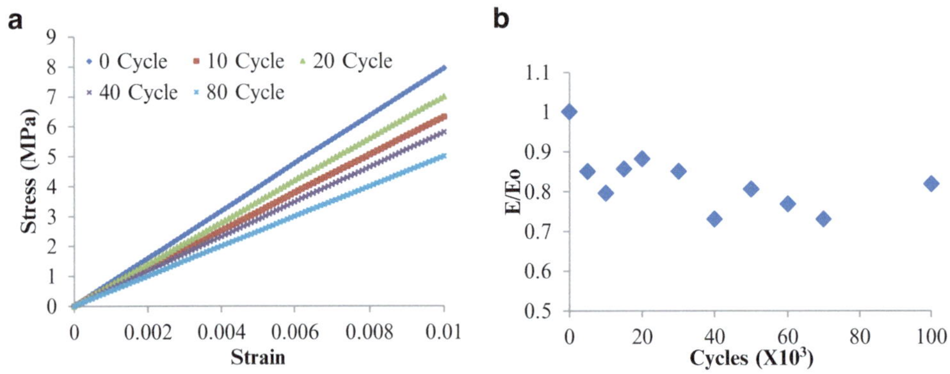

Fig. 12.4 (**a**) Stress-strain plot at different fatigue cycle loading (**b**) normalized modulus of elasticity with cycles

12.4 Results and Discussion

12.4.1 Modulus of Elasticity

The Modulus of Elasticity is measured between consecutive fatigue load intervals. The results for the modulus of Elasticity are shown on Fig. 12.4a, as it can be observed the slope of the stress strain plot has decreased as the loading cycles increased. Stiffness of the material is observed to decrease as number of cycles gets higher, this is taken as indication for the presence of damage within the material.

During compressive fatigue loading of Polymer composites, first the interfaces between the polymer and filler particle start to break and microcracks started to be formed, because the interface is the weakest bond within the material network. Following, since the matrix is under higher deformation than the filler, as the fatigue cycle increases the polymer chains collapses and it starts to rupture. Further fatigue loading starts to crash the nanofiller within the matrix, combination of all this phenomena alters the material network and reduces the stiffness of the material gradually as the cycle grows. In fact, microcracks or damages within the material made discontinuity in material network [6] and this discontinuity introduces interruption on the stress or heat or electric flow within the material. Though, it is difficult to quantify the relation between the number of cycle and reduction in Modulus of Elasticity, it is clearly seen from Fig. 12.4b that rate at which the Modulus

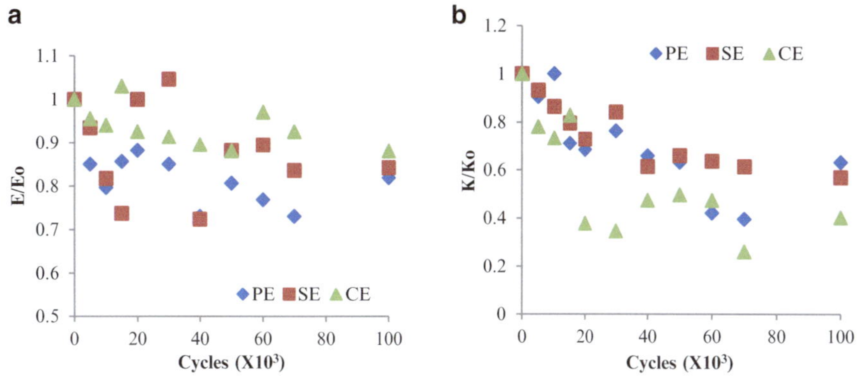

Fig. 12.5 (**a**) Normalized modulus of elasticity with cycles for different samples. (**b**) Normalized thermal conductivity with cycles different samples

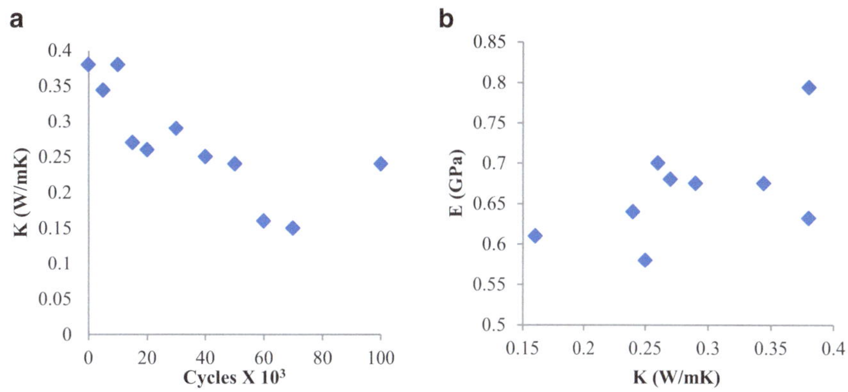

Fig. 12.6 (**a**) Variation of thermal conductivity with cycles and (**b**) modulus of elasticity vs. thermal conductivity

of Elasticity declines is different as the cycle number progress. It decreases with fast rate up to 20×10^3 cycles and it starts to decrease slowly further cycling. Previous studies [1] have shown that there is exponential relation between damage quantity and reduction in stiffness, and from the plots of Fig. 12.4b our result also follow similar trend.

Additionally, from Fig. 12.5b, the scale of degradation for pure Epoxy (PE) and nanocomposites (SE and CE) is different. Relatively higher degradation is observed for PE than the others, this is as a result of nanofiller reinforcement. Since the loading is compressive and nanofillers are the main load carrier, and larger loading cycle is required to damage nanofiller, the modulus of Elasticity is not influenced much for SE and CE as that of PE.

12.4.2 Thermal Conductivity

As in the case of the Modulus of Elasticity, the thermal conductivity of the material between the cycle intervals is measured for different specimen and the results are depicted on Fig. 12.5b. Here the thermal conductivity has shown to decrease as the number of cycle increases. It is believed that the material discontinuities because of microcracks interrupt the continuity of vibration along the chain of the polymer, also between the matrix and filler. This interruption of vibration causes phonon scattering which reduce the efficiency to transfer heat or increases thermal resistance, this situation reduces the thermal conductivity of the material. As it can be seen from Fig. 12.6a the thermal conductivity is decreasing as loading cycle increases in a linear form.

Comparing thermal conductivity degradation for different fillers (Pure Epoxy (PE), Silica-Epoxy (SE) and CNT-Epoxy (CE)) from Fig. 12.4b, the reduction of thermal conductivity as damage grows follows similar trend and the values for different materials are also close to each other for the initial cycles (approximately up to 40×10^3). However, as the loading cycle progresses the nanofilled composites showed slow decrement while the pure epoxy keep reducing almost with the same pace as small loading cycle. From the plot Fig. 12.5b, which shows the modulus of elasticity with thermal conductivity and it is seems that an exponential trend is found between the two quantities.

12.5 Summary

Even though, it is difficult to prepare one sample that can fit for thermal conductivity and cyclic compressive test same time, appreciable result are obtained from our study and it is summarized as;

- Degradation of Modulus of Elasticity is observed, this indicates that damage has occurred within the material. Based on the indication, the extent of damage varies for the pure epoxy and nanofilled epoxy materials.
- The thermal conductivity also has shown reduction as the loading cycle increases, but pace of reduction was not the same for nanofilled sample through the whole cycling intervals, while it almost the same for the pure epoxy.
- The thermal conductivity and Modulus of Elasticity can be related to each other, one can be used to estimate the other implicitly.

Acknowledgement The financial support for this study is funded from NSF under Grant No EEC-1342379 is gratefully acknowledged.

References

1. Yi-Ming, J., Yu-Hsiang, Y.: A study of two-stage cumulative fatigue behavior for CNT/epoxy composites. Procedia Eng. **2**, 2111–2120 (2010)
2. Sevostianov, I., Kachanov, M.: Explicit cross-property correlations for anisotropic two-phase composite materials. J. Mech. Phys. Solids **50**, 253–282 (2002)
3. Liu, S., Nairn, J.A.: Fracture mechanics analysis of composite microcracking: experimental results in fatigue. In: Proceedings of the 5th Technical Conference on Composite Materials, American Society of Composites, East Lansing, 11–14 June 1990
4. Kawaguchi, T., Pearson, R.A.: The effect of particle–matrix adhesion on the mechanical behavior of glass filled epoxies: part 1. A study on yield behavior and cohesive strength. Polymer **44**, 4229–4238 (2003)
5. Kidane, A.: On the failure and fracture of polymer foam containing discontinuities. ISRN Mater. Sci. **2013**, 1–9 (2013)
6. Yu, N., Zhang, Z.H., He, S.Y.: Fracture toughness and fatigue life of MWCNT/epoxy composites. Mater. Sci. Eng. A **494**, 380–384 (2008)
7. Tessema, A., Kidane, A.: The effects of particle size on the thermal conductivity of polymer nanocomposite. In: Tandon, G. (ed.) Composite, Hybrid, and Multifunctional Materials, vol. 4, pp. 151–157. Springer, Cham (2015)
8. Cahill, D.G., Braun, P.V., Chen, G., Clarke, D.R., Fan, S., Goodson, K.E., Keblinski, P., King, W.P., Mahan, G.D., Majumdar, A., Maris, H.J., Phillpot, S.R., Pop, E., Shi, L.: Nanoscale thermal transport. II. 2003–2012. Appl. Phys. Rev. **1**, 011305 (2014)

Chapter 13
Fatigue Tests on Fiber Coated Titanium Implant–Bone Cement Interfaces

M. Khandaker, Y. Li, P. Snow, S. Riahinezhad, and K. Foran

Abstract The goal is developing an efficient bond interface between the implant and the cement by applying micron to nano size fibers to the surface of the implant through an electrospinning process, utilizing biocompatible fibers. Experimental models have been developed to evaluate the forces experienced on a cemented cylinder shape titanium implant through a static and cyclic tests. Finite element analysis (FEA) model for an uncoated cylindrical cemented titanium model was developed and tested under static and fatigue conditions. Our experimental study on cylindrical model found increase of pull out static strength for fiber coated implant (Mean strength = 1.308 MPa) compare to uncoated implant (Mean strength = 1.098 MPa) for 2 samples. Our experimental study also found no noticeable increase of pull out fatigue life for fiber coated implant (Mean fatigue life = 2019 cycles) compare to uncoated implant (Mean fatigue life = 2015 cycles) for 2 samples. Our FEA study on cylindrical model found the design life to be 1690 cycles with element size of 3.0E-3 m under the minimum stress of 112 kPa and maximum stress of 9.71 MPa according to Modified Goodman theorem.

Keywords Titanium • Cement • Interface • PMMA • Polycaprolactone • Fatigue • Implant

13.1 Background

The hip is an important multifunctional joint in the human musculoskeletal system. Since the hip is subject to position change, bending, and extreme force, much wear is imposed on the joint. For these reasons, along with age, health, and weight, a hip replacement is sometimes required and through a total hip replacement (THR) surgery. The most common practice in THR is the use of Titanium (Ti) implants secured in the patients' femur using cement which causes further medical complications including particulates in the blood stream and primarily immobility. The goal of this project is to develop a more efficient bond interface between the implant and the cement by applying micro/nanofibers to the surface of the cylindrical and hip shape implants through an electrospinning process, analyzing results both experimentally and numerically.

Various hip implants have different drawbacks and attributes. Many manufacturers have recalled their hip implants (including Johnson & Johnson, DePuy, and Zimmer Durom) [1]. A patient's age, sex, weight, diagnosis, activity level, surgery condition and implant choice influence the longevity of the device. In the United States of America, five types of total hip replacement devices are currently used with different bearing surfaces [2] which are: metal-on-polyethylene, ceramic-on-polyethylene, metal-on-metal, ceramic-on-ceramic and ceramic-on-metal. The imperfection of the hip implant device can cause pain in the hip or leg, swelling at or near the hip joint, change in walking ability, and popping in the hip joint. A more suitable hip implant device which decreases the number of risks is needed [3].

The interfacial mechanics at the bone-cement or implant-cement interfaces is a critical issue for implant fixation and the filling of tissue defects created by disease [4]. Electro spinning is a process by which fibers with sub-micron diameters can be obtained from an electrostatically driven jet of polymer solution [5]. These fibers have a high surface area to volume ratio, which have numerous engineering applications [6–8]. The present study is based on the hypothesis that the differences of the surface properties at bone/cement interface due to incorporation of micro and sub-micron diameters fiber may have significant influence on the quality of bone/cement union. The objectives of this research were to: (1) develop aligned Poly(ε-caprolactone) PCL fiber coated cylindrical titanium implant, (2) conduct the fatigue tests on without and with fiber coated cylindrical titanium implant, and (3) develop a finite element model to determine the life of the without and with fiber coated cylindrical titanium implant.

M. Khandaker (✉) • Y. Li • P. Snow • S. Riahinezhad • K. Foran
Department of Engineering and Physics, University of Central Oklahoma, Edmond, OK 73034, USA
e-mail: mkhandaker@uco.edu

A.M. Beese et al. (eds.), *Fracture, Fatigue, Failure and Damage Evolution, Volume 8*, Conference Proceedings of the Society for Experimental Mechanics Series, DOI 10.1007/978-3-319-21611-9_13

13.2 Materials and Methods

13.2.1 Materials

Titanium (Ti) bars (6Al-4V ELI, ASTM B 348 standard, grade 23, biocompatible) of dimension (22 × 12 × 2) mm were purchased from Titanium Metal Supply, Inc., Poway, CA. Cobalt™ HV bone cement (Biomet Inc., Warsaw, IN) was used as the polymethyl methacrylate (PMMA) cement. PCL was purchased from Sigma Aldrich. PCL was selected as fiber material since it is biocompatibility, biodegradable and electrostatic. PCL solution was prepared by ultrasonic (Sonics & Materials, Inc., Vibra-cell VCX 130) mixing of 7.69 wt% of PCL beads with acetone. The sonication process was carried out at approximately 80 °C for an hour. The solution was poured into a glass syringe in an infusion pump (Harvard Ins.).

13.2.2 Deposition of Aligned Fiber on Titanium Implant

The cylindrical titanium implant will undergo the electrospinning process called drum collection. This process was chosen because of its capability of high precision controlling deposition of the fibers and producing nano-level fibers. The detail method of preparation of PCL solution for electrospun fiber is appeared in Khandaker et al. [2]. A rotating collecting drum is spun at high speed with a DC motor which is used in conjunction with a Proportional Integral Derivative (PID) control system to control the revolutions of the motor under the electrospinning setup as shown in Fig. 13.1. The implant model will then undergo the same electrospinning process as the cylindrical model in order to coat the model with a layer of micro/nanofibers.

13.2.3 Preparation of Titanium/Cement Samples

The titanium rod is encapsulated by a 3D printed (Dimension elite 3D printer) cylindrical holder in order to cure the rod to the cement without being contaminated. Figure 13.2 shows the fabricated holder for the production of titanium/cement sample.

Fig. 13.1 Electrospin setup for the deposition of aligned fiber on titanium rod

Fig. 13.2 Fabricated Ti/PMMA sample for static and fatigue tests

According to Biomet HV PMMA cement preparation protocol, PMMA cement was prepared by hand mixing 2.2 g of PMMA powder with 1.1 mL of methyl methacrylate (MMA) monomer using powder: monomer ratio of 2:1. PMMA was poured on top of the different titanium samples during doughy phase in the mold. After the cement is allowed to cure the cylinder can be removed and placed in the Test Resources 800LE4 dynamic universal testing machine to be put the titanium/PMMA specimen under cyclic loading for a fatigue test.

13.2.4 Mechanical Testing on Titanium/Cement Samples

Pull out static and cyclic tests were performed on two Ti/PMMA samples with and without using Test Resources universal test system (UTM) (Fig. 13.3). Static tests were conducted to find maximum breaking stress of Ti/PMMA samples at strain rate 0.05 mm/s. The maximum pull-out force will be determined and interfacial shear strength will be calculated by dividing the force at the point of failure by the surface area of the implant in contact with tissue. Cyclic tests (amplitude = 100 N) were conducted to find the fatigue life of Ti/PMMA samples at 1 Hz using stepwise load (100 N) for 1000 cycles starting from 350 N until the failure of the interfaces.

13.2.5 Finite Element Modeling

The Finite Element Analysis (FEA) model was build according to the dimension of the experimental model. The geometric FEA model without micro/nanofiber was modeled and analyzed in ANSYS Workbench 14.0 (Fig. 13.4) with different element size of 5.0E-3, 4.0E-3, 3.0E-3 and 9E-4 m. Materials used in the analysis were titanium alloy (rod), PMMA cement, and abs plastic which corresponds to the 3D printed holder in the experimental model. Table 13.1 shows the properties of materials used for FEA analysis.

Figure 13.5 shows the meshed FE model with the element size of 9.0E-4 m. The fatigue test was processed under static structural analysis. Modified Goodman theory was used for the prediction of life. A downward load of 475N that corresponds to the failure load for the experimental model was applied to the top surface of the titanium rod. A normal stiffness factor of 1 was used for the Ti/PMMA interface, and ANSYS program controlled stiffness factor was used for PMMA/abs plastic interface to model the interface. The outside surface of the abs plastic model were constraint in all direction.

Fig. 13.3 Fatigue test setup with specimen

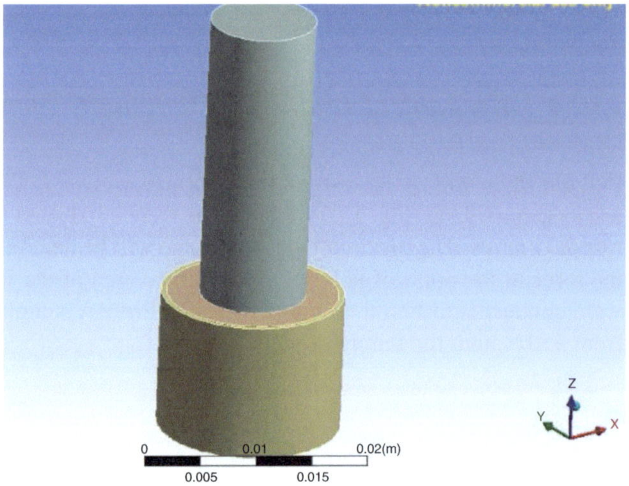

Fig. 13.4 Geometric model of Ti/PMMA for FEA analysis

Table 13.1 Material Properties used for the development of the Ti/PMMA model

Properties	Titanium alloy	PMMA cement	ABS plastic
Young's modulus (GPa)	116	3.3	2.2
Poisson's ratio	0.32	0.39	0.35
Shear modulus (GPa)	43.939	1.1871	0.81481
Density (kg/m^3)	4507	1190	1040
Compressive ultimate strength (MPa)	120	120	50
Compressive yield strength (MPa)	970	120	65
Tensile ultimate strength (MPa)	950	69	40
Tensile yield strength (MPa)	880	70	43

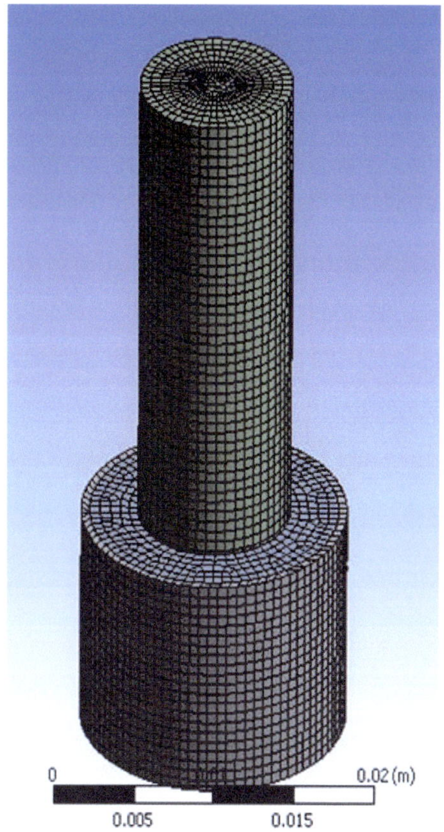

Fig. 13.5 Meshed model of Ti/PMMA samples for fatigue analysis with element size of 9.0E-4 m

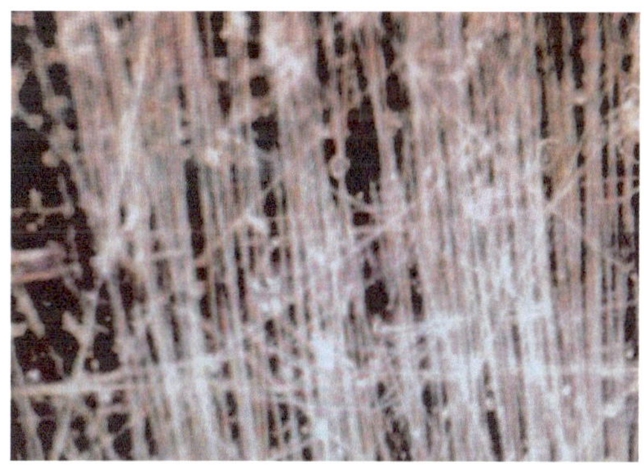

Fig. 13.6 Fiber coated titanium implant

13.3 Results

The study was able to create micro to nano size fiber coated sample. Uni-directional fibers were successfully fabricated using the electrospinning unit having horizontal drum collector method. The fiber collections were viewed in stereomicroscope (Fig. 13.6).

The typical load and displacement plots of a with and without fiber Ti/PMMA specimen during the quasi-static tests can be found in Fig. 13.7. Our studies showed a clear difference of the load-displacement curve among different kinds of titanium cement specimens. The result shows the Ti/cement without fiber has higher stiffness compared to Ti/cement with fiber. The result also shows that Ti/cement incorporated with fiber has the ability to store higher energy before reaching a breaking point than Ti/cement without fiber.

The fatigue life and strength of the cylindrical model with and without fiber were compared as shown in Fig. 13.8. The results showed that the samples with fiber had greater fracture strength and slightly higher fatigue life. The additional surface roughness due to the deposition of the fiber is the main reason for this improvement of the fracture strength.

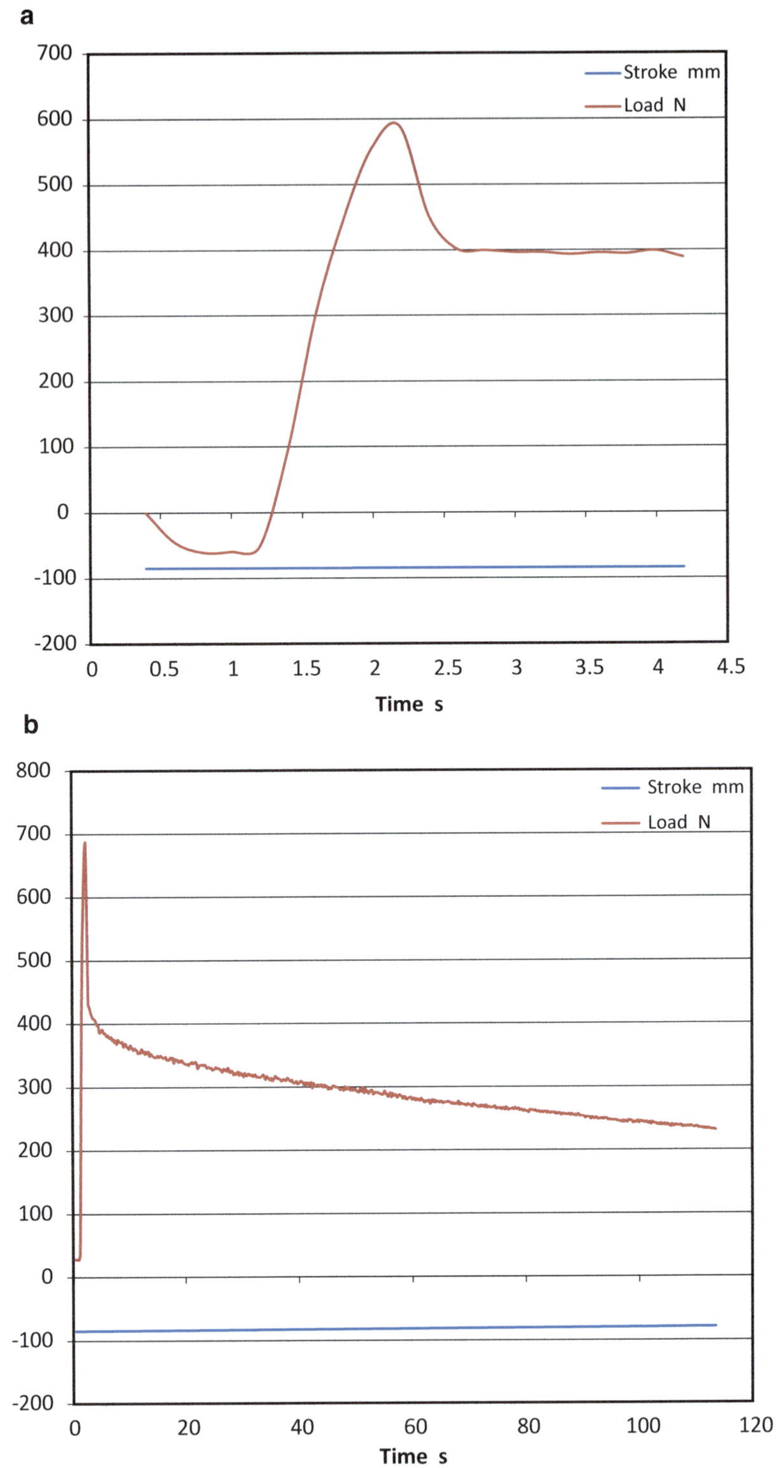

Fig. 13.7 Load and deformation (**a**) with and (**b**) without fiber Ti/PMMA specimen during the quasi-static tests

The FEA contour plots of stress and deformation of the Ti/PMMA model are shown in Fig. 13.9 and Fig. 13.10, respectively. The ANSYS simulation analyzed fatigue test for the cylindrical model under static structural analysis and the design life according to Modified Goodman theorem was found to be 1690 cycles with element size of 3.0E-3 m under the minimum stress of 112 kPa and maximum stress of 9.71 MPa; the life found with element size of 4.0E-3 m was 2367.7 cycles under the minimum stress of 339 kPa and maximum stress of 8.72 MPa; and the life found with element size of 5.0E-3 m was 1905.1 cycles under the minimum stress of 310 kPa and maximum stress of 8.25 MPa The relationship between mesh

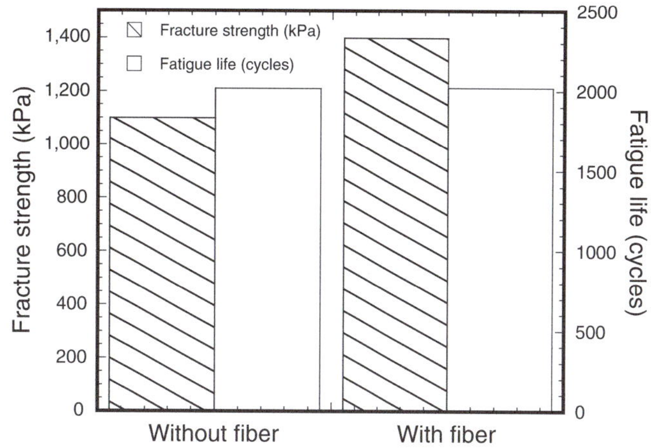

Fig. 13.8 Fracture strength and fatigue life comparison between fiber coated Ti/PMMA and without fiber coated Ti/PMMA samples

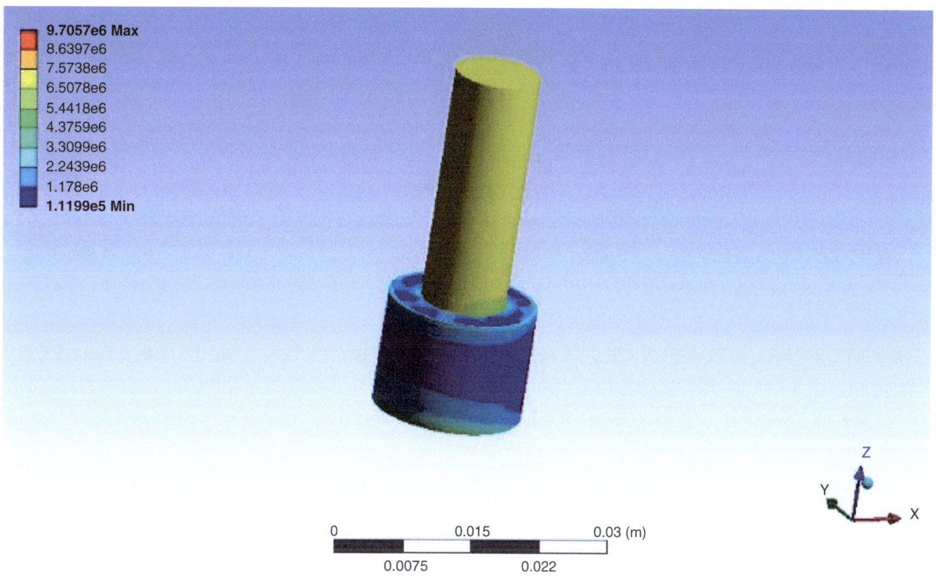

Fig. 13.9 Contour plot of Von-Mises stress

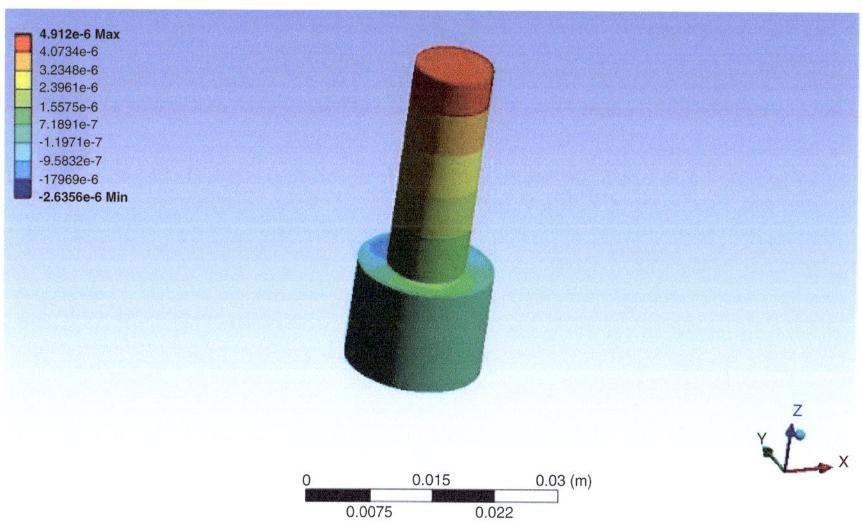

Fig. 13.10 Contour plot of deformation in y-direction

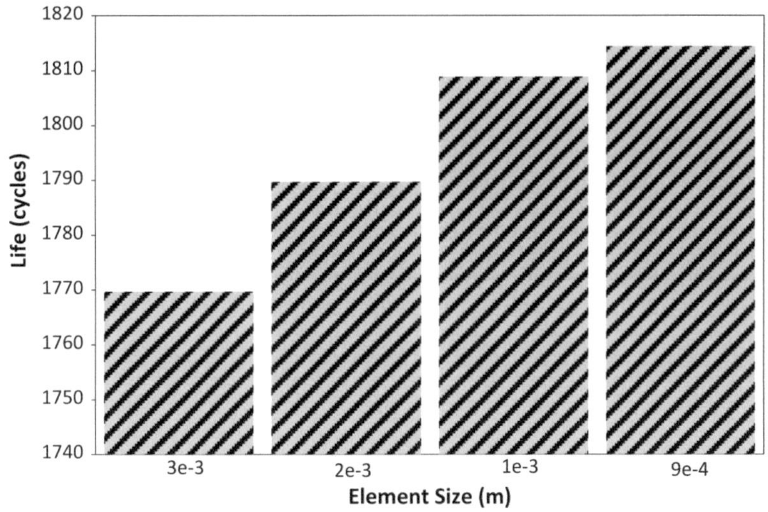

Fig. 13.11 Fatigue life variations with different meshing element sizes

Table 13.2 Convergence tests of the FEA results

				Alternating stress (Pa)			
Element size (m)	#Elements	#Nodes	Life cycles	Minimum stress	Maximum stress	Mean stress	% of difference between iterations
3.00E-03	524	3406	1769.7	1.91E+05	1.16E+07	5.90E+06	
2.00E-03	1244	7491	1789.7	1.22E+04	2.04E+07	1.02E+07	0.561892454
1.00E-03	8296	41760	1808.9	2.89E+04	2.12E+07	1.06E+07	0.533540821
9.00E-04	11607	56298	1814.4	1.44E+04	2.32E+07	1.16E+07	0.151795325

size and fatigue life is presented in Fig. 13.11. The difference between FEA predicted fatigue life and experiment fatigue life for titanium/PMMA model without fiber was found to 17.56 % for the element size of 3 mm.

Table 13.2 shows convergence of FEA results. The mesh size at 0.9 mm is sufficient, since the FEA results with 0.9 mm mesh size differ by less than 0.1 % from those with mesh size of 3 mm.

There are limitations in the current work. Only a small number of samples were tested under static and cyclic compressive loadings, thus it is not possible to present any statistical evaluations, but only a general trend of fracture evolution in titanium-cement interfaces built with and without fiber coated titanium.

13.4 Conclusions

The micro/nanofiber coated titanium/cement interface was designed and constructed by electrospinning process. PCL-collagen was selected for the fiber preparation because of its easy mixture. Fatigue analysis of implant model was done both numerically and experimentally. The predicted life of the developed FEA model was comparable with the experimental fatigue life of titanium/PMMA interface without fiber. Further in vitro and in vivo improvement of fatigue life for titanium/PMMA interface is possible by aligned, uniform and less stiff fiber coating on Ti and adding nanoparticles with PMMA that is sought in our future studies.

Acknowledgements This research was made possible by grant 8P20GM103447 from the US National Institutes of Health and an on-campus faculty grant program from the University of Central Oklahoma Office of Research and Grants.

References

1. FDA U.S. Food and Drug Administration. http://www.fda.gov/medicaldevices/productsandmedicalprocedures/implantsandprosthetics/metalonmetalhipimplants/ucm241770.htm
2. FDA U.S. Food and Drug Administration. http://www.fda.gov/MedicalDevices/ProductsandMedicalProcedures/ImplantsandProsthetics/MetalonMetalHipImplants/ucm241594.htm#risks

3. Ramos, A., Completo, A., Relvas, C., Simões, J.A.: Design process of a novel cemented hip femoral stem concept. Mater. Des. **33**, 313–321 (2012)
4. Khandaker, M., Tarantini, S.: Material mismatch effect on the fracture of a bone-composite cement interface. Adv. Mater. Sci. Appl. **1**(1), 1–8 (2012)
5. Khandaker, M., Utsaha, K.C., Morris, T.: Interfacial fracture toughness of titanium-cement interfaces: effects of fibers and loading angles. Int. J. Nanomedicine **9**(1) (2014)
6. Deitzel, J., Kleinmeyer, J., Hirvonen, J., Beck, T.: Controlled deposition of electrospun poly(ethylene oxide) fibers. Polymer **42**, 8163–8170 (2001)
7. Kwon, I.K., Kidoaki, S., Matsuda, T.: Electrospun nano- to microfiber fabrics made of biodegradable copolyesters: structural characteristics, mechanical properties and cell adhesion potential. Biomaterials **26**(18), 3929–3939 (2005)
8. Wang, H.B., Mullins, M.E., Cregg, J.M., Hurtado, A., Oudega, M., Trombley, M.T., Gilbert, R.J.: Creation of highly aligned electrospun poly-l-lactic acid fibers for nerve regeneration applications. J. Neural Eng. **6**(1), 016001 (2009). doi:10.1088/1741-2560/6/1/016001

Chapter 14
Fatigue Behavior of Carburized Steel at Long Lives

D.V. Nelson and Z. Long

Abstract The fatigue behavior of two carburized steels is investigated with rotating bending tests for lives between approximately 10^5 and 10^8 cycles. Cracks are observed to start at sub-surface inclusions within carburized cases and form "fish eyes" often seen in high cycle fatigue of hardened steels. Optically dark areas (ODAs) are found surrounding the inclusions that acted as crack origins. Fatigue strengths are evaluated by Murakami's $\sqrt{area}$ model, but accounting for the influence of residual stress and hardness at the different depths at which cracks started. Values of stress intensity factor range at the periphery of inclusions and the outer border of ODAs are computed accounting for the influence of residual stress. Possible factors that may influence crack growth within ODAs are discussed.

Keywords Fatigue • Carburized • Inclusion • Fish eye • Residual stress

14.1 Introduction

Numerous mechanical components made of high strength steel are designed to have fatigue lives exceeding a million cycles, a regime in which failures tend to originate at subsurface inclusions. A penny shaped crack typically grows from an inclusion with a distinct appearance known as a "fish-eye." An example is shown in Fig. 14.1. This type of fatigue failure has been the subject of considerable study during the past decade, and is described in detail in a book [1] and number of review articles [2–6]. A zone of special interest surrounding an inclusion is often found within a fish eye. The zone is referred to as an optically dark area (ODA) [1] (or in some cases, a granular bright facet [7], fine granular area [3] or rough surface area [8]). ODAs are believed to start early in fatigue life (first 10 % or so of life) and gradually increase in size, occupying most of the fatigue life [9, 10]. Very little life is typically spent in the growth of a crack from the ODA size to fracture [11]. A sample of an ODA is shown in Fig. 14.2.

Studies of "fish eye" fatigue have primarily used through-hardened specimens. Many mechanical components are case-hardened, producing a profile in hardness and residual stresses as a function of depth below the surface. Carburizing is a widely used case hardening process applied to gears, bearings, camshafts, etc. Not only does it alter subsurface hardness and create residual stresses, but it also leads to hydrogen around internal nonmetallic inclusions [12]. In this paper, the high cycle fatigue behavior of carburized gear steels is explored to assess the influences of residual stresses and hardness on "fish eye" fatigue cracks occurring within carburized cases.

A 1968 study by Funatanai and Noda [13] on the bending fatigue of carburized specimens noted that fish eye cracks initiated at depths approximately equal to case depth, where compressive residual stresses were relatively minimal (less than −100 MPa) compared to bending stresses. Naito et al. [14] tested carburized specimens with either non-martensitic or electropolished surface layers. Surface failures originated in non-martensitic layers even at lives approaching 10^8 cycles (the highest number of cycles tested). In electropolished specimens, several fish eye failures at inclusions were noted for lives of about 5×10^6 cycles and greater, but sizes and depths of inclusions were not reported except in one photo where the depth was approximately 250 μm, which was within the carburized case. Murakami et al. [15] tested carbonitrided specimens with a case depth of approximately 2 mm. Fish eye cracks started at inclusions deeper than the case. Residual stresses vs. depth

D.V. Nelson (✉)
Mechanical Engineering Department, Stanford University, Stanford, CA 94305, USA
e-mail: dnelson@stanford.edu

Z. Long
Department of Mechanical and Vehicular Engineering, Beijing Institute of Technology, Beijing 100081, China

Mechanical Engineering Department, Stanford University, Stanford, CA 94305, USA

A.M. Beese et al. (eds.), *Fracture, Fatigue, Failure and Damage Evolution, Volume 8*, Conference Proceedings of the Society for Experimental Mechanics Series, DOI 10.1007/978-3-319-21611-9_14

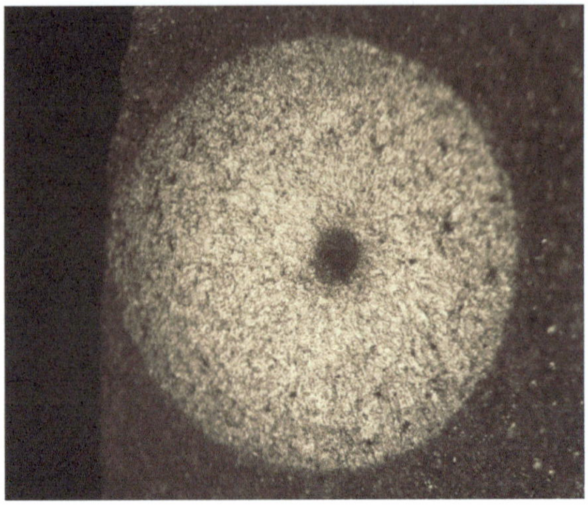

Fig. 14.1 Typical fish eye observed originating at a subsurface inclusion in a high hardness steel. (Specimen surface is shown to the *left* of the fish eye.)

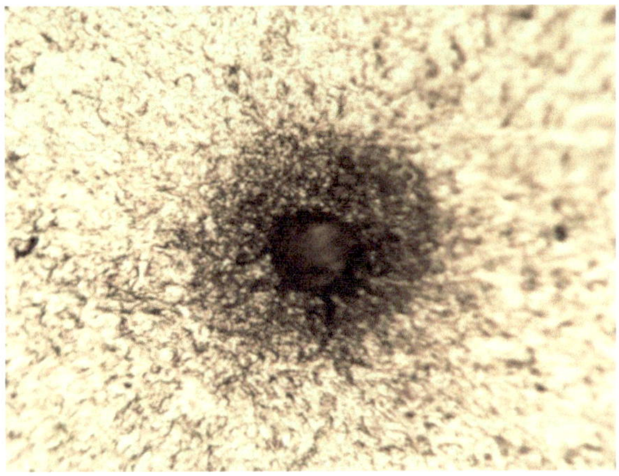

Fig. 14.2 Optically dark area surrounding an inclusion

within the case were not reported, which is understandable since cracks did not originate there. Huang et al. [16] tested carburized specimens with a relatively shallow case (approx. 0.3 mm). Depths of inclusions at which cracks formed and residual stresses vs. depth were not mentioned.

This study differs from those just reviewed by reporting the results of fatigue tests in which all fish eye cracks started within the cases of carburized specimens, at depths where magnitudes of compressive residual stresses were significant relative to applied stresses and where hardness values were much higher than in the core. To date, approaches for predicting fish eye fatigue strength have been developed largely from test data for cracks originating at depths with minimal residual stresses and hardness nearly uniform with depth. This study provides an opportunity to assess the applicability of a prominent predictive approach using data from locations where residual stresses and hardness values much larger than in the core.

14.2 Experimental Procedure

Two Cr-Ni carburized gear steels, 20Cr2Ni4 and 17Cr2Ni2, were tested. Table 14.1 shows their nominal chemical compositions and tensile properties. Specimens were carburized at 910 °C for 20 h with a carbon potential of 1.2 % and then air-cooled and held at 650 °C for 2 h. After that, the specimens were austenitized at 880 °C for 2 h, oil quenched, and tempered at 180 °C for 2 h. The geometry and nominal dimensions of specimens are shown in Fig. 14.3. After carburizing and hardening, specimens were gently ground and polished in the longitudinal direction with abrasive paper (to grade #2000) to achieve a mirror-like finish.

Carburized cases were etched with picric acid to reveal prior austenite grain boundaries. Applying ASTM E112-10 [17], the average grain size was determined to be G8.5 and G9 for steels A and B, respectively, or 19 and 16 μm. Micro-hardness profiles of carburized cases determined according to ASTM E384-10 [18] are reported in Fig. 14.4. The two steels had similar profiles with depth, and the core hardness values were nearly the same, about 450 HV. Longitudinal residual stresses

Table 14.1 Chemical composition (wt.%) and tensile properties of the steels prior to carburizing

Alloy	C	Si	Mn	Cr	Ni	Mo	S_y (MPa)	S_u (MPa)	RA (%)
A	0.17～0.23	0.17～0.37	0.3～0.6	1.25～1.65	3.25～3.65	0.1～0.1 5	1292	1483	57
B	0.14～0.19	0.17～0.37	0.3～0.6	1. 5～1.8	1.4～1.7	0.25～0.35	1150	1400	35

S_y yield strength, S_u ultimate tensile strength, *RA* reduction of area

Fig. 14.3 Geometry of fatigue specimens (dimensions in mm)

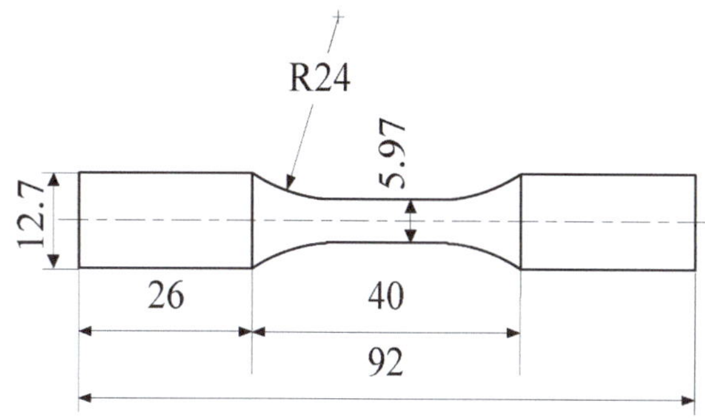

Fig. 14.4 Micro-hardness profiles of carburized steels *A* and *B*

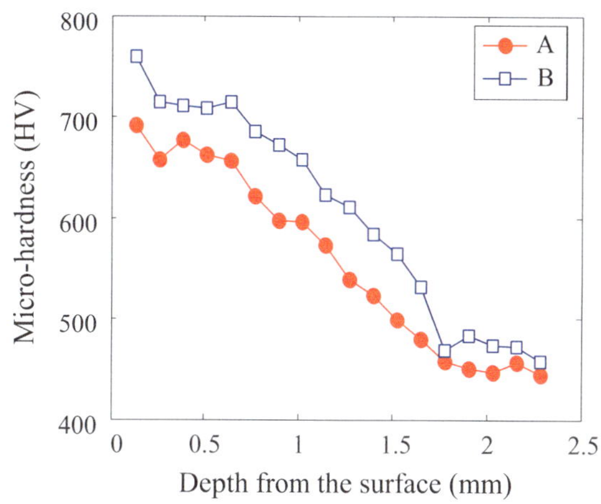

Fig. 14.5 Residual stress profiles for carburized steels *A* and *B*

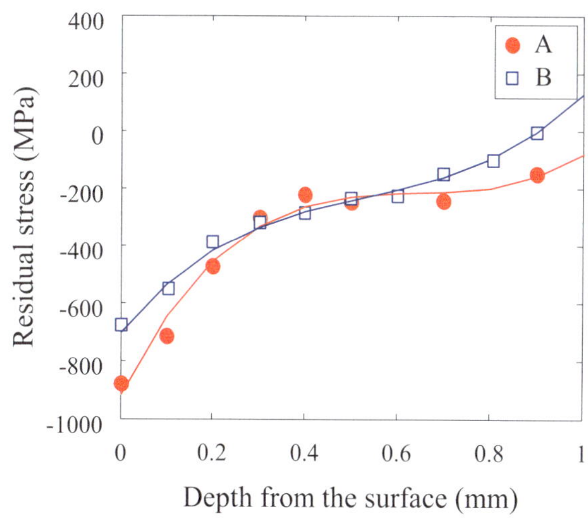

vs. depth were measured by X-ray diffraction using conventional Cr-Ka radiation (2.291 Å wavelength), [211] planes and the $\sin^2\psi$ method combined with layer removal by electropolishing. Stresses were corrected for the effect of layer removal [19]. Residual stresses measured in specimens that had been cycled without failure are shown in Fig. 14.5. The peak value of compressive residual stresses in steel A was −880 MPa, about 200 MPa more compressive than that in steel B. The transition depth from compressive to tensile residual stresses in the steel A was 1.1 mm, 0.2 mm deeper than that for the steel B.

Magnitudes of residual stresses in Fig. 14.5 associated with carburizing (depths greater than about 0.2 mm) are comparable to typical values in the range of −250 to −350 MPa [20] while values near the surface are in line with those approaching −800 MPa reported for carburized and ground gear teeth [21]. The magnitudes of residual stresses measured suggest that little or no residual stress relaxation occurred from fatigue cycling. The yield strength near the surface of specimens was estimated to be approximately 2000–2100 MPa based on correlations between hardness and yield strength for carburized steel [22]. The bending stress amplitude at the surface was 1150 MPa or less in the testing to be described shortly, so that the sum of applied and residual stresses would not be expected to cause significant relaxation of residual stresses by plastic deformation [23].

14.3 Experimental Results

Rotating bending fatigue tests were conducted at room temperature in air. The test frequency was 75 Hz. The results of fatigue testing are shown in Fig. 14.6 where stress amplitude refers to values at the surface. Also included in that plot are recent data from specimens of the same type as in Fig. 14.3, except with a nominal diameter of 5.5 mm. Fatigue tests with specimens of the original diameter required loading levels that unexpectedly caused ball bearings of the test machine to fail from fatigue. A smaller specimen diameter allowed subsequent testing at lower loads. Two B specimens had unexpectedly short lives, for reasons that are not immediately evident.

The fracture surfaces of the failed specimens were examined by optical microscopy as well as a scanning electron microscope (SEM) equipped with an X-ray energy dispersive spectrometer (EDS). Based on the result of EDS, all internal fractures originated at non-metallic inclusions that were either Al_2O_3 or TiN. The sizes of inclusions in terms of a $\sqrt{area}$ parameter ranged between 16 and 49 μm, where *area* is the projected area of a defect or crack on a plane normal to the direction of maximum tensile stress [1]. In the case of a circular area, the parameter $\sqrt{area} = 0.89 \times$ diameter. Inclusions causing fish eye cracks were observed at depths from 10 to 660 μm. Optically dark areas (ODAs) could be always identified at the crack initiation site with sizes in terms of the $\sqrt{area}$ parameter (including the area of the inclusion within an ODA) varying between 27 and 90 μm. Figure 14.7a shows an example of a fish-eye fracture that initiated from an unbroken Al_2O_3 inclusion, while Fig. 14.7b shows a pore left by an inclusion debonded from the matrix, suggesting relatively weak bonding between the inclusion and matrix. Several radial ridges emanated from the crack origin. The length of these ridges ranged from 16 to 100 μm, and many of the short ridges stopped at the outer border of ODAs. Nishijima and Kanazawa [24] have suggested that such ridges indicate shear cracking similar to what might be found for cracks initiated at a surface.

The size of each ODA normalized by the size of the corresponding inclusion from which a fish eye crack started is shown in Fig. 14.8. Although there is significant scatter, the ratio of ODA-to-inclusion size tends to increase at longer lives.

For the remainder of this paper, evaluations will only consider the results from specimens with the original diameter, since data on residual stress and hardness vs. depth for the smaller diameter specimens are not yet available as of this writing. The data are expected to be similar to those in Figs. 14.4 and 14.5, but that needs to be determined. Figure 14.9 shows depths of inclusions from which cracks originated relative to values of hardness and to the sum of applied bending stress amplitude plus compressive residual stress vs. depth. The applied stress accounted for the linear variation with depth and, for the sake of illustration, used bending stress surface values of 1060 and 1120 for steels A and B, respectively. Referring to Fig. 14.5,

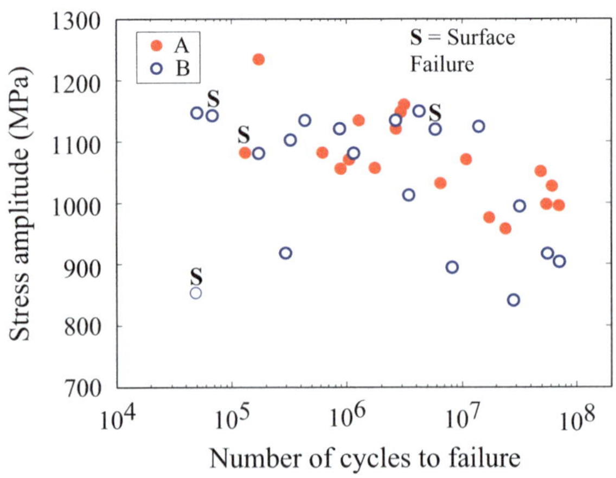

Fig. 14.6 Fully-reversed bending fatigue data for steels *A* and *B*

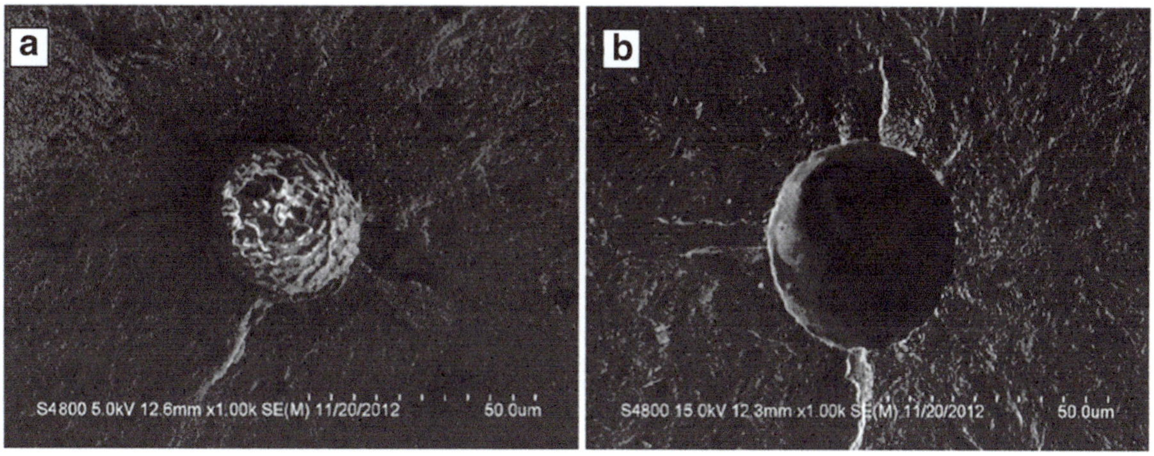

Fig. 14.7 (**a**) Inclusion at the center of a fish eye for a specimen of steel A, $\sigma_a = 1041$ MPa, $N_f = 2{,}651{,}000$ cycles, $\sqrt{area} = 28$ μm, (**b**) pore at the center of a fish-eye for a specimen of steel B, $\sigma_a = 1067$ MPa, $N_f = 1{,}149{,}800$ cycles, $\sqrt{area} = 35$ μm

Fig. 14.8 Ratio of the size of an ODA to that of the inclusion from which fish eye cracking developed

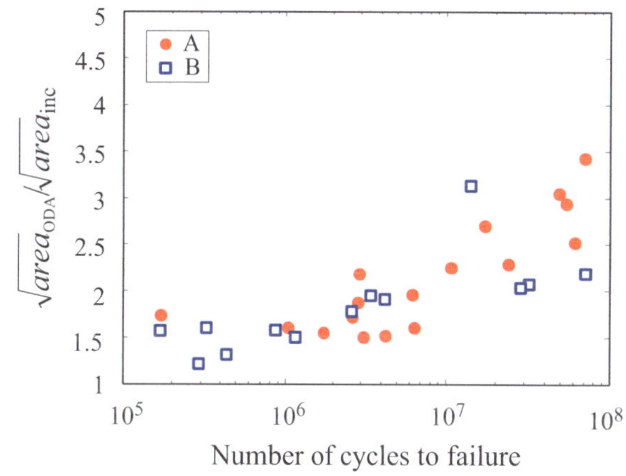

Fig. 14.9 Depths of inclusions (X) from which fish eye cracks started within carburized cases

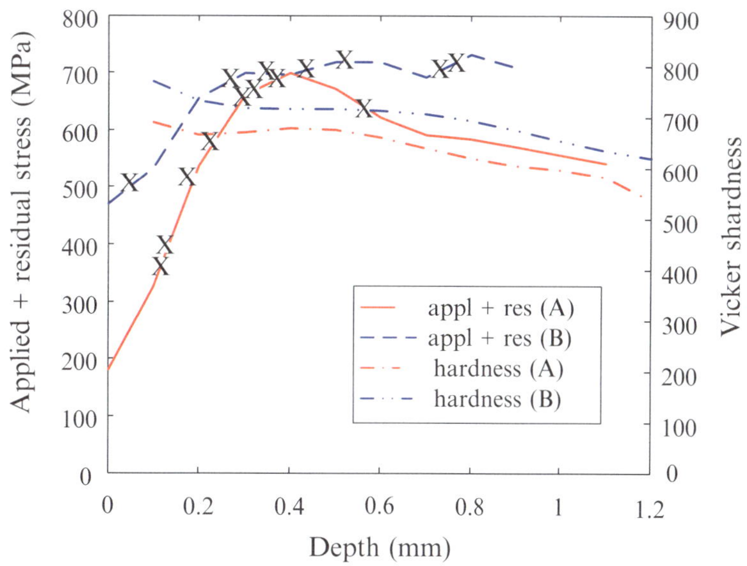

note that many cracks started at depths where compressive residual stresses were substantial. Also, there was no evidence of oxidation or decarburization at any of the locations where cracks started.

The fatigue strength, σ_w (MPa), of each specimen with a fish eye fracture was estimated by Murakami's $\sqrt{area}$ parameter model [1] shown in (14.1), derived by treating a nonmetallic inclusion approximately as a small crack:

$$\sigma_w = \frac{1.56\,(Hv + 120)\,[(1\text{-}R)/2]^{\beta}}{\left(\sqrt{area_{inc}}\right)^{1/6}} \tag{14.1}$$

where: Hv is Vickers hardness, $area_{inc}$ is the projected area (μm) of an inclusion normal to the direction of stress, R is the ratio of minimum to maximum stress in a cycle, and β is given by:

$$\beta = 0.226 + 10^{-4} Hv \tag{14.2}$$

$$R = (\sigma_m - \sigma_w) / (\sigma_m + \sigma_w) \tag{14.3}$$

where σ_m is mean applied stress.

The effectiveness of (14.1) has been demonstrated for high cycle fatigue with different values of applied mean stress [1]. Here, the influence of residual stresses σ_{res} is taken into account as if it was a mean stress σ_m in (14.3). The bending stress amplitude σ_a at the sub-surface location of an inclusion was obtained from

$$\sigma_a = \sigma_{nominal}\left(1 - \frac{d_{inc}}{r_s}\right) \tag{14.4}$$

where $\sigma_{nominal}$ is the bending stress amplitude at the surface, d_{inc} is the depth of an inclusion from the surface, and r_s is the measured radius of the minimum section of specimens. Values of hardness and residual stress present at the depth of a given inclusion were used in (14.1) to (14.3). Since σ_w is a function of R in (14.1), and R is a function of σ_w in (14.3), determination of both can start by taking $\sigma_w = \sigma_a$ to obtain an initial value of R, and then to find an updated value of σ_w, which is used to find an updated of R, and so forth until convergence is obtained.

Figure 14.10a shows the relation between the ratio of the applied stress amplitude to the estimated fatigue strength, (σ_a/σ_w), and the number of cycles to failure, N_f. The ratio (σ_a/σ_w) is conservative and varies between 1.10 and 1.22 for steel A and 1.04 and 1.15 for steel B. The scatter in values of (σ_a/σ_w) in both steels appears comparable. The values of (σ_a/σ_w) provide evidence that Murakami's model in (14.1) is also effective for fatigue strength estimates of carburized and ground steel when fish eye cracks start at depths in carburized cases where residual stresses are significant and fatigue lives are in the range of 10^5 to 10^7 cycles for the data set from specimens of original diameter considered. The relation between (σ_a/σ_w) and the depth of inclusions is shown in Fig. 14.10b. There is no apparent dependence of (σ_a/σ_w) on the depth of inclusions for the data set considered. The use of $\sqrt{area_{ODA}}$ rather than $\sqrt{area_{inc}}$ in (14.1) would provide lower, more conservative estimates of σ_w (approximately 7 % lower on average for the data here).

It has been proposed that there may be two kinds of crack growth thresholds for fish eye failures. A larger one, characterized by the stress intensity factor range ΔK at the outer border of ODAs, represents a threshold for a transition from small crack growth to "Paris law" behavior [25], and a smaller one, characterized by ΔK at the border of inclusions, represents a threshold for crack growth to occur from an inclusion [26]. To account for the effect of compressive residual stress, the stress intensity factor range at the border of an inclusion, ΔK_{inc}, and at the outer border of an ODA, ΔK_{ODA}, was computed using an $\sqrt{area}$ model [1] with superposition of applied and residual stresses:

$$\Delta K_{inc} = 0.5\,(\sigma_a + \sigma_{res})\sqrt{\pi\sqrt{area_{inc}}} \tag{14.5}$$

$$\Delta K_{ODA} = 0.5\,(\sigma_a + \sigma_{res})\sqrt{\pi\sqrt{area_{ODA}}} \tag{14.6}$$

where $\sqrt{area_{inc}}$ and $\sqrt{area_{ODA}}$ are the projected areas of an inclusion and an ODA, respectively. Values of applied stress amplitude, σ_a, and residual stress, σ_{res}, were taken at the sub-surface location of a given inclusion. (Compressive σ_{res} has a negative sign in (14.5) and (14.6).)

Figure 14.11 shows ΔK_{inc} and ΔK_{ODA} values vs. cycles to failure N_f as well as depths of inclusions for steel A. There is no apparent relation with N_f or depth. The values of ΔK_{inc} vary between 2.5 and 3.3 MPa$\sqrt{m}$ with an average of 2.8 MPa$\sqrt{m}$.

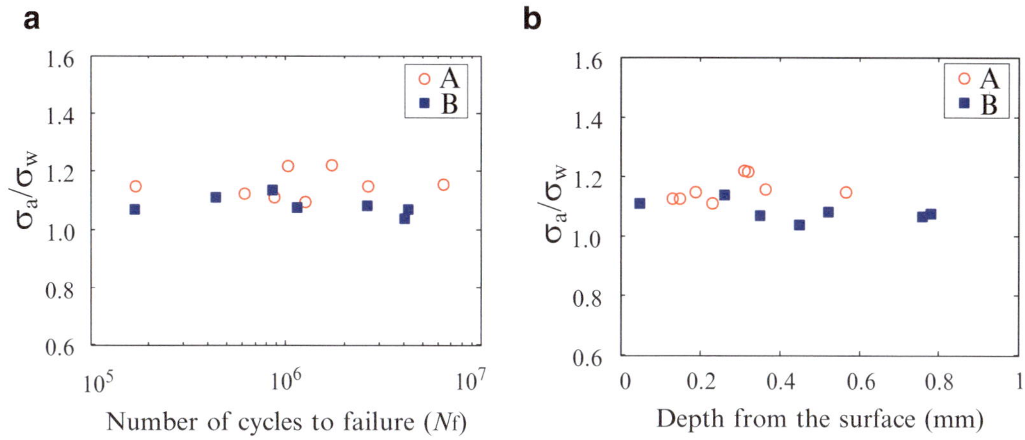

Fig. 14.10 Fish eye failures evaluated using Murakami's $\sqrt{area}$ model (14.1): ratio of applied stress amplitude to estimated fatigue strength vs. (**a**) cycles to failure and (**b**) depth of inclusion

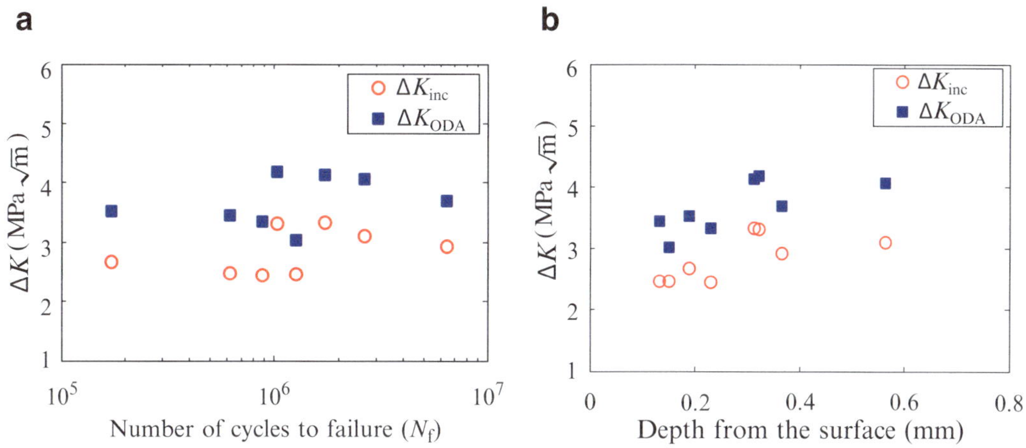

Fig. 14.11 Stress intensity factor range at inclusions and ODAs for steel A vs. (**a**) cycles to failure and (**b**) depths of inclusions

The values of ΔK_{ODA} vary from 3.0 to 4.3 MPa$\sqrt{\mathrm{m}}$ with an average of 3.7 MPa$\sqrt{\mathrm{m}}$. Figure 14.12 shows ΔK_{inc}, ΔK_{ODA} values for steel B. Values of ΔK_{inc} vary between 2.5 and 3.6 MPa$\sqrt{\mathrm{m}}$ with an average of 3.1 MPa$\sqrt{\mathrm{m}}$. The smallest ΔK_{inc} leading to failure occurs at the longest fatigue life. The values of ΔK_{ODA} vary from 3.4 to 4.5 MPa$\sqrt{\mathrm{m}}$ with an average of 4.1 MPa$\sqrt{\mathrm{m}}$. The smallest ΔK_{ODA} occurs at the shallowest inclusion.

The values of ΔK_{inc} for both steels in this study were approximately 3 MPa$\sqrt{\mathrm{m}}$, while values of ΔK_{ODA} were in the range of 4–5 MPa$\sqrt{\mathrm{m}}$. Those values, computed to account for the presence of significant residual stresses, compare well with values in the range of ΔK_{inc} from 2 to 4 MPa$\sqrt{\mathrm{m}}$ for steels with similar martensitic microstructures and hardness but *without* residual stresses from case hardening [26–35]. Likewise, the values of ΔK_{ODA} fall in the range of 3–6 MPa$\sqrt{\mathrm{m}}$ typically reported for such steels [26, 27, 30–37].

If data on residual stress and hardness vs. depth, both within a case and below, are obtained, the evaluation of fatigue strength by the $\sqrt{area}$ model as well by ΔK_{inc} and ΔK_{ODA} offer the potential to assess the risk of fish eye failures as a function of depth for different assumed inclusion sizes. This could provide information useful in design and manufacturing.

14.4 Discussion

The $\sqrt{area}$ approach used to compute stress intensity range ΔK with (14.5) and (14.6) uses "macro" residual stresses measured by X-ray diffraction with a typical beam diameter of about one mm. The actual residual stress distributions in the

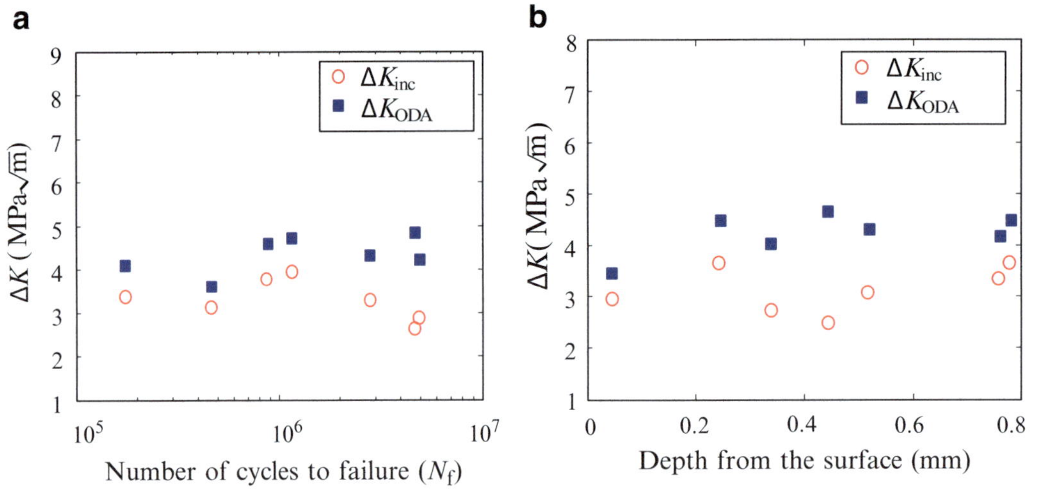

Fig. 14.12 Stress intensity factor range at inclusions and ODAs for steel B vs. (**a**) cycles to failure and (**b**) depth of inclusions

region of an ODA may differ from the "macro" residual stresses, but little appears to be known about such differences. For example, computational modeling [38] suggests that the residual stresses near inclusions the develop from shot peening of hardened steel can differ significantly from stresses measured in typical volumes probed by X-ray diffraction with conventional beam diameters. The distributions of residual stresses that develop locally around inclusions from martensitic transformations, as in case hardening, are difficult to predict with confidence owing to the complexities involved in material behavior at the crystalline size scale. A better undemanding of residual stresses around inclusions may improve prediction of cracking within ODAs.

Different mechanisms have been proposed for development of cracking within ODAs. Murakami et al. [15] suggested that hydrogen played an important role in cracking, perhaps by embrittlement. Shiozawa & Lu [39] proposed that ODAs are formed by carbide decohesion, growth of microcracks along boundaries between carbides and matrix and microcrack coalescence. Larger precipitated carbide particles (approx. 1 μm in size) of Mo, V and tungsten were observed on fracture surfaces within ODAs. Finer particles (e.g., a fraction of a micron) of chromium carbide and iron carbide were also found. Fractography revealed that larger particles ruptured and smaller ones peeled off. Hydrogen is known to have the potential to reduce cohesive strength at carbide-to-matrix interfaces [40]. It seems possible that hydrogen may promote the mechanism suggested by Shiozawa et al. [9], and future study of this potential influence may enhance the understanding of how ODAs develop. Sakai [3] suggested a different mechanism for formation of ODAs in which very fine sub-grains with different crystallographic orientations are formed by micro-scale polygonization and that debonding occurred along their boundaries, along with coalescence of the micro-debondings. Grad et al. [41] have found that cyclic loading at the stress concentration created by inclusions causes significant grain refinement within ODAs. As Li [4] has noted, hydrogen assisted local plasticity may encourage dislocation activity and multi-slip around inclusions, encouraging mechanisms such as polygonization. It may of interest in future studies of fish eye fatigue to investigate how hydrogen content varies with depth within carburized cases using a measurement approach like that applied by Murakami & Matsunaga [12] to hydrogen charged specimens.

14.5 Conclusions

1. Previous studies on fish eye fatigue in carburized specimens in which depths of crack origins were reported revealed that origins were at depths where compressive residual were minimal and hardness values were close to those in the core. By contrast, all fish eye fatigue cracks started at sub-surface inclusions within the case of carburized specimens in this study, at depths where compressive residual stresses were significant and where hardness values were much higher than in the core.
2. Murakami's $\sqrt{area}$ parameter model was found to be effective in predicting fatigue strength when taking into account hardness and compressive residual stresses at the locations of inclusions from which cracks originated.
3. Values of $\triangle K_{inc}$ at the border of inclusions, computed to account for residual stresses by superposition with applied stresses, were similar for both steels tested and in the range of 2.5–3.6, MPa$\sqrt{\text{m}}$ comparable to values in the literature for specimens that had minimal residual stresses at locations where fish eye cracks originated. Values of $\triangle K_{ODA}$ at the outer

border of optically dark areas, also computed to account for residual stresses, were in the range of 3–4.5 MPa$\sqrt{m}$ for both steels tested, again comparable to values from specimens with minimal residual stresses. The ability to account for the influence of residual stresses in computing $\triangle K_{inc}$ and $\triangle K_{ODA}$ simply by superposition is supported by the data generated in this study.

References

1. Murakami, Y.: Metal Fatigue: Effects of Small Defects and Nonmetallic Inclusions. Elsevier Science, Oxford (2002)
2. Chapetti, M., Tagawa, T., Miyata, T.: Ultra-long cycle fatigue of high-strength carbon steels part I: review and analysis of the mechanism of failure. Mater. Sci. Eng. A **356**, 227–235 (2003)
3. Sakai, T.: Review and prospects for current studies on very high cycle fatigue of metallic materials for machine structural use. J. Solid Mech. Mater. Eng. **3**, 425–439 (2009)
4. Li, X.: Effects of inclusions on very high cycle fatigue properties of high strength steels. Int. Mater. Rev. **57**, 92–114 (2012)
5. Bathias, C.: Gigacycle fatigue of bearing steels. Mater. Sci. Technol. **28**, 27–33 (2012)
6. Murakami, Y.: Material defects as the basis of fatigue design. Int. J. Fatigue **41**, 2–10 (2012)
7. Shiozawa, K., Lu, L., Ishihara, S.: S-N curve characteristics and subsurface crack initiation behavior in ultra-long life fatigue of a high carbon steel-chromium bearing steel. Fatigue Fract. Eng. Mater. Struct. **24**, 781–790 (2001)
8. Ochi, Y., Matsumura, T., Masaki, K., Yoshida, S.: High cycle rotating bending fatigue property in very long-life regime of high-strength steels. Fatigue Fract. Eng. Mater. Struct. **25**, 823–830 (2002)
9. Shiozawa, K., Morii, Y., Nishino, S., Lu, L.: Subsurface crack initiation and propagation mechanism in high-strength steel in a very high cycle fatigue regime. Int. J. Fatigue **28**, 1521–1532 (2006)
10. Nakajima, M., Kamiya, N., Itoga, H., Tokaji, K., Ko, H.: Experimental estimation of crack initiation lives and fatigue limit in subsurface fracture of high carbon steel. Int. J. Fatigue **28**, 1540–1546 (2006)
11. Huang, Z., Wagner, D., Wang, Q., Bathias, C.: Effect of carburizing treatment on the "fish eye" crack growth for a low alloyed chromium steel in very high cycle fatigue. Mater. Sci. Eng. A **559**, 790–797 (2013)
12. Murakami, Y., Nomoto, T., Ueda, T., Murakami, Y.: On the mechanism of fatigue failure in the superlong life regime ($N > 10^7$ cycles). Part I: influence of hydrogen trapped by inclusions. Fatigue Fract. Eng. Mater. Struct. **23**, 893–902 (2000)
13. Funatani, K., Noda, F.: The influence of residual stress on the fatigue strength of carburized hardened steel. J. Soc. Mater. Sci. Jpn. **17**, 1124–1128 (1968)
14. Naito, T., Udea, H., Kikuchi, M.: Fatigue behavior of carburized steel with internal oxides and nonmartensitic microstructure near the surface. Metall. Mater. Trans. A **15A**, 1431–1436 (1984)
15. Murakami, Y., Nomoto, T., Ueda, T.: Factors influencing the mechanism of superlong fatigue failure in steels. Fatigue Fract. Eng. Mater. Struct. **22**, 581–590 (1999)
16. Huang, Z., Du, W., Wagner, D., Bathias, C.: Relation between the mechanical behavior of a high strength steel and the microstructure in gigacycle fatigue. Mater. Sci. Forum **636–637**, 1459–1466 (2010)
17. ASTM: Standard E112-12, Standard Test Method for Determining Average Grain Size. American Society for Testing and Materials, East Conshohocken (2012)
18. ASTM: Standard E384-10, Standard Test Method for Knoop and Vickers Hardness of Materials. American Society for Testing and Materials, West Conshohocken (2011)
19. Moore, M., Evans, W.: Mathematical corrections for stress in removed layers in X-ray diffraction residual stress analysis. SAE Trans. **66**, 340–345 (1958)
20. Parrish, G.: Carburizing: Microstructures and Properties. ASM International, Materials Park (1999)
21. Inoue, K., Sonoda, H., Deng, G., Yamanaka, M., Kato, M.: Effects of CBN grinding on the bending fatigue strength of carburized gears. J. Mech. Des. **120**, 606–611 (1998)
22. Pedersen, R., Rice, S.: Case crushing of carburized and hardened gears. Technical paper 610034. SAE International, Warrendale (1961)
23. McClung, R.: A literature survey on the stability and significance of residual stresses during fatigue. Fatigue Fract. Eng. Mater. Struct. **30**, 173–205 (2007)
24. Nishijima, S., Kanazawa, K.: Stepwise S-N curve and fish-eye failure in gigacycle fatigue. Fatigue Fract. Eng. Mater. Struct. **22**, 601–607 (1999)
25. Sakai, T., Sato, Y., Oguma, N.: Characteristic S–N properties of high carbon–chromium bearing steel under axial loading in long life fatigue. Fatigue Fract. Eng. Mater. Struct. **25**, 765–773 (2002)
26. Shimatani, Y., Shiozawa, K., Nakada, T., Yoshimoto, T.: Effect of surface residual stress and inclusion size on fatigue failure mode of matrix HSS in very high cycle regime. Procedia Eng. **2**, 873–882 (2010)
27. Shiozawa, K., Hasegawa, T., Kashiwagi, Y., Lu, L.: Very high cycle fatigue properties of a bearing steel under axial loading condition. Int. J. Fatigue **31**, 880–888 (2009)
28. Qian, G., Hong, Y., Zhou, C.: Investigation of high cycle and very high cycle fatigue behaviour for a structural steel with smooth and notched specimens. Eng. Fail. Anal. **17**, 1517–1525 (2010)
29. Hui, W., Nie, Y., Dong, H., Weng, Y., Wang, C.: High-cycle fatigue fracture behavior of ultrahigh strength steels. J. Mater. Sci. Technol. **24**, 787–792 (2008)
30. Lei, Z., Zhao, A., Xie, J., Sun, C., Hong, Y.: Very high cycle fatigue of GCr15 steel with smooth and hole defect specimens. Theor. Appl. Mech. Lett. **2**, 031003 (2012). doi:10.1063/2.1203103

31. Li, W., Sakai, T., Li, Q., Lu, L., Wang, P.: Reliability evaluation on very high cycle fatigue property of GCr15 bearing steel. Int. J. Fatigue **32**, 1096–1107 (2010)
32. Tanaka, K., Akiniwa, Y.: Fatigue crack propagation behavior derived from S–N data in very high cycle regime. Fatigue Fract. Eng. Mater. Struct. **25**, 775–784 (2002)
33. Zhang, J., Shiozawa, K., Lu, T., Li, W., Zhang, W.: Fatigue fracture behavior of bearing steel GCr15 in very high cycle regime. Adv. Mater. Res. **44–46**, 119–126 (2008)
34. Hong, Y., Lei, Z., Sun, C., Zhao, A.: Propensities of crack interior initiation and early growth for very-high-cycle fatigue of high strength steels. Int. J. Fatigue **58**, 144–151 (2013). doi:10.1016/j.ijfatigue.2013.02.023
35. Akiniwa, Y., Miyamoto, N., Tsuru, H., Tanaka, K.: Notch effect on fatigue strength reduction of bearing steel in the very high cycle regime. Int. J. Fatigue **28**, 1555–1565 (2006)
36. Sun, C., Xie, J., Zhao, A., Lei, Z., Hong, Y.: A cumulative damage model for fatigue life estimation of high-strength steels in high-cycle and very-high-cycle fatigue. Fatigue Fract. Eng. Mater. Struct. **35**, 638–647 (2012)
37. Liu, Y., Yang, Z., Li, Y., Chen, S., Li, S., Hui, W., Weng, Y.: On the formation of GBF of high-strength steel in the very high cycle fatigue regime. Mater. Sci. Eng. A **497**, 408–415 (2008)
38. Zhang, J., Prasannavenkatesan, R., Shenoy, M., McDowell, D.: Modeling fatigue crack nucleation at primary inclusions in carburized and shot-peened martensitic steel. Eng. Fract. Mech. **76**, 315–334 (2009)
39. Shiozawa, K., Lu, L.: Internal fatigue failure mechanism of high strength steels in gigacycle regime. Key Eng. Mater. **378–379**, 65–80 (2008)
40. Novak, P., Yuan, R., Somerday, B., Sofronis, P., Ritchie, R.: A statistical, physical-based, micro-mechanical model of hydrogen-induced intergranular fracture of steel. J. Mech. Phys. Solids **58**, 206–228 (2010)
41. Grad, P., Reuscher, B., Brodyanski, A., Kopnarski, M., Kerscher, E.: Mechanisms of fatigue crack initiation and propagation in the very high cycle fatigue regime of high strength steels. Scr. Mater. **67**, 838–841 (2012)
42. Murakami, Y., Matsunaga, H.: The effect of hydrogen on fatigue properties of steels used for fuel cell system. Int. J. Fatigue **28**, 1509–1520 (2006)

Chapter 15
Fatigue Behavior of Fluid End Crossbore Using a Coupon-Based Approach

Mahdi Kiani, Rayford Forest, Steven Tipton, and Michael W. Keller

Abstract Fracture or mud pumps are known as the heart of the drilling and hydraulic fracturing. Crossbore geometries are central to the design of fluid end module in these positive displacement reciprocating pumps. Intersection between bores emerges as a stress concentrator and because the fluctuating pressure history is extreme, fatigue limits the useful life of the pump. Approaches such as autofrettage are typically used to extend fatigue lives through the imposition of compressive residual stresses at crossbore intersections. Direct investigation of the impact of residual stresses in working pumps is not typically possible. In order to improve understanding of the impact of residual stresses on fatigue life and to optimize the fatigue-strength improvement provided to fluid ends, unique sample geometry was designed to simulate the stresses in the crossbore. These samples are tested on laboratory-based servohydraulic fatigue frames and eliminate the need for complicated in-situ stress analysis on the fluid ends. Using notch strain analysis and modified Smith–Watson–Topper approach a life prediction algorithm was also developed to calculate the fatigue life of the coupon. To optimize the autofrettage load and cyclic loading simulation, elastoplastic FEA was accomplished utilizing a combined nonlinear isotropic/kinematic hardening material model for 4300-series alloy steel.

Keywords Fluid end module • Fatigue failure • Autofrettage • Compressive residual stresses • FEA

15.1 Introduction

Many high-pressure components have intersecting bore geometries, such as in reciprocating pumps that are widely used in gas and oil industries for well stimulation. These intersecting bore geometries are critical regions when designing fluid ends on positive displacement pumps, because the intersection between two bores is a stress concentrator [1–4]. This is especially true as the fluctuating pressure history in a fluid end can be extreme, with pressures alternating between 0 and 15,000 psi at frequencies of up to 5 Hz. At stress concentration points in the fluid ends, the maximum tensile stress can rise up to several times of the fluid pressure in other sections. As a result, fatigue often limits the useful life of the pump. To address these large alternating stresses, compressive residual stresses at crossbore intersections can be introduced to improve fatigue life. These residual stresses can be applied in several ways, such as autofrettage or shot peening [4–6].

The autofrettage approach is one of the most common due to its simplicity and low-cost. During the autofrettage process, the fluid end is pressurized above the design operating pressure, which causes partial yielding. This plastic deformation is concentrated at the crossbore intersection, because of high stress concentration factor in this section. When the autofrettage pressure is released the bulk material elastically relaxes, but the permanent plastic deformation at the crossbore intersection resists. As a result, the bulk elastic material compresses the plastically deformed crossbore intersection leading to imposition of compressive residual stresses exactly where they are most effective, the crossbore intersection. The mean compressive stress generated during autofrettage serves to offset the fluctuating tensile stresses during operation and increase the fatigue life of the fluid end module [4, 7].

To optimize the autofrettage pressure and enhance the compressive residual stresses, understanding the stress-strain response during autofrettage in the crossbore area of the fluid end is critical. Moreover, evaluation of the evolution of the compressive residual stresses after autofrettage is required for fatigue life estimation and design. Understanding the behavior of the imposed residual stresses is critical for evaluating the proper time interval for reapplying the surface treatment process [1, 8, 9].

M. Kiani (✉) • R. Forest • S. Tipton • M.W. Keller
The University of Tulsa, 800 South Tucker Drive, Tulsa, OK 74104, USA
e-mail: mahdi-kiani@utulsa.edu

A.M. Beese et al. (eds.), *Fracture, Fatigue, Failure and Damage Evolution, Volume 8*, Conference Proceedings of the Society for Experimental Mechanics Series, DOI 10.1007/978-3-319-21611-9_15

Table 15.1 Simulation specifications

Part	3D, deformable
Material	4300-Series alloy steel, E = 30,000 ksi, $\nu = 0.3$, D = $0.73 \times 10^{-3} lbfS^2/in^4$
Model	Elastic, isotropic
Section	Solid, homogeneous
Step	Static, general
Element type	Standard, 3D stress, quadratic, Hex (C3D20)
Mesh technique	Structured

However, direct investigation of the impact of residual stresses on the crossbore geometry in a working pump is not typically possible. Thus, in order to improve understanding of the impact of residual stresses on fatigue life and to optimize the fatigue-strength improvement provided to fluid ends through autofrettage, a simple, coupon-based sample was designed to simulate the stresses in the crossbore. These samples can be tested on laboratory-based servohydraulic fatigue frames and eliminate the need for complicated in-situ stress analysis on the fluid ends.

15.2 Approach

The coupon sample geometry was designed to create a surface stress gradient that closely resembles the stress distribution derived from FEA analyses on crossbore intersections. This way, the effectiveness of autofrettage can be assessed experimentally and compared to analytical models quickly and efficiently. A life prediction algorithm was also developed to calculate the required load for the certain fatigue life of the coupon. Notch strain analysis (strain energy density rule) was used to estimate the stress-strain condition for any applied load history and fatigue lives were calculated based on the modified Smith–Watson–Topper approach.

15.2.1 Geometrical Design

FEA models of commercially available triplex and quintuplex fluid end modules were analyzed to obtain the crossbore intersection elastic stress gradient. A one eighth of a cubic crossbore geometry model was also analyzed using a commercial FEA code (Abaqus 6.12) to extract the stress distribution along the line from the maximum stress location. The goal of the specimen design was to match these derived stress gradients in a test coupon as closely as possible. Getting direction from analytical formulas of solid mechanics, a FEA package (Abaqus 6.12) was used for numerical simulation to accomplish the specimen design. Simulation specifications have been summarized in Table 15.1.

Initial specimen design were created using simple analytical approaches, and candidate geometries were selected and then modeled using FEA in order to provide detailed information to match the stress gradient in the fluid end. Comparison among stress gradients of the fluid end, simplified 1/8 model, and final specimen design (Fig. 15.1) can be seen in Fig. 15.2. As it can be observed, the normalized stress distribution of the coupon specimen closely matches those from the crossbores in the initial 0.5 in. of the specimen.

A quarter model of the sample is shown in Fig. 15.3. Pure axial loading of the new sample is achieved by notching the sample from both sides, symmetrically. This creates two identical regions simultaneously, each with a stress gradient similar to that found at a crossbore intersection as shown in Fig. 15.2. The stress gradients (slope of the stress-position curve) simulate the crossbores well. While the stress gradient is shallower than the larger crossbores, the primary interest of this study is the evolution of the stress gradient and the initiation of cracking, not crack propagation.

15.2.2 Fatigue Analysis

Notch strain analysis was applied to evaluate the stress-strain state on the notch's root of the designed specimen. Then, strain-life approach was considered to estimate the fatigue life. To obtain the strain based fatigue parameters, according to ASTM E-606, the cyclic stress strain curve of the material was experimentally obtained, by performing a companion specimen test.

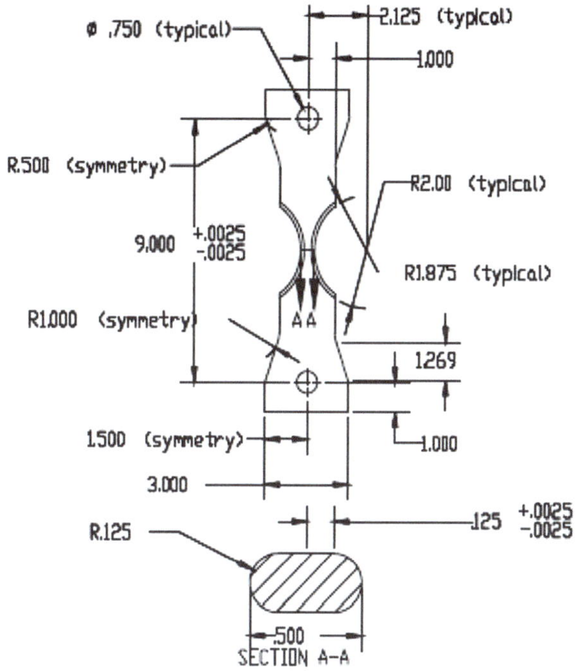

Fig. 15.1 Drawing of the designed notched coupon

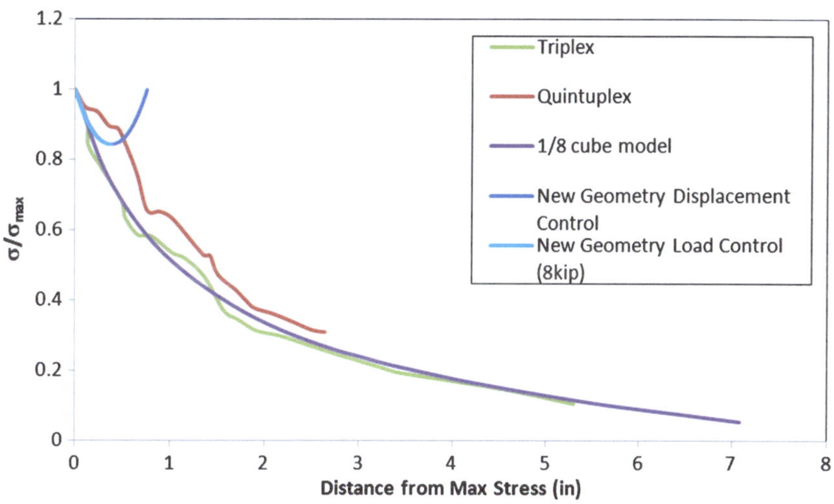

Fig. 15.2 Stress gradient comparison between designed coupon-specimen (new geometry) and different crossbore intersection geometries

Fig. 15.3 FEA stress contour and rendering of the designed coupon-specimen

In addition to cyclic stress-strain curve, strain-life baseline fatigue curve is another required set of information that was developed by low cycle fatigue experimental data. Experimentally obtained stabilized hysteresis loops can be divided into elastic and plastic strain ranges. By plotting these in log-log coordinates as functions of fatigue life, both elastic and plastic components can be approximated by straight lines. In the power equation form, the elastic part can be represented by Basquin's equation (15.1) and Manson–Coffin relationship as shown in (15.2) presents the plastic part. Eventually, combining these two equations, the total strain-life can be introduces by (15.3).

$$\frac{\Delta\sigma}{2} = \sigma_a = \sigma'_f (2N_f)^b \tag{15.1}$$

$$\frac{\Delta\varepsilon_p}{2} = \varepsilon'_f (2N_f)^c \tag{15.2}$$

$$\frac{\Delta\varepsilon}{2} = \varepsilon_a = \frac{\Delta\varepsilon_e}{2} + \frac{\Delta\varepsilon_p}{2} = \frac{\sigma'_f}{E}(2N_f)^b + \varepsilon'_f (2N_f)^c \tag{15.3}$$

where σ_a and ϵ_a are strain and stress amplitudes, σ'_f is fatigue strength coefficient, N_f is number of cycles to failure, b is fatigue strength exponent, ε'_f is fatigue ductility coefficient, c is fatigue ductility exponent, $\Delta\epsilon_p$ is the plastic strain range and $\Delta\epsilon_e$ is the elastic one [10, 11].

The standard strain life equation is formulated based on the fully reversed strain controlled tests. Different models have been developed to account for mean stress effects in the strain life fatigue approach. Smith–Watson–Topper (SWT) relation is the most effective model has been shown to correlate mean stress for a wide range of materials. This model has been developed based on empirical observations that the product of the stabilized maximum stress and the strain range during a cycle is proportional to fatigue life; (15.4) represents the general form of this model [11].

$$\sigma_{max}\varepsilon_a = \frac{\left(\sigma'_f\right)^2}{E}(2N_f)^{2b} + \sigma'_f\varepsilon'_f (2N_f)^{b+c} \tag{15.4}$$

In modified SWT model it is presumed that the product of the maximum stress and the strain amplitude for a given non-fully reversed loading is equal to the maximum stress and strain amplitude of a fully reversed loading, (15.5).

$$\sigma_{max}\varepsilon_a = \sigma_{aR}\varepsilon_{aR} \tag{15.5}$$

where σ_{aR} and ε_{aR} are maximum stress and strain amplitude of the fully reversed loading.

By considering Ramberg–Osgood relationship and combining it to (15.5), (15.6) can be derived.

$$\sigma_{max}\varepsilon_a = \sigma_{aR}\left[\frac{\sigma_{aR}}{E} + \left(\frac{\sigma_{aR}}{K'}\right)^{1/n'}\right] \tag{15.6}$$

where K' is the cyclic strength coefficient and n' is the cyclic strain hardening exponent [11].

Equation (15.6) can iteratively be solved for σ_{aR} and fully reversed strain amplitude can be obtained as shown in (15.7). Eventually, the strain life relation can be reached through (15.8) [11].

$$\varepsilon_{aR} = \frac{\sigma_{aR}}{E} + \left(\frac{\sigma_{aR}}{K'}\right)^{1/n'} \tag{15.7}$$

$$\varepsilon_{aR} = \frac{\sigma'_f}{E}(2N_f)^b + \varepsilon'_f (2N_f)^c \tag{15.8}$$

Applying notch strain analysis and modified SWT fatigue life estimation approach, a Matlab code was developed to calculate the load associated with certain amount of cycles to failure in designed specimen. Considering extracted low cycle fatigue properties, for 50,000 cycles to failure, a load of 12,500 lbs was estimated.

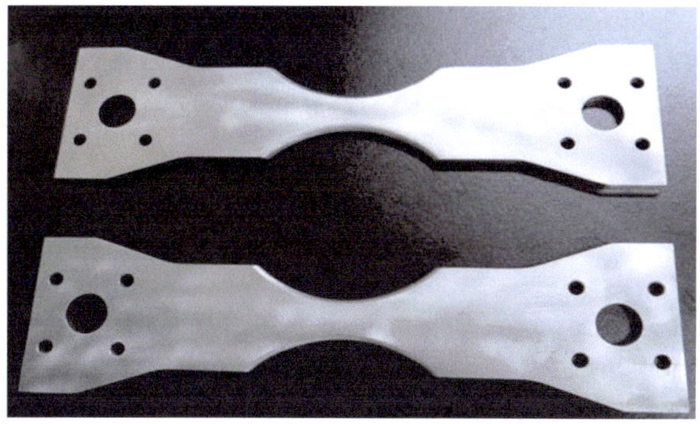

Fig. 15.4 Manufactured notched coupons

Table 15.2 Specifications of notched specimen elastoplastic FEA

Part	3D, deformable
Material	4300-Series alloy steel, E = 30,000 ksi, $\nu = 0.3$, D = 0.73 × $10^{-3} lbfS^2/in^4$
Model	Combined nonlinear isotropic/kinematic hardening, half cycle
Section	Solid, homogeneous
Step	Static, general
Element type	Standard, 3D stress, quadratic, Hex (C3D20)
Mesh technique	Structured

15.2.3 Manufacturing

A scrapped fluid end made of a 4300-series low alloy steel was used to manufacture all of the notched coupons as well as tensile and fatigue specimens for material model calibration. The block was sliced into flat slabs transverse to the bores. After slicing and grinding the slabs to the specified thickness, notched coupons were cut by water jet into the designed shape as shown in Fig. 15.4. Final machining was performed using CNC milling.

15.2.4 Elastoplastic Analysis

To estimate the optimum autofrettage load and evaluate the induced residual stresses the Elastoplastic FEA was applied to the designed notched specimen to simulate the unidirectional overloading and successive unloading which model the autofrettage process. Details of the simulations are presented in Table 15.2.

The combined nonlinear isotopic/kinematic hardening material model was calibrated and applied for elastoplastic FEA. This is the most powerful and accurate model which can be used to simulate the elastoplastic behavior of ductile metals during cyclic loadings. This material model includes the Bauschinger effect and has criterion for unloading. Moreover, cyclic softening/hardening can effectively be simulated by this model [12–15].

Nonlinear isotopic/kinematic hardening material model was composed of two components: nonlinear kinematic hardening and non linear isotropic hardening. Translation of the yield surface in stress space is governed by the first component through considering maximum of 10 back-stresses (α). The isotropic component controls the size of the yield surface (σ^0) as a function of equivalent plastic strain [12].

In Abaqus standard only the Mises yield surface can be applied by this model, however in Abaqus/Explicit both Mises and Hill's yield surface can be used. Equation (15.9) represents the pressure independent yield function associated with this model.

$$F = f(\boldsymbol{\sigma} - \boldsymbol{\alpha}) - \sigma^0 \tag{15.9}$$

where σ^0 is the yield stress, $f(\boldsymbol{\sigma} - \boldsymbol{\alpha})$ is the equivalent Mises stress or Hill's potential and $\boldsymbol{\alpha}$ is the back-stress [12, 13].

Associated plastic flow, as presented in (15.10), is considered as flow rule of this model. As long as microscopic details, such as localized plastic flow due to metal fracture caused by cyclic loading can be ignored, this flow rule is valid for metals undergoing cyclic loads.

$$\dot{\varepsilon}^{pl} = \dot{\bar{\varepsilon}}^{pl} \frac{\partial F}{\partial \boldsymbol{\sigma}} \tag{15.10}$$

where the plastic flow rate has been shown by $\dot{\varepsilon}^{pl}$, and $\dot{\bar{\varepsilon}}^{pl}$ is the rate of equivalent plastic strain [12, 13].

However, in the nonlinear kinematic hardening component of the combined model two terms are effective and the additive combination of these two terms controls the kinematic component. The combination of these two terms for each single back-stress can be seen in (15.11).

$$\dot{\boldsymbol{\alpha}}_k = C_k \frac{1}{\sigma^0} (\boldsymbol{\sigma} - \boldsymbol{\alpha}_k) \dot{\bar{\varepsilon}}^{pl} - \gamma_k \boldsymbol{\alpha}_k \dot{\bar{\varepsilon}}^{pl} \tag{15.11}$$

where C_k presents the initial kinematic hardening modulus and γ_k specifies the decreasing rate of kinematic hardening modulus due to increasing the plastic deformation. These two material parameters can be adjusted by calibration using experimental data [12–15].

The first part of the equation is linear Ziegler law and has purely kinematic effects. The second part which creates the nonlinearity is known as relaxation or recall term. Moreover, several back-stresses can be used to enhance the results. Each different range of strains is controlled by a specific back-stress and large strains are governed by the linear hardening law. The overall back-stress can be calculated using (15.12).

$$\boldsymbol{\alpha} = \sum_{k=1}^{N} \boldsymbol{\alpha}_k \tag{15.12}$$

where N shows the number of back-stresses [12–15].

The isotropic hardening component of the combined model is responsible for changing the yield surface size. This component can be imported by developing a subroutine, arranging a functional table as yield surface size versus equivalent plastic strain or using the exponential law presented in (15.13).

$$\sigma^0 = \sigma_0 + Q_{inf} \left(1 - e^{-b\bar{\varepsilon}^{pl}}\right) \tag{15.13}$$

where $\bar{\varepsilon}^{pl}$ is the equivalent plastic strain, σ_0 is the yield stress at zero plastic strain, Q_{inf} shows the maximum change in size of the yield surface, and b introduces the rate of change in the yield surface size due to plastic deformation. Combined model simplifies to a nonlinear kinematic hardening model if equivalent stress size, σ^0, does not vary and remains equal to the yield stress at zero plastic strain, σ_0 [13–15].

15.3 Results and Discussion

Figure 15.5 shows the residual stress (S22) contour of the notched specimen after elastoplastic simulation of overloading and unloading (autofrettage). Induced residual stresses can be recognized over the thinner section of the specimen, the compressive one has been created on the root of the notch. Cross sectional distribution of the residual stresses over the thinner section can be seen in Fig. 15.6, also, effects of residual stress on the typical elastic loading (8 kip) has been demonstrated in Fig. 15.6. The residual stresses are compressive at the notch root and tensile residual stresses emerge over the middle section to balance them.

Mean stress reduction due to overloading can be observed by comparing loading and reloading data series in Fig. 15.7. Moreover, stress strain response at the notch tip of the specimen subjected to different overloading values has been presented in Fig. 15.8. Analyzing the induced residual stress values associated with different overloading amounts, the load of 30 kip was obtained as the optimum load for overloading (autofrettage).

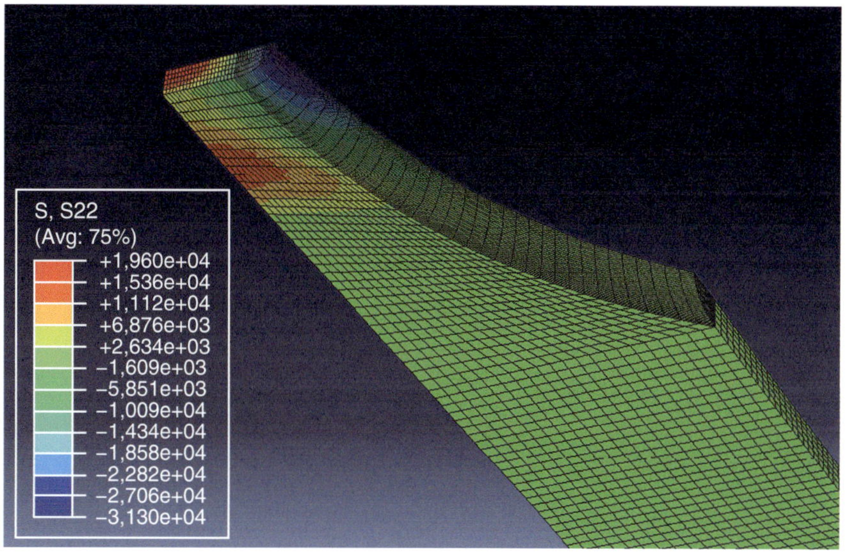

Fig. 15.5 Residual S22 stress contour

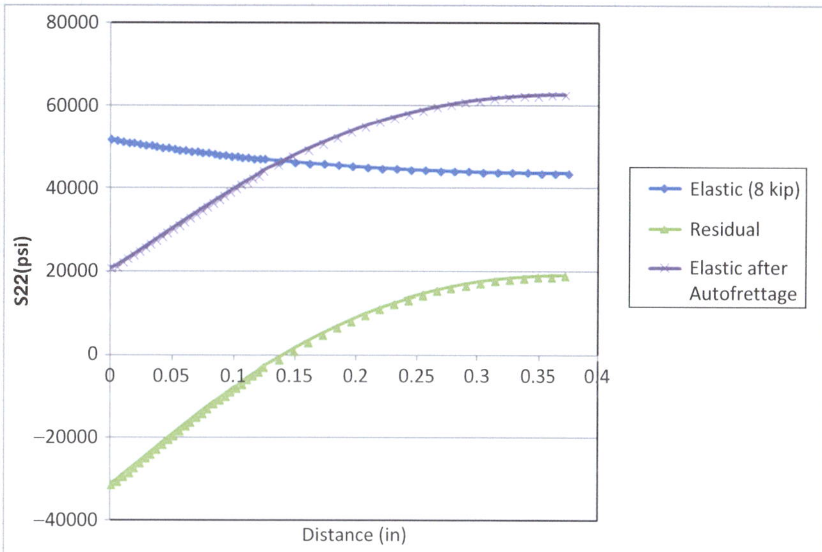

Fig. 15.6 Autofrettage induced residual stresses effects on the elastic loading of notched specimen by considering S22

15.4 Conclusions

1. Unique sample geometry was designed to simulate the stresses in the crossbore section of the fluid end module of fracture pumps. These samples can be tested on laboratory-based servohydraulic fatigue frames and eliminate the need for complicated in-situ stress analysis on the fluid ends.
2. Applying notch strain analysis and modified SWT fatigue life estimation approach, a Matlab code was developed to calculate the load associated with certain amount of cycles to failure in designed specimen. Considering 4300-series alloy steel's extracted low cycle fatigue properties, a load of 12,500 lbs was estimated for 50,000 cycles to failure.
3. The combined nonlinear isotopic/kinematic hardening material model was calibrated for 4300-series alloy steel and applied for elastoplastic FEA on designed notched coupon.
4. Stress strain response at the notch tip of the specimen subjected to different overloading values were analyzed using FEA and evaluating the induced residual stress values associated with different overloading amounts, the load of 30 kip was obtained as the optimum load for overloading (autofrettage).

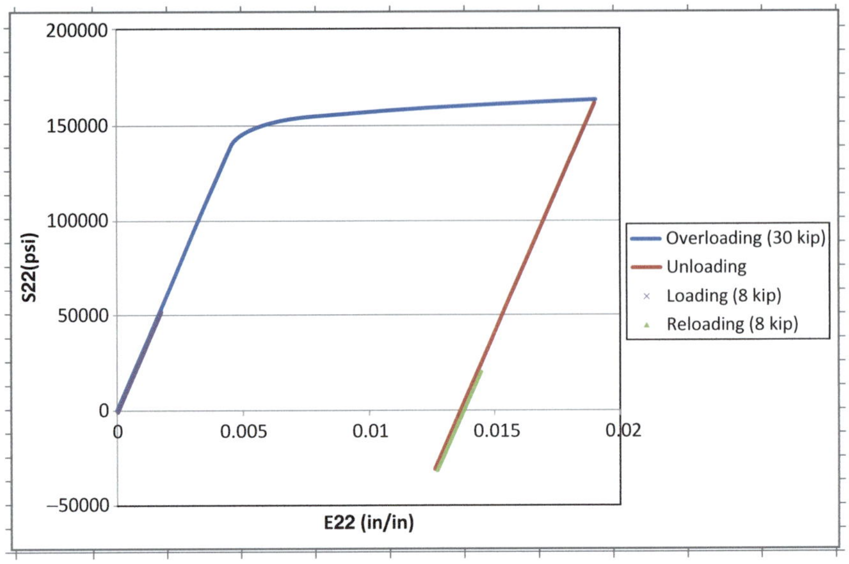

Fig. 15.7 Effects of autofrettage and residual stresses at the notch root during a typical elastic loading

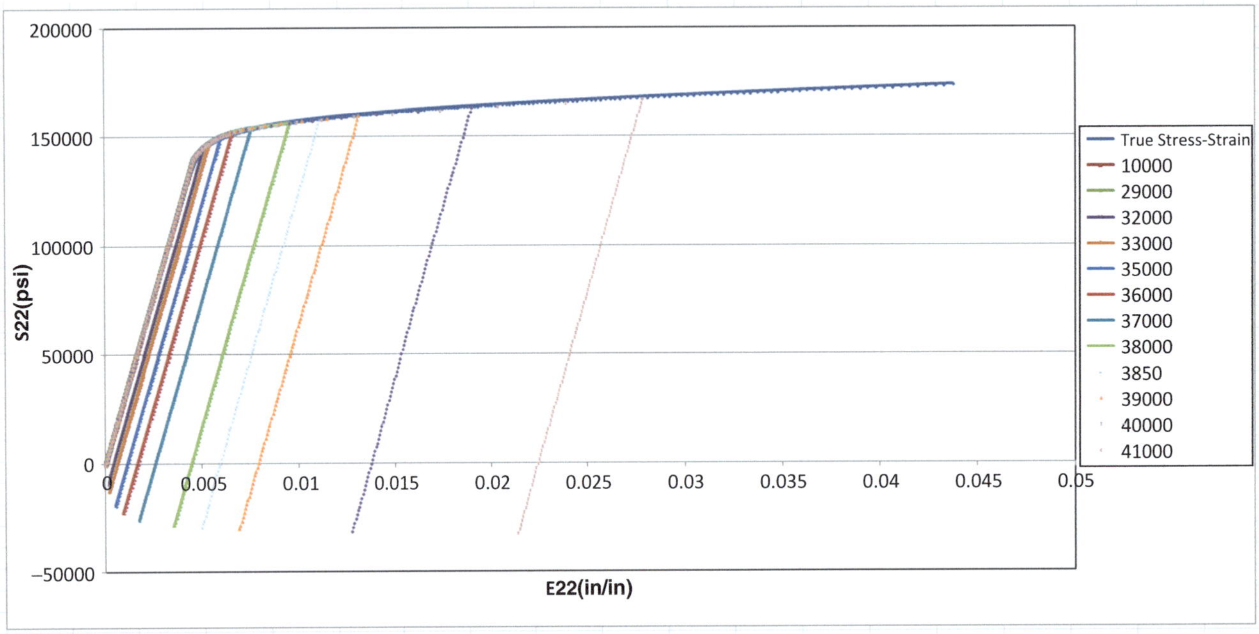

Fig. 15.8 Notch tip stress-strain response under different values of overloading (load values are surface traction force)

References

1. Sorem, J., Shadley, R., Tipton, S.M.: Design curves for maximum stresses in block containing pressurized bore intersections. J. Mech. Des. **113**, 427–431 (1991)
2. Yang, S.: Experimental and analytical investigation of the effects of autofrettage on fatigue damage development in blocks containing crossbores under cyclic internal pressure. Ph.D. Dissertation, University of Tulsa (1995)
3. Badr, E.A.: Estimation of residual stresses induced by autofrettage with and experimental evaluation on the autofrettage process in crossbores of positive displacement pumps. Ph.D. Dissertation, University of Tulsa (1994)
4. Badr, E.A., Sorem, J.R., Tipton, S.M.: Residual stress estimation in crossbore with Bauschinger effect inclusion using FEA and strain energy density. J. Press. Vessel Technol. **121**, 358–363 (1999)
5. Pendleton, G., McPheron, R.: AXON Energy Products, Upstream Pumping, July 2013
6. White, S., White Star Pump Company, Upstream Pumping, July 2011
7. Parker, A.P.: Autofrettage of open-end tubes-pressures, stress, strain and code comparisons. J. Press. Vessel Technol. **123**, 271–281 (2001)
8. Badr, E.A., Yang, S., Sorem, J.R., Tipton, S M.: Development of a cyclic high pressure fatigue test system. High Pressure Technology, PVP-Vol. **281**, ASME (1994)

9. Yang, S., Badr, E.A., Sorem, J.R., Tipton, S.M.: Advantages of sequential crossbore autofrettage of triplex pump fluid end crossbores. High Pressures—Codes, Analysis, and Applications. PVP-Vol. **263**, 81–89 (1993)
10. Dowling, N.E.: Fatigue life and inelastic strain response under complex histories for an alloy steel. J. Test Eval. **1**(4), 271–287 (1973)
11. Stephens, R.I., Fatemi, A., Stephens, R.I., Fuchs, H.Q.: Metal Fatigue in Engineering. Wiley, Toronto (2001)
12. Abaqus documentation. Using a nonlinear isotropic/kinematic cyclic hardening model to define classical metal plasticity
13. Abaqus documentation, 23.2.2 Models for metals subjected to cyclic loading
14. Sarajavi, U., Cronvall, O.: Simulation and analysis of data for enhancing low cycle fatigue test procedures. VTT Technical Research Center of Finland (2006)
15. Sarajavi, U., Cronvall, O.: A procedure to generate input data of cyclic softening and hardening for FEA analysis from constant strain amplitude fatigue tests in LCF regime. VTT Technical Research Center of Finland (2007)

Chapter 16
Notch Strain Analysis of Crossbore Geometry

Mahdi Kiani, Steven Tipton, and Michael W. Keller

Abstract Many high-pressure components have intersecting bore geometries, such as fluid end module of fracture pumps. Imposition of compressive residual stresses at crossbore intersections can extend the fatigue life, thus approaches such as autofrettage are typically used. Understanding the stress–strain response during autofrettage in the crossbore is critical for fatigue life estimation and design. Crossbore geometry is frequently complex and no closed-form analytical solution is available for prediction of residual stresses. As such, numerical methods like FEA are frequently used. Applying FEA to complicated geometries requires extensive parametric studies which are computationally expensive and time consuming, thus notch strain analysis methods are promising. Elastoplastic stress–strain responses due to varying internal pressures in a crossbore geometry were evaluated using FEA and notch strain analysis formulas including both Neuber and Glinka approaches. To define the theoretical elastic stress concentration factor based on ratios between maximum Mises or hoop stress and pressure or nominal stress four different values were calculated and imported into the notch strain analysis formulas. It was observed that the results of Glinka approach match better to the FEA results particularly by applying the ratio between maximum Mises stress and nominal stress as elastic stress concentration factor.

Keywords Crossbore • Residual stresses • Autofrettage • FEA • Notch strain analysis

16.1 Introduction

Fluid end modules intended for fracture pumps undergo intense loading conditions when employed in unconventional drilling. The main reason for failures in fluid end modules is fatigue produced by cyclic loading. Intersecting bore geometries or crossbores are central to the design of many fluid ends on positive displacement pumps. Most failures originate in this region because the intersection between two bores is a stress concentrator. Therefore, techniques for the design and analysis of stress in this location are critical for understanding mechanical behavior and reducing fatigue failures [1–4].

Imposition of compressive residual stresses has been proven to be an effective procedure which can extend the fatigue life of engineering components. In positive displacement pumps, autofrettage is typically used to generate beneficial compressive residual stresses at crossbore intersections. This approach is adopted because it is relatively cheap and simple compared to other methods such as laser-shock-peening or shot peening [5–9].

To optimize the autofrettage pressure and enhance any compressive residual stresses, understanding the stress–strain response during autofrettage in the crossbore area of the fluid end is critical. Moreover, evaluation of the compressive residual stresses after autofrettage is required for fatigue life estimation and design. Crossbore geometry is complex and no closed-form analytical solution is available to analyze stress and strain. As such, numerical methods like Finite Element Analysis (FEA) are typically used to study the fluid end. Applying FEA to obtain the stress–strain response of complicated geometries requires extensive parametric studies which are computationally expensive and time consuming. Thus, several attempts to derive simple formulas that can estimate the stress–strain response during fluctuating loads in the crossbore have been performed [8–13].

Most of these models have used a notch strain analysis approach, such as Neuber's rule or Glinka's rule (also known as strain energy density rule). These relationships are employed to estimate the stress–strain state in the notched components where there is a stress concentration. The general forms of Neuber and Glinka's rules can be expressed as (16.1) and (16.2), respectively.

M. Kiani (✉) • S. Tipton • M.W. Keller
The University of Tulsa, 800 South Tucker Drive, Tulsa, OK 74104, USA
e-mail: mahdi-kiani@utulsa.edu

A.M. Beese et al. (eds.), *Fracture, Fatigue, Failure and Damage Evolution, Volume 8*, Conference Proceedings of the Society for Experimental Mechanics Series, DOI 10.1007/978-3-319-21611-9_16

$$K_t^2\left(\frac{S^2}{E}\right)=\frac{\sigma^2}{E}+\sigma\left(\frac{\sigma}{K}\right)^{\frac{1}{n}} \tag{16.1}$$

$$K_t^2\left(\frac{S^2}{E}\right)=\frac{\sigma^2}{E}+\frac{2}{n+1}\sigma\left(\frac{\sigma}{K}\right)^{\frac{1}{n}} \tag{16.2}$$

where K_t is the theoretical elastic stress concentration factor, S is the nominal elastic stress, E is the elastic modulus, K is the strength coefficient, n is the strain hardening exponent, and σ is the actual stress on the notch root. It has been reported that when the nominal stress exceeds the 80 % of the material yield strength the nominal stress–strain response at the notch root becomes inelastic and as a result the nonlinear Ramberg–Osgood stress–strain relation should be considered for nominal stress–strain relation in the left side of the (16.1) and (16.2) [14, 15].

In a crossbore, the intersection can be considered as a stress concentration and consequently the notch strain analysis approach can be applied. However, direct application of these models to a crossbore is not straightforward because the loading state is not pure uniaxial or biaxial loading. Hence, all stress and strain components should be considered when predicting stress–strain response. Previous researchers have considered the internal pressure as the nominal stress and an elastic pressure concentration factor (16.3) as the stress concentration factor. Using this definition, expressions of Neuber and Glinka's rules can be derived as shown in (16.4) and (16.5), respectively.

$$K_p=\frac{\sigma_h}{P} \tag{16.3}$$

$$K_p^2\left(\frac{P^2}{E}\right)=\frac{\sigma_h^2}{E}+\sigma_h\left(\frac{\sigma_h}{K}\right)^{\frac{1}{n}} \tag{16.4}$$

$$K_p^2\left(\frac{P^2}{E}\right)=\frac{\sigma_h^2}{E}+\frac{2}{n+1}\sigma_h\left(\frac{\sigma_h}{K}\right)^{\frac{1}{n}} \tag{16.5}$$

where P is the internal pressure, and σ_h is the hoop stress. These equations are valid under plane stress condition [14].

However, in the case of a crossbore, considering the equivalent stress is more appropriate as all stress or strain components can be significant. Equations (16.6) and (16.7) represent Neuber and Glinka's rules modified to consider the equivalent stress.

$$2\left(\frac{P^2}{E}\right)\left[\frac{K_P^2}{2}+(1-\nu)+2\nu K_p\right]=\frac{\sigma_e^2}{E}+\sigma_e\left(\frac{\sigma_e}{K}\right)^{\frac{1}{n}} \tag{16.6}$$

$$2\left(\frac{P^2}{E}\right)\left[\frac{K_P^2}{2}+(1-\nu)+2\nu K_p\right]=\frac{\sigma_e^2}{E}+\frac{2}{n+1}\sigma_e\left(\frac{\sigma_e}{K}\right)^{\frac{1}{n}} \tag{16.7}$$

where ν is the Poisson's ratio, and σ_e represents the equivalent stress [8, 13].

In this paper, the stress–strain response of a crossbore geometry subjected to fluctuating internal pressures has been investigated using FEA and compared to notch strain analysis method. The Glinka and Neuber formulas under different conditions have been employed to study the notch strain analysis approach assessing which one has the least deviation of FEA results.

16.1.1 Finite Element Modeling

A cubic crossbore geometry without a fillet at the intersection, shown in Fig. 16.1, was considered as a model geometry to investigate the stress–strain response during fluctuating internal pressures. Eight load histories tabulated in Table 16.1, were considered in this study.

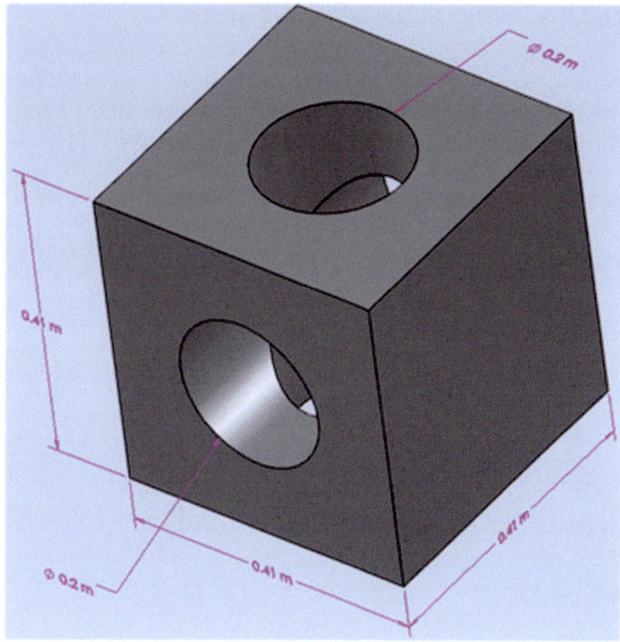

Fig. 16.1 Crossbore geometry

Table 16.1 Applied load histories

Set number	Load history number	Autofrettage pressure (MPa)	Operating pressure (MPa)
1	1	207	103
	2	207	69
2	3	172	103
	4	172	69
3	5	138	103
	6	138	69
4	7	103	103
	8	103	69

Each load history consists of four steps. The first step is same for each two successive load history numbers (known as set number hereafter) and encompasses an overload in order to simulate autofrettage overload. The second step is an unloading and the third step is a loading to the pump operating pressure that leads to yielding for odd load history numbers and is pure elastic for even ones. Finally, the fourth step is unloading to zero pressure.

FEA was performed on one eighth of the crossbore with three symmetry boundary conditions using a commercial FEA code (Abaqus 6.12). Tetrahedral elements with quadratic interpolation were used for meshing. A nonlinear kinematic hardening material model was applied to simulate the elastoplastic behavior of the crossbore. This material model includes the Bauschinger effect and has criterion for unloading. Both capabilities are required to accurately simulate the reverse yielding that occurs during the unloading step. Accurate modeling of reverse yielding is critical in order to precisely predict the residual stresses during fluctuating loading conditions [16, 17].

The material model was obtained from a fit to experimental monotonic stress–strain data for a precipitation hardening stainless steel. The behavior of material during autofrettage process and first cycle of operating pressure can be extracted by performing a monotonic tensile test. Thus, ASTM standard round tensile bars with 12.5 mm diameter gage sections were tested. The experimental stress–strain curve was the average of 3 specimens.

16.1.2 Notch Strain Analysis

To develop notch strain analysis models for a specific material, properties including elastic modulus (E), strength coefficient (K) and strain hardening exponent (n) and elastic stress concentration factor (K_t) are required. Curve fitting technique using Ramberg–Osgood relation was applied to extract the mechanical properties of material (K, n and E) from the same experimental stress–strain curve used for developing the FEA material model.

16.1.2.1 Notch Strain Analysis Equations

As already mentioned, different notch strain analysis formulas have been suggested to analyze the crossbore geometry. In this research, these different approaches were studied for each load history. Each approach had both Glinka and Neuber forms, giving 6 total relationships. Neuber's rule can be presented by (16.8)–(16.10).

$$K_t^2\left(\frac{S^2}{E}\right) = \frac{\sigma_e^2}{E} + \sigma_e\left(\frac{\sigma_e}{K}\right)^{\frac{1}{n}} \tag{16.8}$$

$$K_t^2\left[\frac{S^2}{E} + S\left(\frac{S}{K}\right)^{\frac{1}{n}}\right] = \frac{\sigma_e^2}{E} + \sigma_e\left(\frac{\sigma_e}{K}\right)^{\frac{1}{n}} \tag{16.9}$$

$$2\left(\frac{S^2}{E}\right)\left[\frac{K_t^2}{2} + (1-\nu) + 2\nu K_t\right] = \frac{\sigma_e^2}{E} + \sigma_e\left(\frac{\sigma_e}{K}\right)^{\frac{1}{n}} \tag{16.10}$$

Glinka's approach can be written as (16.11)–(16.13).

$$K_t^2\left(\frac{S^2}{E}\right) = \frac{\sigma_e^2}{E} + \frac{2}{n+1}\sigma_e\left(\frac{\sigma_e}{K}\right)^{\frac{1}{n}} \tag{16.11}$$

$$K_t^2\left[\frac{S^2}{E} + S\left(\frac{S}{K}\right)^{\frac{1}{n}}\right] = \frac{\sigma_e^2}{E} + \frac{2}{n+1}\sigma_e\left(\frac{\sigma_e}{K}\right)^{\frac{1}{n}} \tag{16.12}$$

$$2\left(\frac{S^2}{E}\right)\left[\frac{K_t^2}{2} + (1-\nu) + 2\nu K_t\right] = \frac{\sigma_e^2}{E} + \frac{2}{n+1}\sigma_e\left(\frac{\sigma_e}{K}\right)^{\frac{1}{n}} \tag{16.13}$$

16.1.2.2 Elastic Stress Concentration Factor

Because elastic stress concentration factor is geometry dependent and not load dependent, a single elastic FEA is all that is required to determine K_t for any load history. Four different approaches were considered to describe the theoretical elastic stress concentration factor presented in (16.14)–(16.17).

$$K_t = \frac{\sigma_m}{P} = K_{Pm} \tag{16.14}$$

$$K_t = \frac{\sigma_h}{P} = K_{Ph} \tag{16.15}$$

$$K_t = \frac{\sigma_m}{S_n} = K_{Sm} \tag{16.16}$$

$$K_t = \frac{\sigma_h}{S_n} = K_{Sh} \tag{16.17}$$

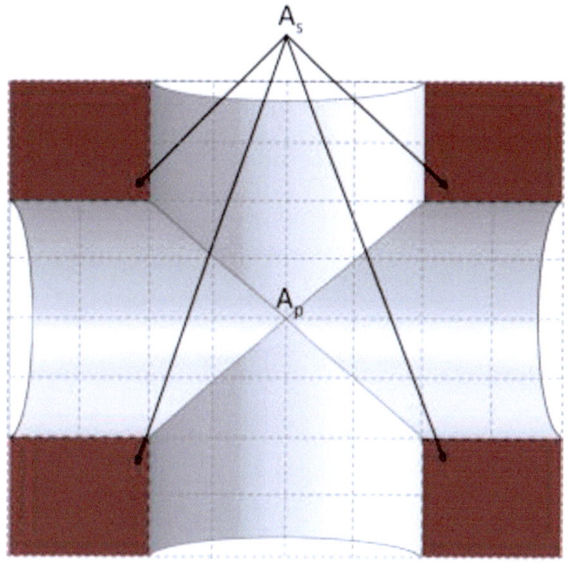

Fig. 16.2 Illustration of A_P and A_S in the half crossbore

where σ_h and σ_m respectively represent the maximum elastic hoop stress and maximum elastic Mises stress on the most critical node, which undergoes the maximum stress (both Mises and hoop) during a pure elastic loading, achieved by running an elastic FEA.

The elastic nominal stress S_n used above can be calculated by (16.18).

$$S_n = \frac{A_P}{A_S} P \tag{16.18}$$

where A_P is the projected area which internal pressure (P) is applied on and A_S is the area where internal stress is supported due to applied internal pressure. Figure 16.2 illustrates these two areas in a half-section of the model crossbore.

Thus, by combining (16.14)–(16.17) with (16.8)–(16.13), four different conditions can be derived for each of (16.8)–(16.13), giving 24 possible relationships, as presented in Table 16.2. To investigate the notch strain analysis, a general computer code was developed that can iteratively solve all 24 formulas for the notch strain analysis problem and plot the stress–strain response on the notch root during an applied load history. Since the results of this analysis involve many variables, a specific notation was defined for each combination. This notation starts with a digit which represents the load history number from Table 16.1. The second digit shows the number of applied equation from (16.8) to (16.10) by using numbers *I*, *II* and *III*, and the third term is a letter of N or G which stands for Neuber or Glinka and the fourth place in each notation is occupied by stress concentration factor symbol from (16.14) to (16.17). Finally, as presented in Table 16.2 a code number from 1 to 24 has been assigned to each notation.

16.2 Results and Discussion

To obtain the stress–strain curve for each simulated load history, Mises equivalent stress and total equivalent strain were extracted from the most critical node. Comparison of the predictions of formulas 1–24 in Table 16.2, shows that some of the equations are insensitive to load history when certain stress concentration forms are used. For example, (16.8) and (16.9) predict the same response for all load histories when either K_{Pm} or K_{Ph} is used (formulas 1, 2, 7 and 8 give identical results to formulas 3, 4, 9, 10, respectively). However, when K_{Sm} or K_{Sh} is substituted, (16.8) and (16.9) predict the same response for load histories 8, 7, 6 and 5 but predict different behavior for load histories 1, 2, 3, and 4. If we compare the results of (16.8) when K_{Pm} or K_{Sm} is substituted, the prediction is the same for all load histories (formulas 1 and 2 lead to same results as formulas 13 and 14, respectively). Inserting K_{Pm} or K_{Sm} into (16.9) leads to identical behavior for load histories 7 and 8, but results diverge for other load histories. There is similar behavior when K_{Ph} and K_{Sh} are used. We will now discuss the accuracy of selected analytical forms from Table 16.2. Since the loading consists of four critical steps, we will discuss each step below.

Table 16.2 Notch strain analysis formulas

Code number	Notation definition	Formula
1	i/I/N/K_{Pm}	$K_{Pm}^2\left(\frac{P^2}{E}\right) = \frac{\sigma_e^2}{E} + \sigma_e\left(\frac{\sigma_e}{K}\right)^{\frac{1}{n}}$
2	i/I/G/K_{Pm}	$K_{Pm}^2\left(\frac{P^2}{E}\right) = \frac{\sigma_e^2}{E} + \frac{2}{n+1}\sigma_e\left(\frac{\sigma_e}{K}\right)^{\frac{1}{n}}$
3	i/II/N/K_{Pm}	$K_{Pm}^2\left[\frac{P^2}{E} + P\left(\frac{P}{K}\right)^{\frac{1}{n}}\right] = \frac{\sigma_e^2}{E} + \sigma_e\left(\frac{\sigma_e}{K}\right)^{\frac{1}{n}}$
4	i/II/G/K_{Pm}	$K_{Pm}^2\left[\frac{P^2}{E} + P\left(\frac{P}{K}\right)^{\frac{1}{n}}\right] = \frac{\sigma_e^2}{E} + \frac{2}{n+1}\sigma_e\left(\frac{\sigma_e}{K}\right)^{\frac{1}{n}}$
5	i/III/N/K_{Pm}	$2\left(\frac{P^2}{E}\right)\left[\frac{K_{Pm}^2}{2} + (1-\nu) + 2\nu K_{Pm}\right] = \frac{\sigma_e^2}{E} + \sigma_e\left(\frac{\sigma_e}{K}\right)^{\frac{1}{n}}$
6	i/III/G/K_{Pm}	$2\left(\frac{P^2}{E}\right)\left[\frac{K_{Pm}^2}{2} + (1-\nu) + 2\nu K_{Pm}\right] = \frac{\sigma_e^2}{E} + \frac{2}{n+1}\sigma_e\left(\frac{\sigma_e}{K}\right)^{\frac{1}{n}}$
7	i/I/N/K_{Ph}	$K_{Ph}^2\left(\frac{P^2}{E}\right) = \frac{\sigma_e^2}{E} + \sigma_e\left(\frac{\sigma_e}{K}\right)^{\frac{1}{n}}$
8	i/I/G/K_{Ph}	$K_{Ph}^2\left(\frac{P^2}{E}\right) = \frac{\sigma_e^2}{E} + \frac{2}{n+1}\sigma_e\left(\frac{\sigma_e}{K}\right)^{\frac{1}{n}}$
9	i/II/N/K_{Ph}	$K_{Ph}^2\left[\frac{P^2}{E} + P\left(\frac{P}{K}\right)^{\frac{1}{n}}\right] = \frac{\sigma_e^2}{E} + \sigma_e\left(\frac{\sigma_e}{K}\right)^{\frac{1}{n}}$
10	i/II/G/K_{Ph}	$K_{Ph}^2\left[\frac{P^2}{E} + P\left(\frac{P}{K}\right)^{\frac{1}{n}}\right] = \frac{\sigma_e^2}{E} + \frac{2}{n+1}\sigma_e\left(\frac{\sigma_e}{K}\right)^{\frac{1}{n}}$
11	i/III/N/K_{Ph}	$2\left(\frac{P^2}{E}\right)\left[\frac{K_{Ph}^2}{2} + (1-\nu) + 2\nu K_{Ph}\right] = \frac{\sigma_e^2}{E} + \sigma_e\left(\frac{\sigma_e}{K}\right)^{\frac{1}{n}}$
12	i/III/G/K_{Ph}	$2\left(\frac{P^2}{E}\right)\left[\frac{K_{Ph}^2}{2} + (1-\nu) + 2\nu K_{Ph}\right] = \frac{\sigma_e^2}{E} + \frac{2}{n+1}\sigma_e\left(\frac{\sigma_e}{K}\right)^{\frac{1}{n}}$
13	i/I/N/K_{Sm}	$K_{Sm}^2\left(\frac{S^2}{E}\right) = \frac{\sigma_e^2}{E} + \sigma_e\left(\frac{\sigma_e}{K}\right)^{\frac{1}{n}}$
14	i/I/G/K_{Sm}	$K_{Sm}^2\left(\frac{S^2}{E}\right) = \frac{\sigma_e^2}{E} + \frac{2}{n+1}\sigma_e\left(\frac{\sigma_e}{K}\right)^{\frac{1}{n}}$
15	i/II/N/K_{Sm}	$K_{Sm}^2\left[\frac{S^2}{E} + S\left(\frac{S}{K}\right)^{\frac{1}{n}}\right] = \frac{\sigma_e^2}{E} + \sigma_e\left(\frac{\sigma_e}{K}\right)^{\frac{1}{n}}$
16	i/II/G/K_{Sm}	$K_{Sm}^2\left[\frac{S^2}{E} + S\left(\frac{S}{K}\right)^{\frac{1}{n}}\right] = \frac{\sigma_e^2}{E} + \frac{2}{n+1}\sigma_e\left(\frac{\sigma_e}{K}\right)^{\frac{1}{n}}$
17	i/III/N/K_{Sm}	$2\left(\frac{S^2}{E}\right)\left[\frac{K_{Sm}^2}{2} + (1-\nu) + 2\nu K_{Sm}\right] = \frac{\sigma_e^2}{E} + \sigma_e\left(\frac{\sigma_e}{K}\right)^{\frac{1}{n}}$
18	i/III/G/K_{Sm}	$2\left(\frac{S^2}{E}\right)\left[\frac{K_{Sm}^2}{2} + (1-\nu) + 2\nu K_{Sm}\right] = \frac{\sigma_e^2}{E} + \frac{2}{n+1}\sigma_e\left(\frac{\sigma_e}{K}\right)^{\frac{1}{n}}$
19	i/I/N/K_{Sh}	$K_{Sh}^2\left(\frac{S^2}{E}\right) = \frac{\sigma_e^2}{E} + \sigma_e\left(\frac{\sigma_e}{K}\right)^{\frac{1}{n}}$
20	i/I/G/K_{Sh}	$K_{Sh}^2\left(\frac{S^2}{E}\right) = \frac{\sigma_e^2}{E} + \frac{2}{n+1}\sigma_e\left(\frac{\sigma_e}{K}\right)^{\frac{1}{n}}$
21	i/II/N/K_{Sh}	$K_{Sh}^2\left[\frac{S^2}{E} + S\left(\frac{S}{K}\right)^{\frac{1}{n}}\right] = \frac{\sigma_e^2}{E} + \sigma_e\left(\frac{\sigma_e}{K}\right)^{\frac{1}{n}}$
22	i/II/G/K_{Sh}	$K_{Sh}^2\left[\frac{S^2}{E} + S\left(\frac{S}{K}\right)^{\frac{1}{n}}\right] = \frac{\sigma_e^2}{E} + \frac{2}{n+1}\sigma_e\left(\frac{\sigma_e}{K}\right)^{\frac{1}{n}}$
23	i/III/N/K_{Sh}	$2\left(\frac{S^2}{E}\right)\left[\frac{K_{Sh}^2}{2} + (1-\nu) + 2\nu K_{Sh}\right] = \frac{\sigma_e^2}{E} + \sigma_e\left(\frac{\sigma_e}{K}\right)^{\frac{1}{n}}$
24	i/III/G/K_{Sh}	$2\left(\frac{S^2}{E}\right)\left[\frac{K_{Sh}^2}{2} + (1-\nu) + 2\nu K_{Sh}\right] = \frac{\sigma_e^2}{E} + \frac{2}{n+1}\sigma_e\left(\frac{\sigma_e}{K}\right)^{\frac{1}{n}}$

16.2.1 Autofrettage Overload

For all load histories the Neuber approach leads to larger overestimations in the stress component when compared to the Glinka approach as shown in Fig. 16.3 for load history 2. For load set 1 (load histories 1 and 2), all of the 24 formulas overestimate the amount of stress with the lowest overestimation being 2 %, obtained by 1,2/I/G/K_{Ph}, 1,2/II/G/K_{Ph} and 1,2/I/G/K_{Sh}. There is similar observation for load histories 3 and 4 for the lowest overestimation of 3 %. For load histories 5 and 6, as well as previous formulas which lead to the lowest overestimation for load histories 1–4, 5,6/II/G/K_{Sh} also shows the lowest overestimation of 2 %. For load histories 7 and 8 formulas 8, 10, 20 and 22 underestimate the stress component and the lowest absolute value of the percentage difference is obtained by 7,8/III/G/K_{Ph} which is 0.2 %.

In the case of strain components, predictions from both approaches are less accurate when compared to the predictions for the stress states. For example, stress predictions using 3,4/III/N/K_{Pm} are 10 % different from the FEA results but the strain difference is 62 %. To evaluate which notch strain analysis formula better predicts the strain component of the autofrettage overload Fig. 16.4 can be used. This figure shows the percentage difference between Abaqus and notch strain analysis for strain component during autofrettage. By reviewing Fig. 16.4 it is obvious that there is no decisive trend in strain values to decide which notch strain analysis formula has the least deviation. Thus, average of absolute values of percentage differences was considered as criteria to compare different notch strain analysis formulas. The lowest average value for strain difference is 4 % for i/II/G/K_{Sm}. To evaluate the prediction of notch strain analysis formulas for autofrettage overload's both components (stress and strain) a certain criteria is required. Therefore, average of absolute values of percentage difference of strain was multiplied by stress one. The lowest multiplication value, 13 %, is also for i/II/G/K_{Sm}, thus it can be concluded i/II/G/K_{Sm} predicts the autofrettage point more accurately than other formulas.

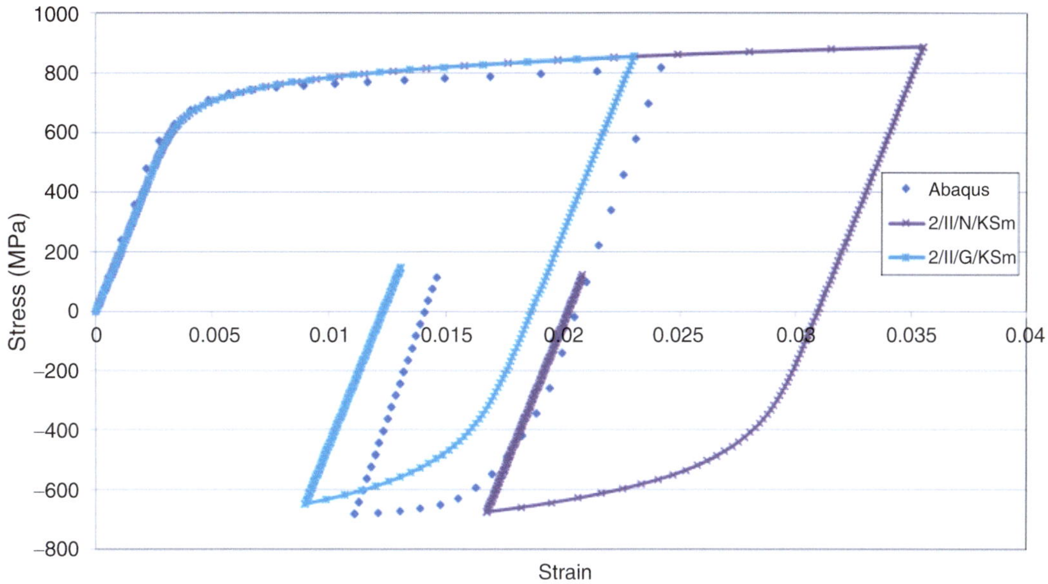

Fig. 16.3 Comparison between Abaqus, Neuber and Glinka

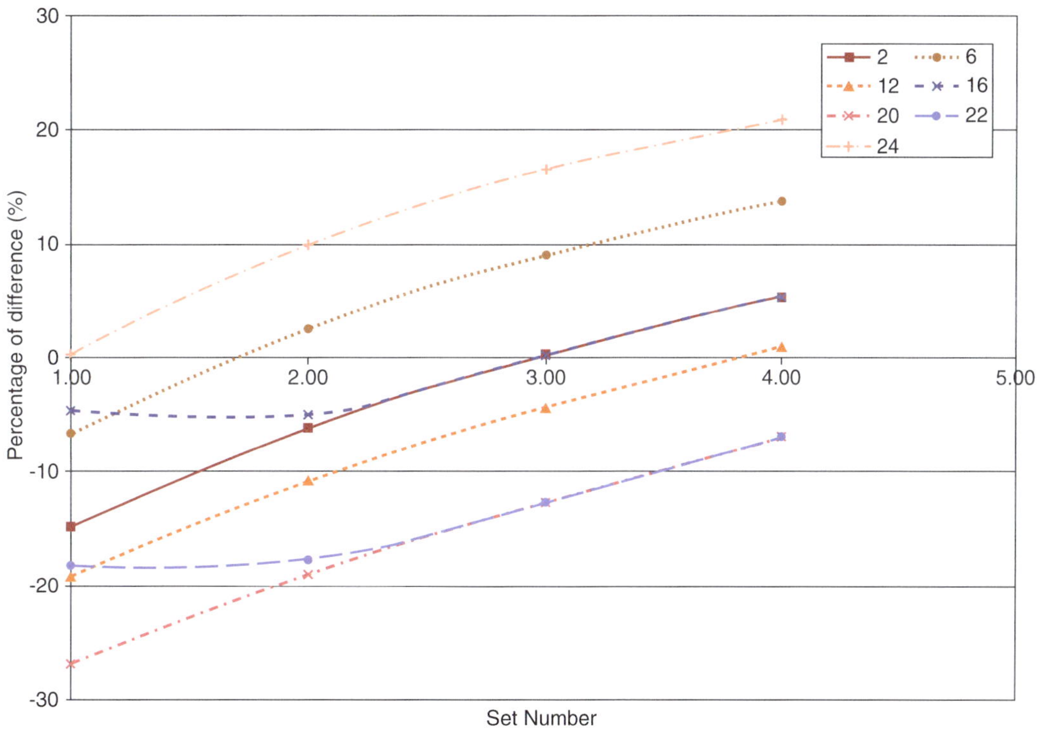

Fig. 16.4 Difference between Abaqus results and notch strain analysis for strain component of autofrettage overload

16.2.2 Autofrettage Unload

The stress and strain components at full unload are governed by the Bauschinger effect and the overload history. Figure 16.5 shows the percentage difference between Abaqus and notch strain predictions for the stress component. As it can be seen, the largest difference between Abaqus and notch strain analysis happens for set 4 which has the lowest autofrettage load, thus it can be stated that as autofrettage load increases, the notch strain analysis predictions for unloaded stress state begin to convergeto FEA predictions.

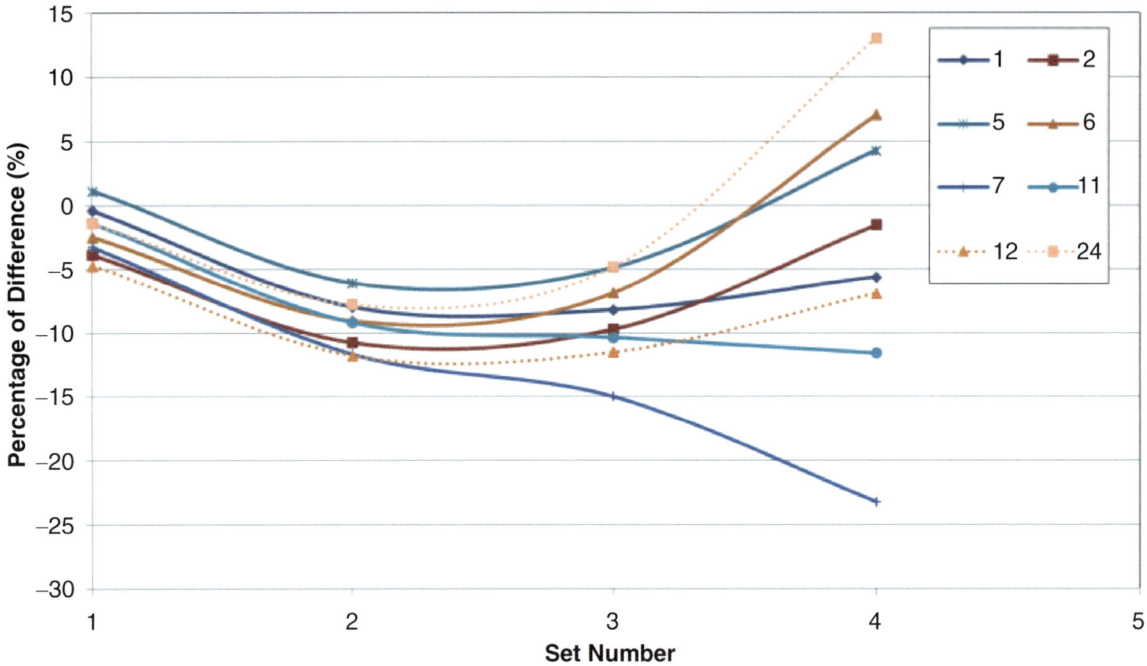

Fig. 16.5 Difference between Abaqus results and notch strain analysis for stress component of autofrettage unload

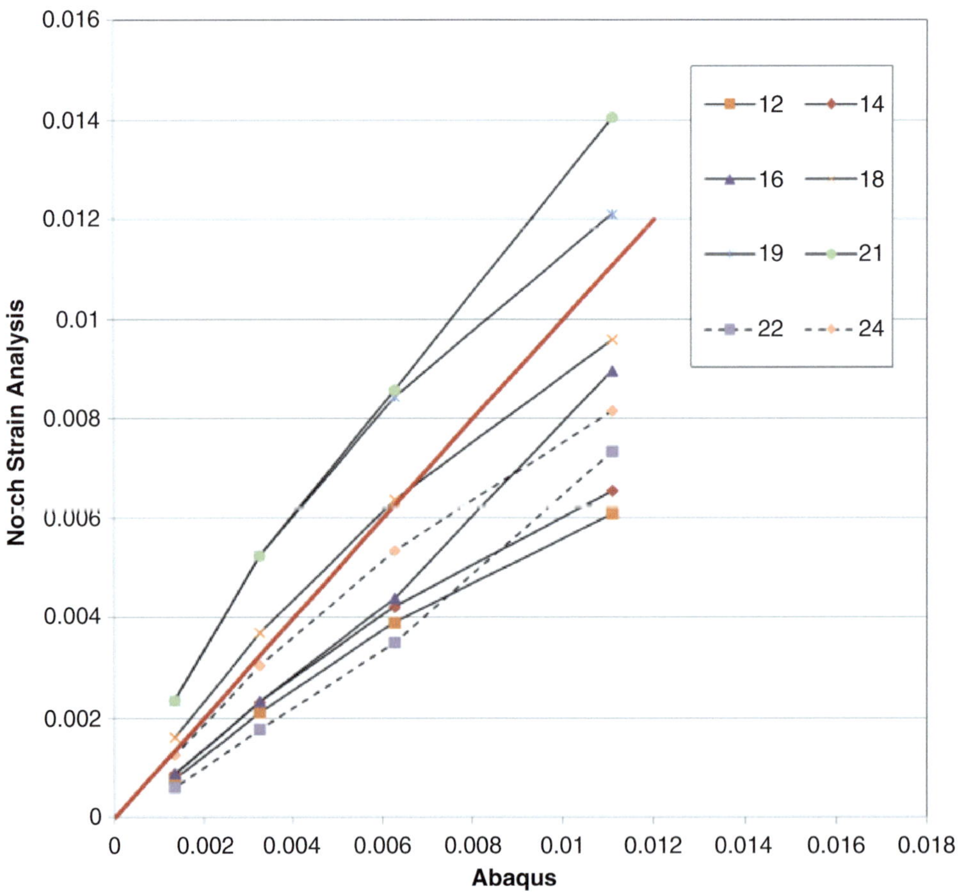

Fig. 16.6 Notch strain analysis versus Abaqus for strain component of autofrettage unload

Figure 16.6 shows strain components of the autofrettage unload in such a way that notch strain analysis results have been plotted versus Abaqus results. It is obvious that formula 18, i/III/G/K_{Sm}, has the lowest deviation from abaqus results.

16.2.3 Operating Pressure Cycles

The maximum and minimum pressures after the autofrettage cycles define the mean stress and strain amplitude. Both mean stress and strain amplitude depend on loading condition and material history and have significant effects on fatigue life estimation approaches. Calculating mean stress values based on different approaches showed the difference depends on load history. Thus, by considering the average of absolute values of percentage difference it can be concluded that i/I/G/K_{Pm}, i/II/G/K_{Pm}, i/III/N/K_{Pm}, i/I/G/K_{Sm} and i/II/G/K_{Sm} have the lowest deviation from FEA results. Analyzing strain amplitudes obtained through 24 notch strain analysis formulas as well as FEA ones for each load history demonstrated formulas 8, 10, 20 and 22 have the minimum average of percentage difference of approximately 6 %.

16.3 Conclusions

1. Stress–strain response of a crossbore geometry subjected to 8 different internal pressure load histories was studied considering both FEA and different notch strain analysis approaches.
2. Equation (16.16) of Table 16.2, i/II/G/K_{Sm}, estimates the stress and strain components of the autofrettage overload more accurately comparing to other notch strain analysis formulas.
3. Regarding the stress component of the autofrettage unload formula 5 of Table 16.2, i/III/N/K_{Pm}, has the lowest average of percentage difference absolute values, 4 %.
4. Equation (16.18) of Table 16.2, i/III/G/K_{Sm}, has the lowest deviation from abaqus results for strain component of the autofrettage unload.

References

1. Bayyouk, J.: Weir oil and gas, long-life fluid ends, Upstream Pumping, February 2013
2. Foote, E.: Southwest Oilfield Products, Inc, Upstream Pumping, March 2013
3. Pendleton, G., McPheron, R.: AXON Energy Products, Upstream Pumping, July 2013
4. White, S.: White Star Pump Company, Upstream Pumping, July 2011
5. Rybicki, E.F., Stonesifer, R.B., Olson, R.J.: Stress intensity factors due to residual stresses in thin-walled girth welded pipes. J. Press. Vessel Technol. **103**, 60–75 (1981)
6. Brinksmeier, E., Cammett, J.T., Konig, W., Leskovar, P., Peters, J., Tonshoff, H.K.: Residual stresses—measurement and causes in machining processes. CIRP Ann. Manuf. Technol. **31**(2), 491–510 (1982)
7. Yang, S., Badr, E.A., Sorem, J.R., Tipton, S.M.: Advantages of sequential crossbore autofrettage of triplex pump fluid end crossbores. High Pressures—Codes, Analysis, and Applications, PVP-Vol. **263**, ASME, pp. 81–89 (1993)
8. Badr, E.A., Sorem, J.R., Tipton, S.M.: Residual stress estimation in crossbore with Bauschinger effect inclusion using FEM and strain energy density. J. Press. Vessel Technol. **121**, 358–363 (1999)
9. Parker, A.P.: Autofrettage of open-end tubes-pressures, stresses, strain and code comparisons. J. Press. Vessel Technol. **123**, 271–281 (2001)
10. Sorem, J., Shadley, R., Tipton, S.M.: Design curves for maximum stresses in block containing pressurized bore intersections. J. Mech. Des. **113**, 427–431 (1991)
11. Stephens, R.I., Fatemi, A., Stephens, R.I., Fuchs, H.Q.: Metalic Fatigue in Engineering. Wily, Toronto (2001)
12. Yang, S.: Experimental and analytical investigation of the effects of autofrettage on fatigue damage development in blocks containing crossbores under cyclic internal pressure. Ph.D. Dissertation, University of Tulsa (1995)
13. Badr, E.A.: Estimation of residual stresses induced by autofrettage with and experimental evaluation on the autofrettage process in crossbores of positive displacement pumps. Ph.D. Dissertation, University of Tulsa (1994)
14. Zeng, Z., Fatemi, A.: Elasto-plastic stress and strain behavior at notch roots under monotonic and cyclic loading. J. Strain Anal. Eng. Des. **36**(3), 287–300 (2001)
15. Hoffman, M., Seeger, T.: A generalized method for estimating multiaxial elastic-plastic notch stresses and strains part 1: theory. J. Eng. Mater. Technol. **107**, 250–254 (1985)
16. Abaqus documentation, 23.2.2 Models for metals subjected to cyclic loading
17. Sarajavi, U., Cronvall, O.: Simulation and analysis of data for enhancing low cycle fatigue test procedures. VTT Technical Research Center of Finland (2006)

Chapter 17
Opto-acoustic and Neutron Emissions from Fracture and Earthquakes

Alberto Carpinteri

Abstract TeraHertz phonons are produced in condensed matter by mechanical instabilities at the nano-scale (fracture, turbulence, buckling). They present a frequency that is close to the resonance frequency of the atomic lattices and an energy that is close to that of thermal neutrons. A series of fracture experiments on natural rocks has recently demonstrated that the TeraHertz phonons are able to induce fission reactions on medium weight elements with neutron and/or alpha particle emissions. The same phenomenon appears to have occurred in several different situations and to explain puzzles related to the history of our planet, like the ocean formation or the primordial carbon pollution, as well as scientific mysteries, like the so-called "cold nuclear fusion" or the correct radio-carbon dating of organic materials.

Very important applications to earthquake precursors, climate change, energy production, and cell biology can not be excluded.

Keywords Fracture • Turbulence • Buckling • TeraHertz phonons • Ultrasonic pressure waves • Piezonuclear fission reactions • Neutron emissions • Alpha particle emissions • Acoustic emissions • Electromagnetic emissions • Compositional changes • Great Oxidation Event • Carbon pollution • Earthquake precursors • Chemical evolution • Great Red Spot of Jupiter • Cold nuclear fusion • Turin Shroud • Cell mechanotransduction • Protein folding

17.1 Fracture and Acoustic Emission: From Hertz to TeraHertz Pressure Wave Frequencies

When you cut a stretched rubber band, it remains subject to rapid fluctuations for a few moments. The same phenomenon occurs in any solid body when it breaks in a brittle way, even if only partially. In the case of the formation or propagation of micro-cracks, such dynamic phenomenon appears under the form of longitudinal waves of expansion/contraction (tension/compression), in addition to transverse or shear waves. These are generally said pressure waves, or phonons when their particle nature is emphasized, and travel at a speed which is characteristic of the medium, and, for most of the solids and fluids, presents an order of magnitude of 10^3 m/s. On the other hand, the wavelength of pressure waves emitted by forming or propagating cracks appears to be of the same order of magnitude of crack size or crack advancement length. The wavelength can not therefore exceed the maximum size of the body in which the crack is contained and may vary from the nanometre scale (10^{-9} m), for defects in crystal lattices such as vacancies and dislocations, up to the kilometre, in the case of Earth's Crust faults. Applying the well-known relationship: frequency = speed/wavelength, one obtains the two extreme cases corresponding to the frequency of pressure waves: 10^{12} oscillations/s (TeraHertz), in the case of the formation of nano-cracks, as well as one oscillation/s (Hertz), in the case of large-scale tectonic dynamics (Fig. 17.1).

In fact in solids, whatever their size, the cracks that are formed or propagate are of different lengths, sometimes belonging to different orders of magnitude. In particular, in the nanoscopic bodies the only frequency present should be the TeraHertz, since the higher frequencies would imply defects at the atomic or subatomic scales. On the contrary, in the Earth's Crust and during an earthquake, cracking is a multi-scale phenomenon as well as the frequencies of pressure waves are spread over a broad spectrum (Fig. 17.1). Moreover, while at the early stages of the seismic event mainly small cracks will be present and active and therefore high frequencies, so at the end large cracks and low frequencies will prevail, the latter typically in the audible field.

A. Carpinteri (✉)
Department of Structural, Geotechnical and Building Engineering, Politecnico di Torino, Corso Duca degli Abruzzi 24, Torino 10129, Italy
e-mail: alberto.carpinteri@polito.it

A.M. Beese et al. (eds.), *Fracture, Fatigue, Failure and Damage Evolution, Volume 8*, Conference Proceedings of the Society for Experimental Mechanics Series, DOI 10.1007/978-3-319-21611-9_17

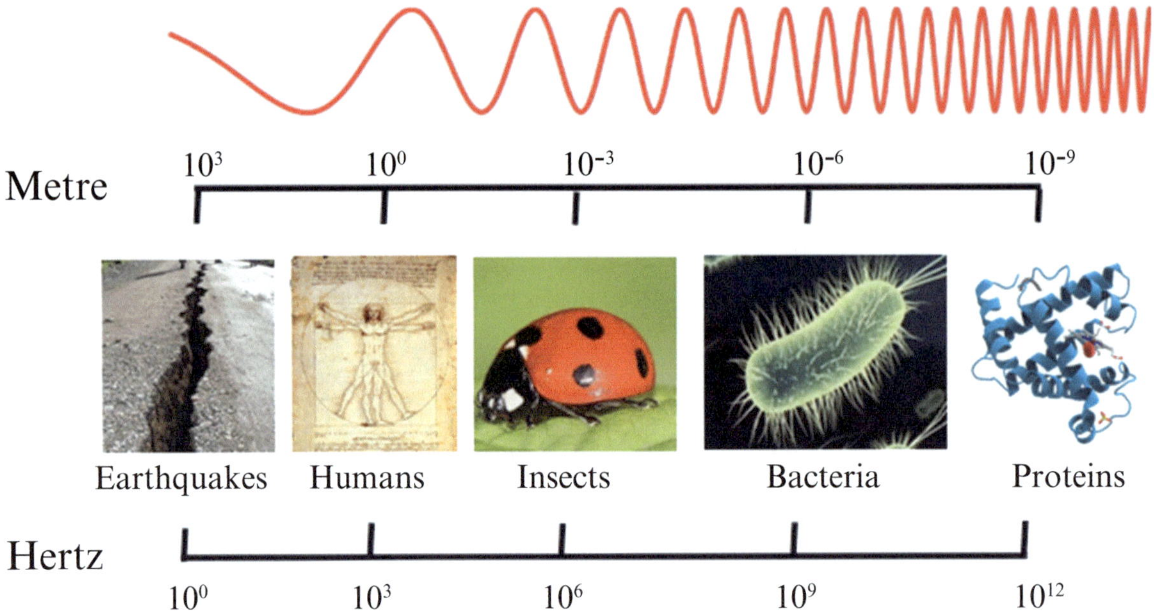

Fig. 17.1 Correlation between wavelength scale and frequency scale by assuming a constant pressure wave speed

17.2 Seismic Precursors: Acoustic, Electromagnetic, Neutron Emissions

Further considering the very important case of earthquakes, you can complete the picture by stating that, as fracture at the nanoscale (10^{-9} m) emits phonons at the frequency scale of TeraHertz (10^{12} Hz), so fracture at the microscale (10^{-6} m) emits phonons at the frequency scale of GigaHertz (10^9 Hz), at the scale of millimetre emits phonons at the scale of MegaHertz (10^6 Hz), at the scale of metre emits phonons at the scale of kiloHertz (10^3 Hz), and eventually faults at the kilometre scale emit phonons at the scale of the simple Hertz, which is the typical and most likely frequency of seismic oscillations (Fig. 17.1) [1].

The animals with sensitive hearing in the ultrasonic field (frequency >20 kHz) "feel" the earthquake up to 1 day in advance, when the active cracks are still below the metre scale. Ultrasounds are in fact a well-known seismic precursor [2, 3]. With frequencies between Mega- and GigaHertz, and therefore cracks between the micron and the millimetre scale, phonons can generate electromagnetic waves of the same frequency, which turn out to be even a more advanced seismic precursor (up to a few days before) [4, 5].

When phonons show frequencies between Giga- and TeraHertz, and then with cracks below the micron scale, we are witnessing a phenomenon partially unexpected: phonons resonate with the crystal lattices and, through a complex cascade of events (acceleration of electrons, bremsstrahlung gamma radiation, photo-fission, etc.), may produce nuclear fission reactions [6–14]. It can be shown experimentally how such fission reactions can emit neutrons [15–17] like in the well known case of uranium-235 but without gamma radiation and radioactive wastes. Note that the Debye frequency, i.e., the fundamental frequency of free vibration of crystal lattices, is around the TeraHertz, and this is not a coincidence, since it is simply due to the fact that the inter-atomic distance is just around the nanometre, as indeed the minimum size of the lattice defects. As the chain reactions are sustained by thermal neutrons in a nuclear power plant, so the piezonuclear reactions are triggered by phonons that have a frequency close to the resonance frequency of the crystal lattice and an energy close to that of thermal neutrons. Neutrons therefore appear to be as the most advanced earthquake precursor (up to 3 weeks before) [18–23].

17.3 Chemical Evolution of Our Planet and Its Reproduction in the Fracture Mechanics Laboratory

The piezonuclear fission reactions appear then to be induced by pressure waves at very high frequencies (TeraHertz). They are often accompanied and revealed by the emission of neutrons and/or alpha particles. However, gamma rays and radioactive wastes appear to be absent in the experiments. Ultrasonic pressure waves may in turn be produced by the most common mechanical instabilities, such as fracture in solids and turbulence in fluids. Both are hierarchical, multi-fractal, and dissipative phenomena, where cracks and vortexes, respectively, are present at the different scales.

After the early experiments conducted at the National Research Council of Italy (CNR) [24, 25], soliciting with ultrasounds aqueous solutions of iron salts, the research group of the Politecnico di Torino has conducted fracture experiments on solid samples, using iron-rich rocks like granite [26–36], basalt and magnetite [37, 38], and then marble [39], mortar [40], and steel [41]. Different types of detectors have demonstrated the presence of significant neutron emissions, in some cases by different orders of magnitude higher than the usual environmental background (up to 10 times from granitic rocks, up to 100 times from basalt, up to 1000 times from magnetite).

The neutron flux was found to depend, besides on the iron content, on the size of the specimen through the well-known brittleness size effect [42–45]: larger sizes imply a higher brittleness, i.e. a more relevant strain energy release, and therefore more neutrons.

These studies have also been able to give an answer to some puzzles related to the history of our planet. It has been shown how the piezonuclear reactions that would have occurred between 3.8 and 2.5 billion years ago, during the period of formation and most intense activity of tectonic plates, have resulted in the splitting of atoms of certain elements, which were so transformed into other lighter ones. Since the product-elements, i.e., the fragments of the fissions, appear to be stable isotopes, all the excess neutrons are therefore emitted. Several of the most abundant chemical elements have been involved in similar transformations, like a part of magnesium that transformed into carbon, forming the dense atmospheres of the primordial terrestrial eras [46, 47]. In a similar way, calcium depletion contributed to the formation of oceans as a result of fracture phenomena in limestone rocks.

Considering the entire life of our planet and all the most abundant chemical elements [48–50], it can be seen how ferrous elements have dramatically decreased in the Earth's Crust (−12 %), as well as at the same time aluminum and silicon have increased (+8.8 %). An increment in magnesium (+3.2 %), which then transformed into carbon, has been assumed as the origin of carbon-rich primordial atmospheres. Similarly, alkaline-earth elements have strongly decreased (−8.7 %), whereas alkaline elements (+5.4 %) and oxygen (+3.3 %) have increased. The appearance of a 3.3 % oxygen represents the well-known Great Oxidation Event, a phenomenon that led to the formation of oceans and the origin of life on our planet.

These transformations, that have lasted for billion years in the Earth's Crust, have been reproduced in the laboratory in a fraction of a second by crushing different rock samples. We were able to confirm, through advanced micro-chemical analyses, the most relevant compositional variations described above at the geological and planetary scales: the transformation of iron into aluminum, or into magnesium and silicon (in iron-rich natural rocks [37]), as well as the transformation of calcium and magnesium into other lighter elements including carbon (in the samples of marble [39]). Such variations are shown to be not modest at all. The iron decrement in magnetite was found to be of 27.9 %, compared to an overall increment of 27.7 % in lighter elements. So in marble, carbon has increased by 13 %, compared to an exactly equivalent overall decrement in heavier elements.

Since the natural carbon production of the primordial eras, although at a slower rate, is going on even today, due to the seismic activity, the monitoring of carbon dioxide in relation to major earthquakes can be considered as a potential earthquake precursor [47], in addition to acoustic, electromagnetic, and neutron emissions.

17.4 Chemical Evolution in the Solar System

Even in the case of the other planets of the Solar System we are witnessing a series of experimental evidences that can be interpreted in the light of piezonuclear fission reactions [51]. In particular, the data coming from different surveys on the crust of planet Mars, made available by the NASA space missions over the past 15 years, suggest that the increase in certain elements (iron, chlorine, and argon) and the concomitant decrease in others (nickel and potassium), together with the emission of neutrons from the major fault lines in the planet, should all be considered as phenomena directly correlated. These data provide a clear confirmation that seismic activity has contributed to the chemical evolution of the Red Planet. Similar experimental evidences are concerning Mercury, Jupiter (with its relevant emission of neutrons from the Great Red Spot), and the Sun itself. The piezonuclear phenomena are triggered by earthquakes in rocky planets and by storms in gaseous planets. In the Sun, for example, the drastic decrease in lithium appears to be due to the fission of the same lithium into helium and hydrogen.

17.5 A Plausible Explanation of the So-Called Cold Nuclear Fusion

Several evidences have been observed during the last 25 years of anomalous nuclear reactions in electrolytic experiments. The purpose was that of providing an explanation to the phenomena related to the so-called "cold nuclear fusion", and of evaluating the possibility of a heat generation from electrolytic cells [52–54]. Despite the large amount of positive experimental results, the understanding of these phenomena is still unsatisfactory. On the other hand, as reported in most of the articles on cold nuclear fusion, the appearance of micro-cracks on the surface of the electrodes used in the experiments is one of the most common observations. It is therefore possible to give an explanation of a mechanical nature which takes into account the hydrogen embrittlement of the metallic electrodes.

In our earlier experiments [55, 56], electrolytic phenomena were produced by means of an anode of a nickel-iron alloy and a cathode of a cobalt-chromium alloy, immersed in an aqueous solution of potassium carbonate. During these experiments, emissions of neutrons and alpha particles were revealed. Furthermore, the composition of the electrodes was analyzed before and after the experiments, allowing to identify piezonuclear fissions occurred in the electrodes. The primary process appears to be a symmetric fission of the atom of nickel into two atoms of silicon, or two atoms of magnesium. In the latter case, additional fragments were found to be constituted of alpha particles.

In our later experiments [57, 58], where a palladium electrode was used, the primary process appears to be the non-symmetric fission of palladium into iron and calcium, whereas the secondary processes appear to be the further fissions of both such products into oxygen atoms and alpha particles.

17.6 A Catastrophic Earthquake Behind the Mystery of the Shroud

A neutron radiation, produced by the historical earthquake occurred around the year AD 33, may have caused the erroneous radiocarbon dating of the Shroud of Turin in 1988 [59, 60]. Neutron radiation could have also caused the image formation of a crucified man, who many believe to be Jesus Christ, on the linen cloth. The Shroud has attracted a large interest since Secondo Pia took the first photograph in 1898: in fact, one wonders if this is really the shroud of Jesus Christ, investigating his true age and the manner in which the image was produced. According to carbon-14 dating, the cloth would approximately be only 750 years old. From 1988 onwards, several researchers have instead argued that the Shroud would be much older and that the process of dating would have been wrong because of neutron radiation, so as to form new isotopes of carbon from nitrogen atoms. However, so far, any plausible physical reason has not yet identified that can justify the origin of this neutron radiation [61]. The mechanical and chemical experiments described in the previous sections allow, on the other hand, to hypothesize that high-frequency pressure waves, generated in the Earth's Crust by the historical earthquake of AD 33, which took place in old Jerusalem with a magnitude between 8 and 9 on the Richter Scale, may have produced a neutron radiography on linen fibers and seemingly rejuvenated the same fabric. Let us consider that, although the calculated integral flux of 10^{13} neutrons per square centimetre is 10 times greater than the cancer therapy dose, nevertheless it is 100 times smaller than the lethal dose.

17.7 Future Applications Also to Biology?

Regarding the living cells, the piezonuclear reactions could explain the mechanism that governs the so-called "sodium-potassium pump" and, more in general, the metabolic processes. In the case of the ionic pump, the ions of sodium and potassium would be subject to a continuous and recurrent transformation of one into the other, losing and regaining an oxygen atom at each passage through the cell membrane. As cells are microscopic objects, so proteins are nanoscopic and the typical mechanisms of "folding", which make the passage of ions through the cell membrane possible, are accompanied by vibrational phenomena of resonance at the frequency of TeraHertz [62, 63] (mechanotransduction [64]). More precisely, the folding changes of configuration in the proteins should be interpreted as a dynamic nano-buckling (with snap-through) of thin complex-shaped shells. As in the case of acoustic emission from cracks or vortexes at the nano-scale, also the resonance frequency of nano-structures may be evaluated in the TeraHertz range.

Analogous reasons could explain also the "digestion" of radioactive isotopes intended as their transformation into stable isotopes of chemical elements which are essentials for the vital activity of microbial cultures [65].

Acknowledgements I would like to thank my two colleagues, Prof. Giuseppe Lacidogna and Dr. Amedeo Manuello, for their untiring efforts during the last 6 years in carrying out with me this difficult and cutting-edge research work. Thanks are also due to Dr. Oscar Borla for his specific expertise and contribution in environment neutron measurements.

References

1. Ashcroft, N.W., Mermin, D.N.: Solid state physics. Cengage Learning, Delhi (2013)
2. Lockner, D.A., et al.: Quasi-static fault growth and shear fracture energy in granite. Nature **350**, 39–42 (1991)
3. Carpinteri, A., Lacidogna, G., Niccolini, G.: Acoustic emission monitoring of medieval towers considered as sensitive earthquake receptors. Nat. Hazards Earth Syst. Sci. **7**, 251–261 (2007)
4. Rabinovitch, A., Frid, V., Bahat, D.: Surface oscillations. A possible source of fracture induced electromagnetic oscillations. Tectonophysics **431**, 15–21 (2007)
5. Carpinteri, A., et al.: Mechanical and electromagnetic emissions related to stress-induced cracks. Exp. Tech. **36**, 53–64 (2012)
6. Bridgman, P.W.: The breakdown of atoms at high pressures. Phys. Rev. **29**, 188–191 (1927)
7. Batzel, R.E., Seaborg, G.T.: Fission of medium weight elements. Phys. Rev. **82**, 607–615 (1951)
8. Fulmer, C.B., et al.: Evidence for photofission of iron. Phys. Rev. Lett. **19**, 522–523 (1967)
9. Widom, A., Swain, J., Srivastava, Y.N.: Neutron production from the fracture of piezoelectric rocks. J. Phys. G: Nucl. Part. Phys. **40**(15006), 1–8 (2013)
10. Widom, A., Swain, J., Srivastava, Y.N.: Photo-disintegration of the iron nucleus in fractured magnetite rocks with magnetostriction. Meccanica **50**, 1205–1216 (2015). doi:10.1007/s11012-014-0007-x
11. Hagelstein, P.L., Letts, D., Cravens, D.: Terahertz difference frequency response of PdD in two-laser experiments. J. Cond. Mat. Nucl. Sci. **3**, 59 (2010)
12. Hagelstein, P.L., Chaudhary, I.U.: Anomalies in fracture experiments and energy exchange between vibrations and nuclei. Meccanica **50**, 1189–1203 (2014). doi:10.1007/s11012-014-9988-8
13. Cook, N.D.: Models of the atomic nucleus, 2nd edn. Springer, Dordrecht (2010)
14. Cook, N.D., Manuello, A., Veneziano, D., Carpinteri, A.: Chapter 15: Piezonuclear fission reactions simulated by the lattice model of the atomic nucleous. In: Carpinteri, A., Lacidogna, G., Bertetto, A.M. (eds.) Acoustic, electromagnetic, neutron emissions from fracture and earthquakes. Springer, Heidelberg (2015)
15. Diebner, K.: Fusionsprozesse mit hilfe konvergenter stosswellen—einige aeltere und neuere versuche und ueberlegungen. Kerntechnik **3**, 89–93 (1962)
16. Derjaguin, B.V., et al.: Titanium fracture yields neutrons? Nature **34**, 492 (1989)
17. Fujii, M.F., et al.: Neutron emission from fracture of piezoelectric materials in deuterium atmosphere. Jpn. J. Appl. Phys. **41**(Pt.1), 2115–2119 (2002)
18. Sobolev, G.A., Shestopalov, I.P., Kharin, E.P.: Implications of solar flares for the seismic activity of the Earth. Izvestiya, Phys. Solid Earth **34**, 603–607 (1998)
19. Volodichev, N.N., et al.: Lunar periodicity of the neutron radiation burst and seismic activity on the Earth. In: Proceedings of the 26th International Cosmic Ray Conference, Salt Lake City, 17–25 August 1999
20. Kuzhevskij, M., Nechaev, O.Y., Sigaeva, E.A.: Distribution of neutrons near the Earth's surface. Nat. Hazards Earth Syst. Sci. **3**, 255–262 (2003)
21. Kuzhevskij, M.: Neutron flux variations near the Earth's crust. A possible tectonic activity detection. Nat. Hazards Earth Syst. Sci. **3**, 637–645 (2003)
22. Sigaeva, E.A., et al.: Thermal neutrons' observations before the Sumatra earthquake. Geophys. Res. Abstr. **8**, 00435 (2006)
23. Borla, O., Lacidogna, G., Carpinteri, A.: Chapter 10: Piezonuclear neutron emissions from earthquakes and volcanic eruptions. In: Carpinteri, A., Lacidogna, G., Bertetto, A.M. (eds.) Acoustic, electromagnetic, neutron emissions from fracture and earthquakes. Springer, Heidelberg (2015)
24. Cardone, F., Mignani, R.: Deformed spacetime. Springer, Dordrecht (2007). Chapters 16 and 17
25. Cardone, F., Cherubini, G., Petrucci, A.: Piezonuclear neutrons. Phys. Lett. A **373**, 862–866 (2009)
26. Carpinteri, A., Cardone, F., Lacidogna, G.: Energy emissions from failure phenomena: mechanical, electromagnetic, nuclear. Exp. Mech. **50**, 1235–1243 (2010)
27. Carpinteri, A., Lacidogna, G., Manuello, A., Borla, O.: Piezonuclear fission reactions from earthquakes and brittle rocks failure: evidence of neutron emission and nonradioactive product elements. Exp. Mech. **53**(3), 345–365 (2013)
28. Carpinteri, A., Cardone, F., Lacidogna, G.: Piezonuclear neutrons from brittle fracture: early results of mechanical compression tests. Strain **45**, 332–339 (2009)
29. Carpinteri, A., Chiodoni, A., Manuello, A., Sandrone, R.: Compositional and microchemical evidence of piezonuclear fission reactions in rock specimens subjected to compression tests. Strain **47**(2), 267–281 (2011)
30. Carpinteri, A., Manuello, A.: Geomechanical and geochemical evidence of piezonuclear fission reactions in the Earth's crust. Strain **47**(2), 282–292 (2011)
31. Carpinteri, A., Borla, O., Lacidogna, G., Manuello, A.: Neutron emissions in brittle rocks during compression tests: monotonic vs. cyclic loading. Phys. Mesomech. **13**, 264–274 (2010)
32. Carpinteri, A., Manuello, A.: An indirect evidence of piezonuclear fission reactions: geomechanical and geochemical evolution in the Earth's crust. Phys. Mesomech. **15**, 14–23 (2012)
33. Cardone, F., Carpinteri, A., Lacidogna, G.: Piezonuclear neutrons from fracturing of inert solids. Phys. Lett. A **373**, 4158–4163 (2009)

34. Carpinteri, A., Lacidogna, G., Manuello, A., Borla, O.: Energy emissions from brittle fracture: neutron measurements and geological evidences of piezonuclear reactions. Strength, Fract. Complex. **7**, 13–31 (2011)
35. Carpinteri, A., Lacidogna, G., Manuello, A., Borla, O.: Piezonuclear fission reactions in rocks: evidences from microchemical analysis, neutron emission, and geological transformation. Rock Mech. Rock. Eng. **45**(4), 445–459 (2012)
36. Carpinteri, A., Lacidogna, G., Borla, O., Manuello, A., Niccolini, G.: Electromagnetic and neutron emissions from brittle rocks failure: experimental evidence and geological implications. Sadhana **37**(1), 59–78 (2012)
37. Manuello, A., et al.: Chapter 3: Neutron emissions and compositional changes at the compression failure of iron-rich natural rocks. In: Carpinteri, A., Lacidogna, G., Bertetto, A.M. (eds.) Acoustic, electromagnetic, neutron emissions from fracture and earthquakes. Springer, Heidelberg (2015)
38. Carpinteri, A., et al.: Chapter 4: Frequency-dependent neutron emissions during fatigue tests on iron-rich natural rocks. In: Carpinteri, A., Lacidogna, G., Bertetto, A.M. (eds.) Acoustic, electromagnetic, neutron emissions from fracture and earthquakes. Springer, Heidelberg (2015)
39. Carpinteri, A., Lacidogna, G., Borla, O.: Chapter 5: Alpha particle emissions from Carrara marble specimens crushed in compression and X-ray photoelectron spectroscopy of correlated nuclear transmutations. In: Carpinteri, A., Lacidogna, G., Bertetto, A.M. (eds.) Acoustic, electromagnetic, neutron emissions from fracture and earthquakes. Springer, Heidelberg (2015)
40. Carpinteri, A., Borla, O., Lacidogna, G.: Chapter 6: Elemental content variations in crushed mortar specimens measured by instrumental neutron activation analysis (INAA). In: Carpinteri, A., Lacidogna, G., Bertetto, A.M. (eds.) Acoustic, electromagnetic, neutron emissions from fracture and earthquakes. Springer, Heidelberg (2015)
41. Invernizzi, S., et al.: Chapter 7: Piezonuclear evidences from tensile and compression tests on steel. In: Carpinteri, A., Lacidogna, G., Bertetto, A.M. (eds.) Acoustic, electromagnetic, neutron emissions from fracture and earthquakes. Springer, Heidelberg (2015)
42. Hudson, J.A., Crouch, S.L., Fairhurst, C.: Soft, stiff and servo-controlled testing machines: a review with reference to rock failure. Eng. Geol. **6**, 155–189 (1972)
43. Carpinteri, A.: Cusp catastrophe interpretation of fracture instability. J. Mech. Phys. Solids **37**, 567–582 (1989)
44. Carpinteri, A., Pugno, N.: Are scaling laws of strength of solids related to mechanics or to geometry? Nat. Mater. **4**, 421–423 (2005)
45. Carpinteri, A., Corrado, M.: An extended (fractal) overlapping crack model to describe crushing size-scale effects in compression. Eng. Fail. Anal. **16**, 2530–2540 (2009)
46. Liu, L.: The inception of the oceans and CO2-atmosphere in the early history of the Earth. Earth Planet. Sci. Lett. **227**, 179–184 (2004)
47. Padron, E., et al.: Changes on diffuse CO2 emission and relation to seismic activity in and around El Hierro, Canary Islands. Pure Appl. Geophys. **165**, 95–114 (2008)
48. Taylor, S.R., McLennan, S.M.: Planetary crusts: their composition, origin and evolution. Cambridge University Press, Cambridge (2009)
49. Carpinteri, A., Manuello, A.: Chapter 12: Evolution and fate of chemical elements in the Earth's crust, ocean, and atmosphere. In: Carpinteri, A., Lacidogna, G., Bertetto, A.M. (eds.) Acoustic, electromagnetic, neutron emissions from fracture and earthquakes. Springer, Heidelberg (2015)
50. Carpinteri, A., Manuello, A., Negri, L.: Chapter 13: Chemical evolution in the Earth's mantle and its explanation based on piezonuclear fission reactions. In: Carpinteri, A., Lacidogna, G., Bertetto, A.M. (eds.) Acoustic, electromagnetic, neutron emissions from fracture and earthquakes. Springer, Heidelberg (2015)
51. Carpinteri, A., Manuello, A., Negri, L.: Chapter 14: Piezonuclear fission reactions triggered by fracture and turbulence in the rocky and gaseous planets of the solar system. In: Carpinteri, A., Lacidogna, G., Bertetto, A.M. (eds.) Acoustic, electromagnetic, neutron emissions from fracture and earthquakes. Springer, Heidelberg (2015)
52. Fleischmann, M., Pons, S.: Electrochemically induced nuclear fusion of deuterium. J. Electroanal. Chem. **261**, 301–308 (1989)
53. Preparata, G.: A new look at solid-state fractures, particle emissions and «cold» nuclear fusion. Il Nuovo Cimento **104A**, 1259–1263 (1991)
54. Mizuno, T.: Nuclear transmutation: the reality of cold fusion. Infinite Energy Press, Concord (1998)
55. Carpinteri A. et al.: Mechanical conjectures explaining cold nuclear fusion. In: Conference & Exposition on Experimental and Applied Mechanics (SEM), vol. 3, pp. 353–367. Lombard, (2013)
56. Carpinteri, A., et al.: Chapter 8: Cold nuclear fusion explained by hydrogen embrittlement and piezonuclear fissions at the metallic electrodes—part I: Ni-Fe and Co-Cr electrodes. In: Carpinteri, A., Lacidogna, G., Bertetto, A.M. (eds.) Acoustic, electromagnetic, neutron emissions from fracture and earthquakes. Springer, Heidelberg (2015)
57. Carpinteri A. et al.: Hydrogen embrittlement and "cold fusion" effects in palladium during electrolysis experiments. In: Conference & Exposition on Experimental and Applied Mechanics (SEM), vol. 6, pp. 37–47. Greenville, (2014)
58. Carpinteri, A., et al.: Chapter 9: Cold nuclear fusion explained by hydrogen embrittlement and piezonuclear fissions at the metallic electrodes—part II: Pd and Ni electrodes. In: Carpinteri, A., Lacidogna, G., Bertetto, A.M. (eds.) Acoustic, electromagnetic, neutron emissions from fracture and earthquakes. Springer, Heidelberg (2015)
59. Carpinteri, A., Lacidogna, G., Manuello, A., Borla, O.: Piezonuclear neutrons from earthquakes as a hypothesis for the image formation and the radiocarbon dating of the Turin shroud. Sci. Res. Essays **7**(29), 2603–2612 (2012)
60. Carpinteri, A., Lacidogna, G., Borla, O.: Chapter 11: Is the Shroud of Turin in relation to the Old Jerusalem historical earthquake? In: Carpinteri, A., Lacidogna, G., Bertetto, A.M. (eds.) Acoustic, electromagnetic, neutron emissions from fracture and earthquakes. Springer, Heidelberg (2015)
61. Phillips, T.J., Hedges, R.E.M.: Shroud irradiated with neutrons? Nature **337**, 594 (1989)
62. Acbas, G., et al.: Optical measurements of long-range protein vibrations. Nature Comm. **5**, 3076 (2014). doi:10.1038/ncomms4076
63. http://www.buffalo.edu/news/releases/2014/01/012.html
64. Mofard, M.R.K., Kamm, R.D.: Cellular mechanotransduction. Cambridge University Press, Cambridge (2010)
65. Vysotskii, V.I., Kornilova, A.A.: Nuclear transmutation of stable and radioactive isotopes in biological systems. Pentagon Press, New Delhi (2010)

Chapter 18
Field Theoretical Description of Shear Bands

Sanichiro Yoshida and Tomohiro Sasaki

Abstract Spatiotemporal behaviors of the shear band have been analyzed. Based on a recent field theory of deformation and fracture, it has been hypothesized that (a) a shear band is formed along the boundary of opposite rotational displacements of a specimen. (b) When the propagation velocity of movable dislocations along the front of a shear band is higher than the phase velocity of the rotation wave the shear band appears continuously whereas when the velocity of the dislocations is lower than the rotation wave the shear band travels at the same velocity as the rotational wave appearing intermittently. Electronic-Speckle-Pattern Interferometric setup has been used to monitor the formation and movement of shear bands under various tensile strain rates. The observed spatiotemporal characteristics of the shear bands have been found to support the above hypotheses.

Keywords Shear band • Lüders band • Portevin–Le Chatelier effect • Electronic Speckle-Pattern Interferometry • Plastic deformation

18.1 Introduction

Metal specimens under tensile loads exhibit a localized, band-structured strain known as the shear band. Generally, shear bands are observed in the plastic regime of stress-strain characteristics. Depending on the material and the dynamic behavior, the same type of localized strain is referred to as a Lüders band or a Portevin–Le Chatelier (PLC) band. Normally, the word Lüders band is used when a shear band moves continuously at the yield plateau of steels. On the other hand, the PLC bands are classified as type A–C bands as follows. Type A bands are characterized as continuously propagating over the entire specimen with small stress drops. They are often seen at a high strain rate. Type B bands are sometimes called hopping bands, and characterized as appearing in a more discontinuous fashion than type A bands. While they appear discontinuously, the locations where series of type B bands appear are somewhat correlated. Type B bands are often observed at a medium to high strain rate. Type C bands are observed at a low strain rate and characterized as randomly appearing static bands with large stress drops. From the spatiotemporal characteristics, the Lüders band can be classified as a PLC band of type A.

A number of tensile analyses with an optical interferometry known as the Electronic Speckle-Pattern Interferometry (ESPI) reveal an optical band-structured pattern that can be interpreted as a shear band [1, 2]. These band-patterns are observed in interferometric fringe patterns that represent displacement contours (call the optical band-pattern hereafter.) One of these previous studies on carbon steel specimens specifically compares optical band-patterns and Lüders band via simultaneous observations, and demonstrates that they have identical spatiotemporal behaviors [3]. Many other similar experiments with aluminum alloys indicate that the optical band-patterns show the same spatiotemporal behaviors as type A–C PLC bands. Most importantly, the optical band patterns become stationary at the end of the tensile experiment at the location where the specimen breaks. This strongly indicates the importance of the shear-band as a phenomenon fundamentally related to the mechanism of fracture.

On the theoretical side, a recent field theory of deformation and fracture [4], called the field theory hereafter, explains the shear-band in association with deformation dynamics. Based on a fundamental physical principle known as local symmetry [5], the field theory can describe all stages of deformation and fracture on the same theoretical foundation without relying on any phenomenology. Using the Lagrangian formalism, it derives field equations of deformation dynamics. These field

S. Yoshida (✉)
Department of Chemistry and Physics, Southeastern Louisiana University, SLU 10878, Hammond, LA 70402, USA
e-mail: syoshida@selu.edu

T. Sasaki
Department of Mechanical Engineering, Niigata University, IkarashinNinocho 8050, Nishi-ku, Niigata-shi, Niigata 950-2181, Japan

A.M. Beese et al. (eds.), *Fracture, Fatigue, Failure and Damage Evolution, Volume 8*, Conference Proceedings of the Society for Experimental Mechanics Series, DOI 10.1007/978-3-319-21611-9_18

equations yield wave solutions that describe spatiotemporal characteristics of translational and rotational displacement fields. The strain represented by a shear band can be interpreted as an entity causing energy dissipation when it drifts under the influence of the translational displacement [6]. The shear stress has a local maximum along the boundary of a pair of opposite rotations; it is naturally interpreted that a shear-band is generated along such a boundary. A recent study indicates that the shear band interfaces the field theory and the dislocation theory [7]. The field theoretical wave dynamics of deformation and the interconnection between the shear band and dislocations lead to the following hypothesis. (a) A shear band is formed when dislocations moving along the line of the maximum shear stress reach the other side of the specimen. (It is said hereafter that dislocations bridge the specimen.) (b) If the propagation velocity of these dislocations is higher than the phase velocity of the rotational displacement wave that generates the shear stress, the shear band moves continuously (the Lüders band or type A PLC band), whereas if the dislocations propagate more slowly than the displacement wave, the shear band appears as type B or C.

The aim of this paper is to discuss the shear band based on the field theory, and provide experimental findings that support the above hypothesis. Spatiotemporal behaviors of the optical band-pattern are presented and discussed based on the field theoretical interpretation. With the above-mentioned experimental observation that the optical band-pattern, hence the shear-band, is closely related to the fracture mechanism, it is expected that the present study is useful for better understanding of transition from the final stage of plastic deformation to fracture.

18.2 Theoretical

18.2.1 Plastic Deformation Wave

Details of the field theory can be found elsewhere [4]. In short, this theory is based on a fundamental physical principle known as local symmetry. Here the word symmetry means that a certain physical theory (physics law) is invariant under a transformation; in other words, when a certain transformation is operated on a physical system, the law of physics can be written in the same form as before the transformation. A simple example is that when we rotate a coordinate system (coordinate transformation around an axis of rotation), we know that all physics laws can be written in the same form. Local symmetry means that the physics law is symmetric even if the transformation matrix is coordinate dependent. In the context of deformation dynamics, say a line element vector η undergoes linear transformation U as $\overline{\boldsymbol{\eta}} = U\boldsymbol{\eta}$. Here $\overline{\boldsymbol{\eta}}$ denotes the line element vector after the transformation. If the transformation U is independent of the coordinates, the differential of $\boldsymbol{\eta}$ after the transformation is $d\overline{\boldsymbol{\eta}} - Ud\boldsymbol{\eta}$; it is transformed in the same fashion as $\boldsymbol{\eta}$. However, if U depends on the coordinate variables it follows that $d\overline{\boldsymbol{\eta}} = dU\boldsymbol{\eta} + Ud\boldsymbol{\eta}$; apparently the differential is not transformed in the same way as the case where $\boldsymbol{\eta}$ is independent of the coordinates. Normally, a physics law contains derivatives. Therefore, the physics law after the transformation cannot be written in the same fashion as before. Request of local symmetry means that even if the transformation matrix is coordinate dependent, the physics law after the transformation is written in the same fashion as before. Obviously, it is necessary to compensate the additional term $dU\boldsymbol{\eta}$. This compensation is made by introduction of an additional field called the compensation field. The dynamics associated with the compensation field describes the effect of coordinate dependence. In the present case, this additional dynamics represents the nonlinearity of deformation.

With the use of the Lagrangian formalism, the following field equations can be derived.

$$\nabla \cdot \boldsymbol{v} = j_0 \tag{18.1}$$

$$\nabla \times \boldsymbol{v} = \frac{\partial \boldsymbol{\omega}}{\partial t} \tag{18.2}$$

$$G\nabla \times \boldsymbol{\omega} = -\rho\frac{\partial \mathbf{v}}{\partial \mathbf{t}} - G\mathbf{j} \tag{18.3}$$

$$\nabla \cdot \boldsymbol{\omega} = 0 \tag{18.4}$$

Here $\boldsymbol{v}$ is the rate of the displacement or the local velocity of the particle, $\boldsymbol{\omega}$ is the rotation, G is the shear modulus, ρ is the density of the medium , j_0 and $\mathbf{j}$ are the temporal and spatial components of charge of symmetry [5]. $G\mathbf{j}$ represents the longitudinal resisting force [6] exerted by the medium. In the elastic case, this is the elastic force proportional to the local stretch. In the plastic case, it is a velocity damping force in the following form.

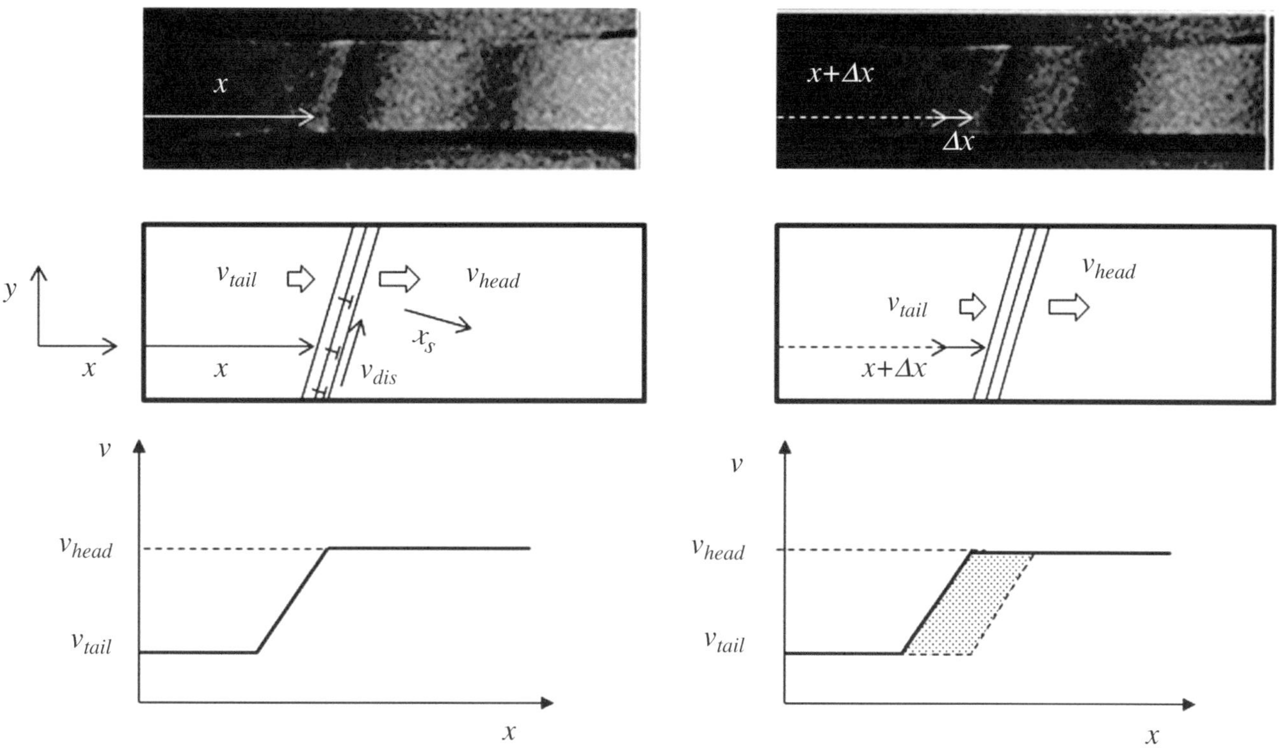

Fig. 18.1 Example of one-dimensional $(\nabla \cdot \boldsymbol{v})$

$$G\mathbf{j} = \mathbf{W}_d \rho (\nabla \cdot \boldsymbol{v}) = \sigma_0 \rho (\nabla \cdot \boldsymbol{v}) \boldsymbol{v} = \sigma_c \boldsymbol{v} \tag{18.5}$$

Here, $\mathbf{W}_d$ is the drift velocity of the quantity $(\nabla \cdot \boldsymbol{v})$, σ_0 is the dimensionless parameter expressing how much faster the drift velocity is than the particle velocity, and $\sigma_c = \sigma_0 \rho (\nabla \cdot \boldsymbol{v})$ as defined in (18.5). In xyz coordinates $(\nabla \cdot \boldsymbol{v}) = \partial v_x/\partial x + \partial v_y/\partial y + \partial v_z/\partial z$; it essentially represents a volume expansion rate.

Figure 18.1 shows a one-dimensional case where $(\nabla \cdot \boldsymbol{v})$ has only dependence on a coordinate axis x_s as

$$(\nabla \cdot \boldsymbol{v}) = \partial v_x/\partial x_y + \partial v_y/\partial y = \partial v_s/\partial x_s. \tag{18.6}$$

Here, the coordinate axes are as shown in Fig. 18.1. The top images in Fig. 18.1 are optical interferometric fringe patterns representing the velocity contours corresponding to deformation in the duration of two time-steps in a tensile experiment with a constant pulling rate. The images are formed by an in-plane displacement sensitive Electronic Speckle-Pattern Interferometry (ESPI) setup. The bright slant banded pattern represents $\partial v_s/\partial x_s$ where the velocity contours inside the banded region are so dense that the image cannot resolve. The middle illustrations indicate the situation where the leading contour of the $\partial v_s/\partial x_s$ pattern v_{head} is higher than the tailing contour v_{tail}, or the pattern essentially represents a normal stretch. The bottom drawing illustrates that as the passage of $\partial v_s/\partial x_s$ the particles behind it lose the momentum as the velocity decreases from v_{head} to v_{tail}. For the entire system, it works as energy loss. A previous experiment proves it as the temperature rise (heat loss) accompanied by the motion of a $\partial v_s/\partial x_s$ pattern [8]. The interpretation that $G\mathbf{j}$ in the plastic case is a velocity damping force will be justified later in this paper.

Elimination of $\boldsymbol{v}$ from (18.2) and (18.3) yields the following wave equation of $\boldsymbol{\omega}$

$$\rho \frac{\partial^2 \boldsymbol{\omega}}{\partial t^2} - G \nabla^2 \boldsymbol{\omega} + \sigma_c \frac{\partial \boldsymbol{\omega}}{\partial t} = 0 \tag{18.7}$$

The solution can be put in the following form in general.

$$\boldsymbol{\omega}(t, \mathbf{r}) = \Omega_0 e^{-\frac{\sigma_c}{2\rho} t} \cos\left(\left(\frac{G}{\rho} k^2 - \frac{\sigma_c{}^2}{4\rho^2} \right)^{1/2} t - \mathbf{k} \cdot \mathbf{r} \right) \tag{18.8}$$

Solution (18.8) indicates that the $\boldsymbol{\omega}$ wave is a decaying sinusoidal wave where Ω_0 denotes the amplitude and $\mathbf{k}$ is the propagation vector. Here $\left(\frac{G}{\rho}k^2 - \frac{\sigma_c{}^2}{4\rho^2}\right)^{1/2}$ and $\frac{\sigma_c}{2\rho}$ can be interpreted as the real and imaginary part of the complex frequency $\omega_f = \omega_{f0} + j\tilde{\omega}_f$ as

$$\omega_{f0}{}^2 = \left(\frac{G}{\rho}k^2 - \frac{\sigma_c{}^2}{4\rho^2}\right) = \frac{G}{\rho}k^2 - \tilde{\omega}_f^2 \tag{18.9a}$$

$$\tilde{\omega}_f = \frac{\sigma_c}{2\rho} \tag{18.9b}$$

The relation between ω_{f0} and $\tilde{\omega}_f$ expressed by (18.9a) can be interpreted as Kramers–Kronig relations. The phase velocity of the $\boldsymbol{\omega}$ wave is given by

$$v_\omega = \frac{\omega_{f0}}{k} = \sqrt{\frac{G}{\rho} - \frac{\tilde{\omega}_f^2}{k^2}} = \sqrt{\frac{G}{\rho} - \frac{\sigma_c{}^2}{(2\rho k)^2}} \tag{18.10}$$

18.2.2 Shear Band Formation

While the dynamics of shear band has not been fully understood, prevailing theories in common ascribe it to interactions of moving dislocations. Dislocation theory [7] says the propagation velocity of dislocation is proportional to the strain rate.

$$v_{dis}(\tau) = \frac{\dot{\gamma}_{ext}}{bn} \tag{18.11}$$

where τ is the shear force, $\dot{\gamma}_{ext}$ is the strain rate, b is the magnitude of the Burgers vector and n is the dislocation density. The force acting on dislocations is

$$Bv_{dis} = b\tau \tag{18.12}$$

Apparently, being proportional to the velocity this force is a velocity damping force with the damping coefficient B. B is given by the sum of the damping coefficient B_e due to conduction electrons and B_p due to phonons.

$$B = B_e + B_p \tag{18.13}$$

18.2.3 Hypothesis

In the plastic case, the medium does not resist in the form of elastic force proportional to the local stretch (differential displacement.) Instead, the propagation of dislocations is the main reaction. With the interpretation that the longitudinal force in the plastic case is a velocity-damping force, the $\boldsymbol{\omega}$ wave travels according to (18.8). The field equation (18.4) indicates that unlike $\boldsymbol{\omega}$ always pair. Figure 18.2 schematically illustrates the situation where a pair of developed $\boldsymbol{\omega}$ are formed in a specimen under a tensile load, and the spatiotemporal variation travels to the right as a decaying wave. Here, the center of the counterclockwise rotation is the crest of the $\boldsymbol{\omega}$ wave and that of the clockwise rotation is the trough (or the other way around depending on the definition of the coordinate axes.) A local maximum of the shear stress appears along the boundary of a $\boldsymbol{\omega}$ pair, and travels at the velocity of the $\boldsymbol{\omega}$ wave. This indicates that the force on dislocations presented by (18.12) is higher along these boundaries than the surrounding part of the specimen, and consequently, the velocity of the dislocations is higher as well.[1] When these dislocations reach the other side of the specimen (bridge the specimen), the stress will drop as the material slips along the border of the $\boldsymbol{\omega}$ pair. If the dislocation velocity is higher than the phase velocity of the $\boldsymbol{\omega}$ wave, the bridging takes place as the $\boldsymbol{\omega}$ wave travels, giving no time for the stress to recover. If the dislocation velocity is

[1] According to the dislocation theory, dislocations move faster at a higher shear stress [7].

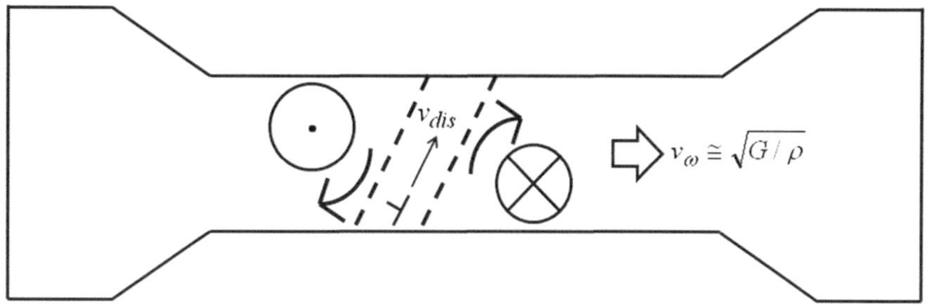

Fig. 18.2 Shear band formed as dislocations propagate along the boundary of unlike rotations

lower than the **ω** wave velocity, the next bridging will not occur immediately after one takes place. The stress will recover as the **ω** wave resumes traveling, and it drops when the next bridging occurs. Below we discuss the shear band dynamics for the cases where the dislocation velocity is higher (Case 1) and lower (Case 2) than the **ω** wave velocity.

Case 1: $v_{dis} > v_\omega$

Here v_ω denotes the **ω** wave velocity expressed by (18.10). When the stress reaches the yield stress, a **ω** wave is generated in accordance with (18.7) and (18.8). At a certain point, movable dislocations bridge the specimen along the boundary of a **ω** pair where the shear stress is locally highest. This accompanies the stress drop. Once this happens, since $v_{dis} > v_\omega$, new bridging occurs one after another at the front of the shear band, keeping the stress at the dropped level. (18.11) indicates that the dislocation velocity v_{dis} decreases as the dislocation density increases. This means that v_{dis} is higher in the early stage of the plastic regime than later stages. This explains why Lüders bands appear in the yield plateau, and type A PLC bands are more likely to appear in an early stage of plastic deformation. It is interesting to note that usually the appearance of a Lüders band accompanies a sharp stress drop and that once the band begins to move the stress remains more or less the same. This is consistent with the mechanism that when $v_{dis} > v_\omega$ the dislocations keep bridging the specimen faster than the **ω** wave travels.

Case 2: $v_{dis} < v_\omega$

This case can be differentiated from Case 1 in two regards. First, after each occurrence of the bridging by dislocations the stress recovers as the **ω** wave resumes. The dislocation velocity becomes lower as the dislocation density increases. This explains why type B and type C tend to appear in later stages of plastic regime accompanying serrations in stress. Second, the shear band reappears at a new location where the **ω** wave creates a local maximum of shear stress when bridging occurs at that location. Between bridging events, a shear band is not generated but since the bridging always takes place at the location where the shear stress is the highest at that moment, the velocity of the intermittent shear band is the same as the **ω** wave. Since the bridging does not occur one after another like Case 1, the shear bands do not appear continuously.

The decrease in the propagation velocity of shear bands with the development of plastic deformation can be discussed along the same line of argument. Under the condition where the **ω** wave travels with a moderate decay constant, the relation $G/\rho >> \sigma_c{}^2/(2\rho k)^2$ holds in the expression of the **ω** wave velocity (18.10); otherwise, the decay rate is so high that the life of each **ω** wave is too short and it does not exist stably. Under this condition, the wave velocity is approximately proportional to the square root of the shear modulus as $v_\omega \cong \sqrt{G/\rho}$. As the plastic deformation develops, the increase in the dislocation density lowers G, hence the wave velocity v_ω. This lowers the velocity of type B and C PLC bands, explaining why their propagation velocity is lower than type A bands. Towards the end of the plastic deformation, the dislocation density becomes so high that their propagation velocity is low making the bridging event less probable. This explains why type C bands appear more irregularly than type B bands.

The irreversibility of plastic deformation can also be explained in conjunction with the above argument. As (18.12) indicates, the force exerted by the medium on dislocations is a velocity-damping force. Every time dislocations bridge the specimen, a strain jump occurs across the banded region as the front end is displaced faster than the tailing end. This pattern drifts at the velocity $\mathbf{W}_d$ as defined by (18.5). This velocity is higher than the particle velocity, $\sigma_0 > 1$ in (18.5), because the motion of the front end is not carried by the particles but is determined by how fast the dislocations bridge the specimen in front of the band. According to (18.11), the dislocation velocity is proportional to the strain rate $\dot{\gamma}_{ext}$, which is proportional to the pulling rate and hence the particle velocity. This explains why $\mathbf{W}_d$ is proportional to the particle velocity and the plastic longitudinal force $G\mathbf{j} = \mathbf{W}_d \rho (\nabla \cdot \mathbf{v})$ is energy dissipative. The longitudinal force $G\mathbf{j}$ is energy dissipative whenever a positive/negative normal strain drifts in the same/opposite direction to the particle velocity. It is unnecessary that the dislocations bridge the specimen (or a band structure is formed.)

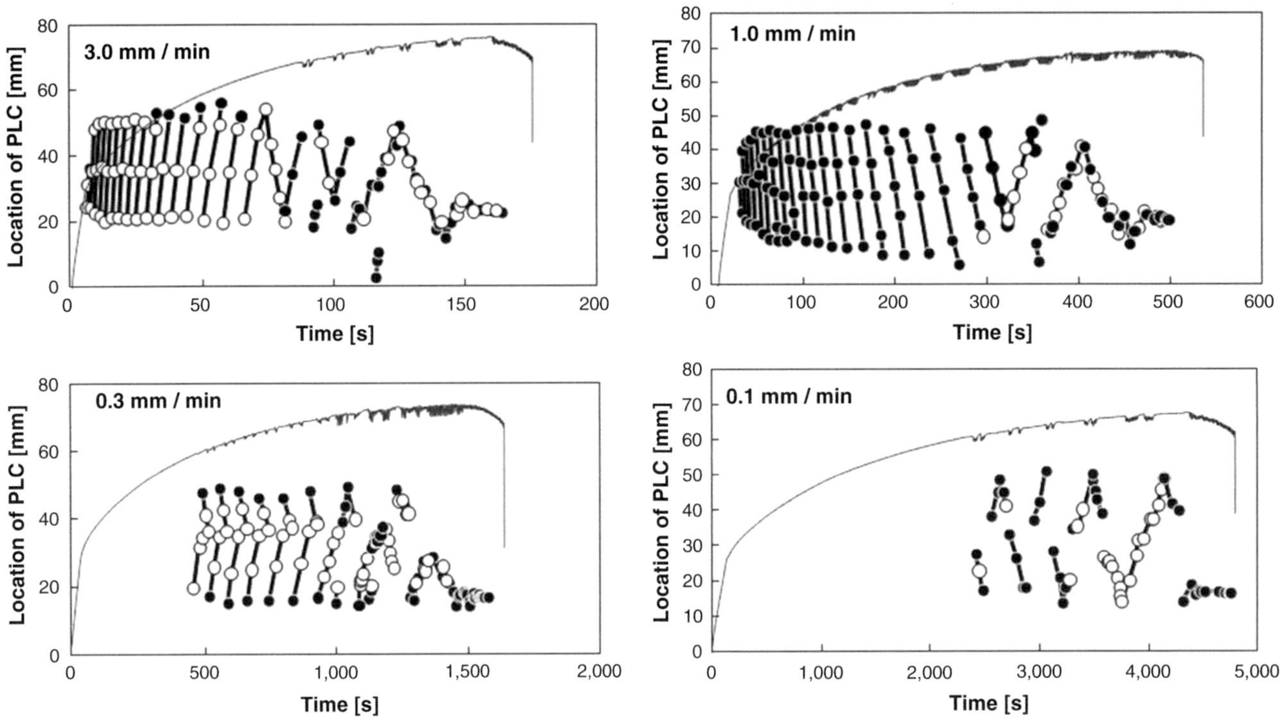

Fig. 18.3 Loading curve and band's motions for various pulling rates

18.3 Experimental

A number of ESPI based tensile experiments support the above argument regarding the spatiotemporal behavior of shear bands. Figure 18.3 plots the appearance of shear bands in tensile tests of aluminum alloy A5083 for four different constant pulling rates ranging from 3 mm/min (highest) to 0.1 mm/min (lowest) along with the loading curves. The higher the pulling rate, the earlier the first shear band appears. This is consistent with the above argument that since the dislocation velocity is proportional to the strain rate and hence the pulling rate, the higher pulling rate makes easier for the dislocations to bridge the specimen. Figure 18.1 also clearly indicates that the earlier the first shear band appears the band moves more continuously (type A PLC band) with less stress drop. In the case of the lowest pulling rate, the appearance of the shear band is irregular and sparse. In all the cases, the shear band stops moving at the end. This situation corresponds to $\mathbf{W}_d = 0$ and $G\mathbf{j} \neq 0$. Here the condition $G\mathbf{j} \neq 0$ is true because the medium exerts reaction to the tensile force provided by the test machine. From (18.5) it is clear that if both conditions are true it follows that $\nabla \cdot \boldsymbol{v} \to \infty$, or particles come out from the unit volume at an infinite rate. This necessarily creates discontinuity in the medium, which can be interpreted as the fracture. Indeed the specimen fractures at the end of the $\mathbf{W}_d = 0$ trend.

Figure 18.4 plots the velocity of the shear band ($\mathbf{W}_d$) as a function of the elapsed time from the beginning of the tensile load. All the cases overlap one another converging to a single line, indicating that the shear band velocity solely depends on the time. As discussed above, the shear modulus is expected to decrease with the development of plastic deformation. This observation that the shear band velocity overlaps one another with different pulling rates strongly support the above argued hypothesis that in the case of type B or C PLC bands, the shear band velocity is determined by the $\boldsymbol{\omega}$ wave velocity approximated as $v_\omega = \sqrt{G/\rho - {\sigma_c}^2/(2\rho k)^2} \cong \sqrt{G/\rho}$. To examine this preposition, the square root of the shear modulus is plotted in the same graph as the shear band velocity in Fig. 18.5. Here, the shear modulus was evaluated from the slope of the loading curve and normalized by a constant so that the maximum value at the beginning is similar to that of the shear band velocity. The two curves overlap fairly well, proving that the shear band velocity decreases in proportion to the $\boldsymbol{\omega}$ wave velocity.

Figure 18.6 shows the decrease in the band width with the passage of time. According to (18.5) and (18.6), $\sigma_c = \sigma_0 \rho \left(\partial v_s / \partial x_s\right)$, and the narrower the band (i.e., higher the $\partial v_s / \partial x_s$), the higher the σ_c and the system is more energy dissipative. According to the field theory, type A PLC bands are explained as a solitary wave [9], whose driving force (the mechanism that type A bands travel) is longitudinal elasticity. From the field theoretical viewpoint, it is explained that

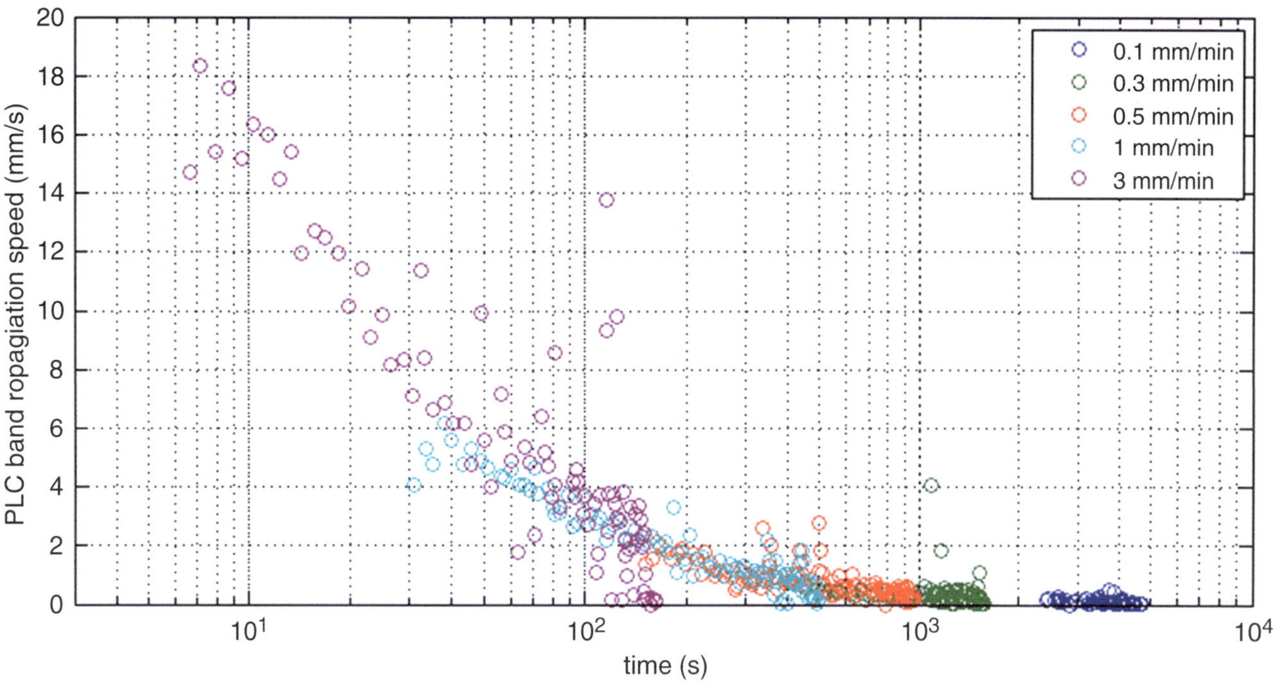

Fig. 18.4 Time vs. band velocity

Fig. 18.5 Square root of shear stress and band velocity as a function of time

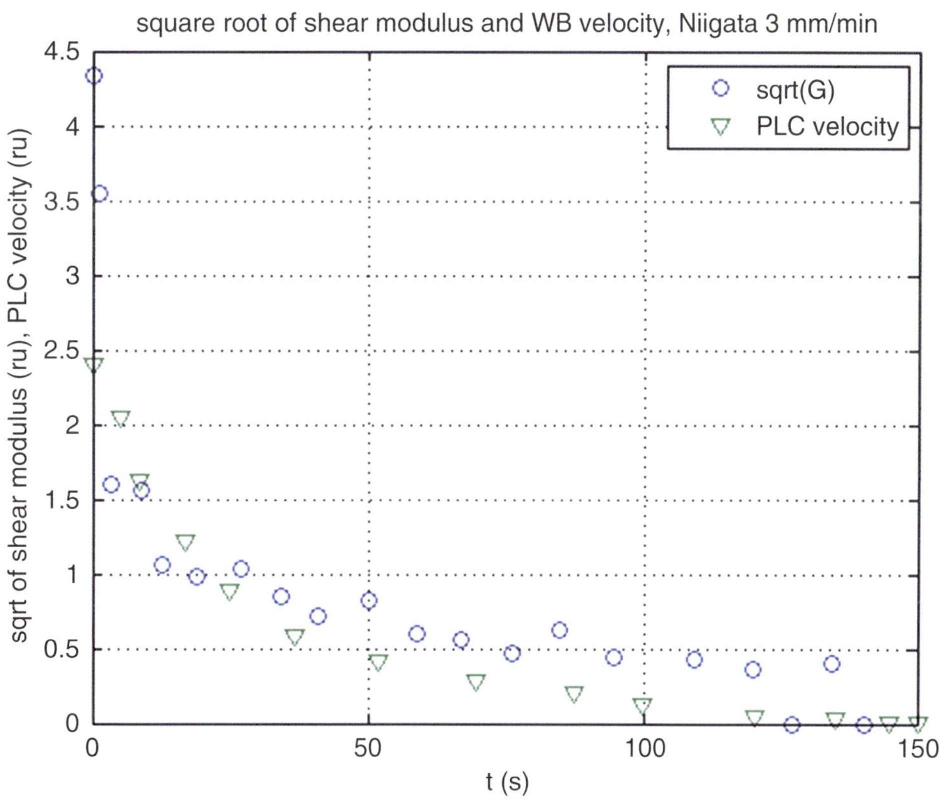

type B and C bands appearing in late stages of plastic regime cannot move continuously because the material does not have sufficient longitudinal elasticity in those stages. Asymptotically, $\partial v_s/\partial x_s$ approaches infinity when the shear band becomes stationary whereas the material needs to exert resistive force in the form of $G\mathbf{j} = \mathbf{W}_d\rho(\nabla \cdot \mathbf{v})$ [eq. (18.5)]; $\mathbf{W}_d = 0$ and $G\mathbf{j} \neq 0$ hence $\nabla \cdot \boldsymbol{v} = \partial v_s/\partial x_s = \infty$. This causes an infinite strain or fracture. This is the field theoretical definition of fracture.

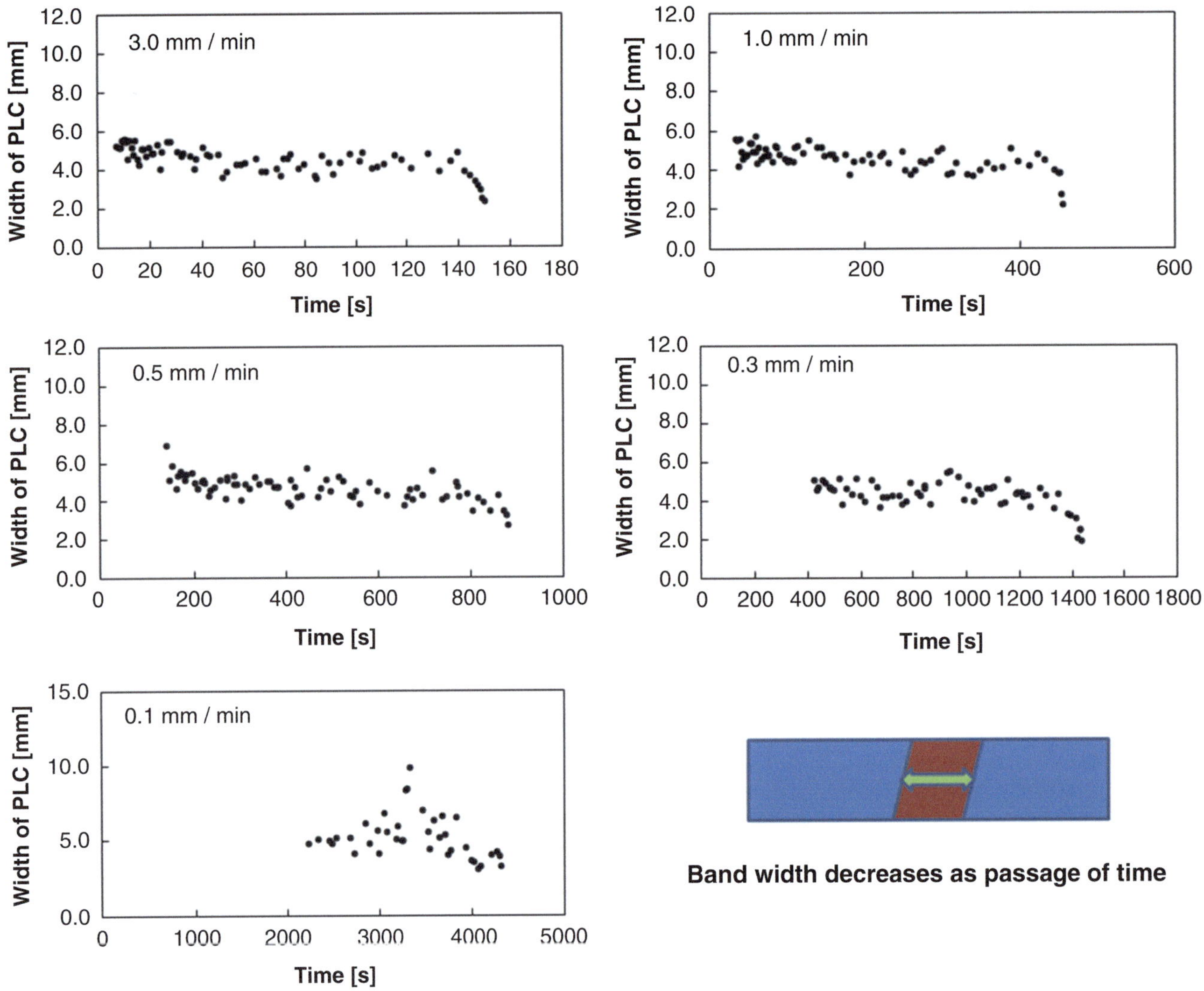

Fig. 18.6 Change in width of shear band

18.4 Conclusions

Dynamic behaviors of the shear band have been discussed based on a field theory of deformation and fracture and dislocation theory. It has been hypothesized that a shear band is formed along the boundary of opposite rotations that travels as a decaying sinusoidal wave and that depending on the propagation velocity of moveable displacement relative to the rotational wave velocity, the shear band can appear continuously with practically no stress rise or it can appear intermittently accompanied by stress serrations. Experimental observations that support the hypotheses have been presented.

References

1. Yoshida, S., Suprapedi, R., Widiastuti, R., Pardede, M., Hutagalung, S., Marpaung, J.S., Muhardy, A.F., Kusnowo, A.: Direct observation of developed plastic deformation and its application to nondestructive testing. Jpn. J. Appl. Phys. **35**, 854–857 (1996)
2. Yoshida, S., Muhamad, I., Widiastuti, R., Kusnowo, A.: Optical interferometric technique for deformation analysis. Opt. Exp. **2**, 516–530 (1998)
3. Yoshida, S., Ishii, H., Ichinose, K., Gomi, K., Taniuchi, K.: An optical interferometric band as an indicator of plastic deformation front. J. Appl. Mech. **72**(5), 792–794 (2005)
4. Yoshida, S.: Deformation and fracture of solid-state materials. Springer, New York (2015)
5. Aitchson, I.J.R., Hay, A.J.G.: Gauge theories in particle physics. Adam Hilger, Bristol (1989)

6. Yoshida, S.: Deformation and fracture of solid-state materials. Springer, New York (2015). Chap. 5
7. Suzuki, T., Takeuchi, S., Yoshinaga, H.: Dislocation dynamics and plasticity. Springer series in material science, vol. 12. Springer, Berlin (1991)
8. Yoshida, S., Gaffney, G.A., Schneider, C.W., Rourks, R.L.: Field Theoretical Approach to Deformation and Fracture. In: Proceedings of SEM XI International Congress, pp. 144–149. Orlando, 2–5 June 2008
9. Yoshida, S.: Wave nature in deformation of solids and comprehensive description of deformation dynamics. Presented at IUTAM Symposium GA 12-08, "Complexity of Nonlinear Waves", Tallinn, 8–12 Sept 2014

Chapter 19
Measuring the Effective Fracture Toughness of Heterogeneous Materials

Chun-Jen Hsueh, Guruswami Ravichandran, and Kaushik Bhattacharya

Abstract Heterogeneous materials where the scale of the heterogeneities is small compared to the scale of applications are common in nature. These materials are also engineered synthetically with the aim of improving performance. The overall properties of heterogeneous materials can be different from those of its constituents; however, it is challenging to characterize effective fracture toughness of these materials. We present a new method of experimentally determining the effective fracture toughness. The key idea is to impose a steady process at the macroscale while allowing the fracture process to freely explore at the level of microstructure. We apply a time-dependent displacement boundary condition called the surfing boundary condition that corresponds to a steadily propagating macroscopic crack opening displacement. We then measure the full-field displacement using digital image correlation (DIC) method, and use it to obtain the macroscopic energy release rate. In particular, we develop a global approach to extract information from DIC. The effective toughness is obtained at the peak of the energy release rate. Finally, the full field images also provide us insight into the role of the microstructure in determining effective toughness.

Keywords Fracture toughness • Heterogeneous material • Surfing boundary conditions • Digital image correlation • Brittle material

19.1 Introduction

There are several experimental methods for measuring the fracture toughness of homogeneous materials including single edge notch tension (SENT), center cracked tension (CCT), ASTM standard compact tension (CT), and single edge notch bend (SENB) [1]. However, general methods for measuring the toughness of heterogeneous materials remains less well developed though there are well-established methods for specific materials such as the double cantilever beam (DCB) method for the laminar material [2] or r-curve measurement of ceramics [3]. Similarly, a systematic understanding of the effective toughness and how it depends on microstructure also remains a topic of active research [4–8]. Recently, Hossain et al. [9] proposed an approach to defining the effective toughness of brittle heterogeneous materials and used it to study the role of microstructure in determining overall toughness. Here, we report on preliminary results of a new experimental method to measure the toughness of heterogeneous materials based on the theoretical work of Hossain et al. The key idea is to enforce a macroscopically steady crack growth while allowing crack deflection, pinning, nucleation of distal cracks etc. at the microscopic scale while measuring the macroscopic energy release rate.

19.2 Experiment Configuration

For this study, an experimental configuration is developed that enables for a steady and controlled crack growth at the macroscopic scale but imposes no constraints on the microscopic scale. The configuration is shown in Fig. 19.1b. We first cut the specimen from 1/8″ thick Homalite H-911 sheets using a laser cutter (shown in Fig. 19.1a). We then insert rods to the specimen and attach these rods to a specifically shaped pair of rail with rollers. The first rod is connected to another roller with cables for self-alignment and the rail is attached on a linear stage. The linear stage pulls the rail downwards while the

C.-J. Hsueh • G. Ravichandran • K. Bhattacharya (✉)
Division of Engineering and Applied Science, California Institute of Technology, 1200 E California Blvd, Pasadena, CA 91125, USA
e-mail: chhsueh@caltech.edu; bhatta@caltech.edu

A.M. Beese et al. (eds.), *Fracture, Fatigue, Failure and Damage Evolution, Volume 8*, Conference Proceedings of the Society for Experimental Mechanics Series, DOI 10.1007/978-3-319-21611-9_19

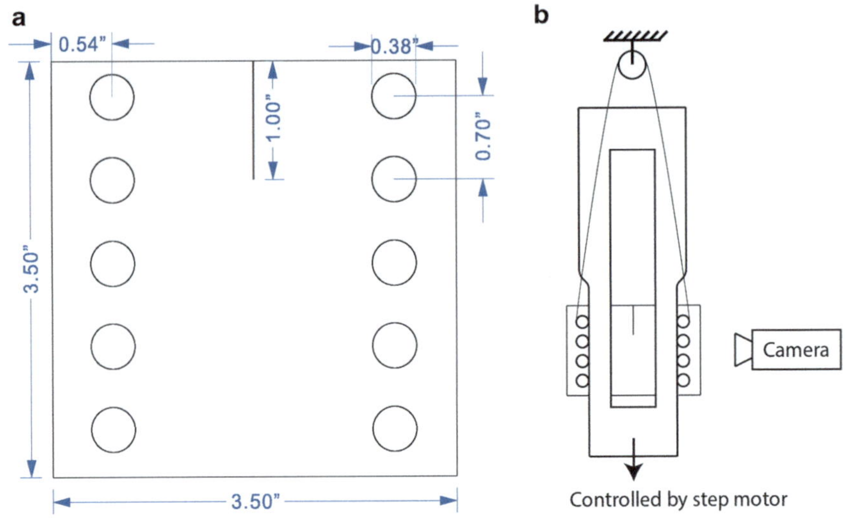

Fig. 19.1 (**a**) (*left*) Homogeneous specimens dimension (**b**) (*right*) experiment configurations

specimen is held fixed by the cables. As a result of the shape of the rails, when the linear stage pulls the rail down and the specimen slides along the rail and the rail imposes a smoothly translating crack-opening displacement that approximates the surfing boundary conditions proposed by Hossain et al. [9].

We determine the macroscopic energy release rate by measuring the macroscopic stress intensity factor. More precisely, we assume the existence of a K-dominant region and we use the digital image correlation (DIC). A random fine speckle pattern is applied on the specimens. When the specimen is deformed, a CCD camera observes the deformation of the pattern. We then use a global data analysis method to calculate the stress intensity factor.

19.3 Data Analysis

The grey level in the deformed image $g(x)$ is related to the grey level in the reference image $f(x)$ by

$$f(x) = g\left(x + u(x)\right) \tag{19.1}$$

where $u(x)$ is the displacement field in reference configuration [10]. We also know that the mode-I asymptotic displacement field is

$$u\left(x; K_I, x_0\right) = \frac{K_I}{2\mu}\sqrt{\frac{r}{2\pi}}U\left(\theta; \kappa\right) \tag{19.2}$$

where μ, κ corresponds to the material properties, r, θ is the polar coordinate with origin coincide with crack tip position x_0, $\theta = 0$ coincide with crack propagate direction, and K_I is the stress intensity factor [1]. Thus we are able to calculate the deformed position of each pixel for a reference image $f(x)$ with given stress intensity factor K_I and crack tip position x_0. We then use linear interpolation to establish the grey value for these points that are enclosed in the deformed meshes (the red dots in Fig. 19.2) and use background value for these points that are not enclosed in the deformed meshes (the blue dots in Fig. 19.2). This way, we are able to calculate the deformed reference image $g(x; K_I, x_0)$.

We define the cost function as the difference between the experiment observation and the deformed reference image, as is shown in the following equation:

$$\int_{\Omega} \left|\left| G(x) - g\left(x; K_I, x_0\right) \right|\right|^2 d\Omega \tag{19.3}$$

where $G(x)$ is the experiment observation and $g(x; K_I, x_0)$ is the deformed reference image with given stress intensity factor and crack tip position. Minimizing the cost function above with respect to K1 and x0 gives us the optimal stress intensity factor and the optimal crack tip position.

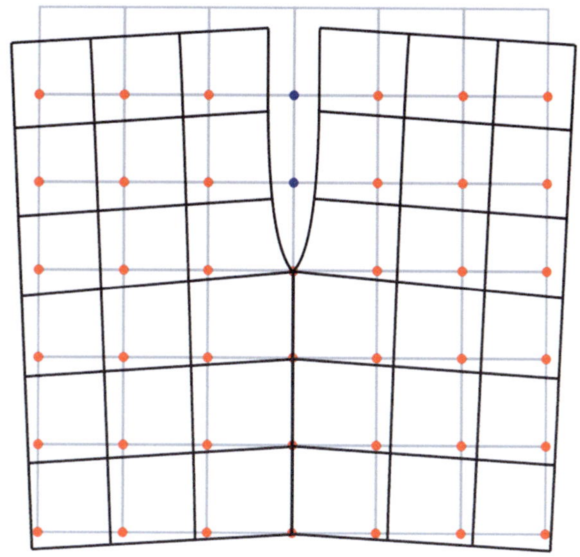

Fig. 19.2 Reference meshes (*grey*) and deformed meshes (*black*)

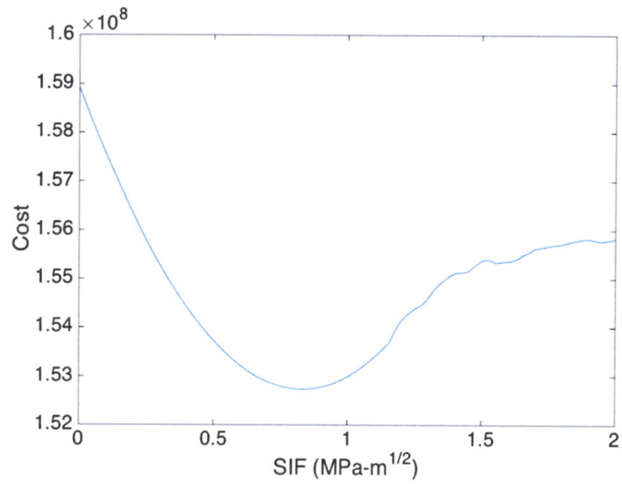

Fig. 19.3 Cost vs. stress intensity factor with fixed crack tip position

19.4 Results

We know that the following optimization problem is not convex in general.

$$\min_{x_0, K_I} \int_{\Omega} ||\, G(x) - g\,(x; K_I, x_0)\, ||^2 d\Omega \tag{19.4}$$

However, when we plot the cost function versus the stress intensity for fixed crack tip position x_0 in Fig. 19.3, we can see the function is convex when the stress intensity factor is smaller than a certain critical value. Thus it is necessary to establish the minimum value for given crack tip position when the initial guess is small enough. In the following sections, we have used $0.5MPa - \sqrt{m}$ as our initial guess.

To calculate the optimal pair (K_I, x_0), we compared the deformed reference image and the experiment observation image and made an initial guess $\tilde{x}_0$ for the crack tip position. We then constructed a region A with initial guess $\tilde{x}_0$ as the center. For every crack tip position contained in region A, we calculated the corresponding optimal stress intensity factor and the cost function. On the basis of these calculations, the optimal pair with the minimum cost function was chosen.

We applied this procedure to the images of crack propagate in a homogeneous Homalite specimen and plotted the stress intensity factor in Fig. 19.4.

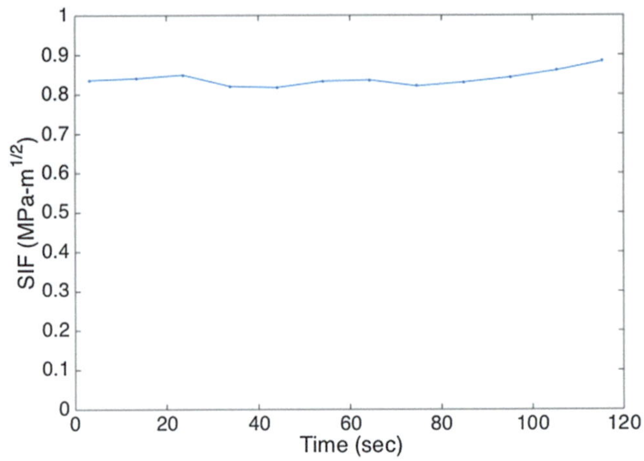

Fig. 19.4 Stress intensity factor for homogeneous Homalite specimen

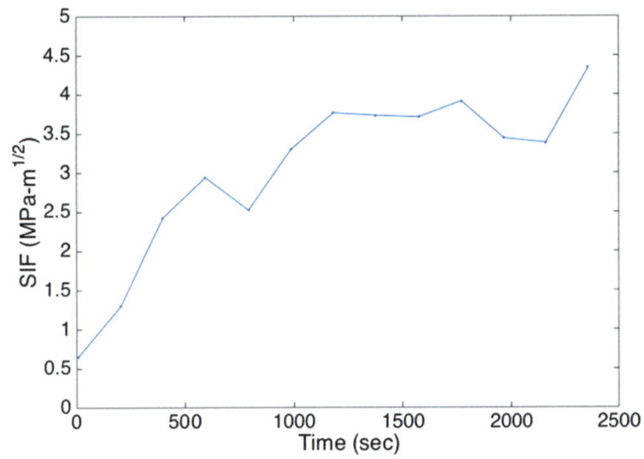

Fig. 19.5 Stress intensity factor for the crack tip when trapped in a hole

As Fig. 19.4 shows, we can see the stress intensity factor is quite stable in homogeneous material as the crack propagates steadily through the specimen. This result is in agreement with findings regarding the properties of linear elastic fracture mechanics (LEFM), $K_I = \sqrt{G_c E}$ (which is a constant).

We also applied the described procedure to specimens with holes (heterogeneity). We observed that when the crack tip is trapped in a hole, the driving force (or stress intensity factor) is increased dramatically as shown in Fig. 19.5. More energy needs to be applied to the crack tip for the crack to propagate. Hence, the effective fracture toughness of this heterogeneous material is higher than that of the homogeneous material.

19.5 Conclusion

In this work, we have proposed an experiment configuration that gives us steady macroscopic crack growth while letting the crack evolve freely microscopically. This method provides a good foundation to investigate fracture propagation in heterogeneous materials. We have also shown that the effective fracture toughness is increased by the holes (or heterogeneity) in the specimens. Heterogeneity thus has a profound effect on effective fracture toughness. We are currently further investigating the effect of heterogeneity in fracture mechanics and the results of which will be presented in due course.

Acknowledgement We gratefully acknowledge the financial support of the U.S. National Science Foundation (Grant No. CMMI-1201102).

References

1. Zehnder, A.T.: Fracture mechanics. Lecture notes in applied and computational mechanics. Springer, London (2012)
2. Liang, F., et al.: Fracture behavior of flock fiber reinforced laminar composite. J. Mater. Sci. Eng. B **1**(1), 1–13 (2011)
3. Kruzic, J.J., et al.: The utility of R-curves for understanding fracture toughness-strength relations in bridging ceramics. J. Am. Ceram. Soc. **91**(6), 1986–1994 (2008)
4. Faber, K.T., Evans, A.G.: Crack deflection processes—I. Theory. Acta Metall. **31**(4), 565–576 (1983)
5. Gao, H., Rice, J.R.: A first-order perturbation analysis of crack trapping by arrays of obstacles. J. Appl. Mech. **56**(4), 828–836 (1989)
6. Bower, A.F., Ortiz, M.: A three-dimensional analysis of crack trapping and bridging by tough particles. J. Mech. Phys. Solids **39**(6), 815–858 (1991)
7. Hutchinson, J.W., Suo, Z.: Mixed mode cracking in layered materials. Adv. Appl. Mech. **29**, 63–191 (1991)
8. Cox, B., Yang, Q.: In quest of virtual tests for structural composites. Science **314**(5802), 1102–1107 (2006)
9. Hossain, M.Z., et al.: Effective toughness of heterogeneous media. J. Mech. Phys. Solids **71**, 15–32 (2014)
10. Mathieu, F., Hild, F., Roux, S.: Identification of a crack propagation law by digital image correlation. Int. J. Fatigue **36**(1), 146–154 (2012)

Chapter 20
Local Strain Analysis of Nitinol During Cyclic Loading

Kenneth Perry, Alex Teiche, and Izak McGieson

Abstract Various approaches for obtaining a cyclic stress-strain curve for superelastic nitinol are evaluated. Smooth test specimens made from seamless drawn tubing and equivalently processed as medical device materials are tested according to one of three methods. The technique of phase shifted moiré interferometry is used to provide in situ strain data with automated data analysis using spatially correlated fringe patterns during the cyclic stress-strain testing. Selected results demonstrate the exceptional spatial and temporal resolution of the measurement system over a wide range of deformations and time scales. General trends in the testing and observations on the feasibility and application of the various methods to medical device design are discussed.

Keywords Moiré interferometry • Phase shifting • Local strain • Fatigue • Nitinol • Medical • Device • Modeling • Design

20.1 Introduction

Many medical implants experience one or more large amplitude loading cycles early in their life and some applications expose devices to relatively large amplitude loading cycles in vivo. In some cases, the degree of non-recoverable deformation, or plasticity is enough to warrant consideration of material models that incorporate such behavior. Generating appropriate material characterization data and employing a methodology to accurately relate alternating strain data to material fatigue limits is a crucial aspect in the design, development and engineering of medical devices.

Efforts to characterize and model cyclic inelasticity in superelastic nitinol include the work of Lagoudas and Entchev [1] and Lubliner and Auricchio [2]. The former is based on a transformation induced plasticity formulation while the latter follows the concept of generalized plasticity. These approaches assume different mechanisms and each accurately captures some aspects of the cyclic inelastic behavior of superelastic nitinol.

The purpose of our work is to develop unique experimental capabilities, methodology and data for computational modeling of superelastic nitinol to include cyclic and loading history dependent inelastic effects. A key element of our approach is to observe and quantify local inelastic strains under isothermal loading conditions using a non-contact, full-field, measurement system optimized for studying the superelastic nitinol material phase transformation. In this work we further extend the technique of phase shifted moiré interferometry [3, 4] to obtain in situ, local strain data during controlled and instrumented cyclic loading.

K. Perry (✉)
ECHOBIO LLC, 11588 Arrow Point Drive, Bainbridge Island, WA 98110, USA
e-mail: kperry@echobio.com

A. Teiche
California Polytechnic State University, San Luis Obispo, CA, USA

I. McGieson
Knox College, Galesburg, IL, USA

A.M. Beese et al. (eds.), *Fracture, Fatigue, Failure and Damage Evolution, Volume 8*, Conference Proceedings of the Society for Experimental Mechanics Series, DOI 10.1007/978-3-319-21611-9_20

A direct objective of the present work is to evaluate three different methods for obtaining cyclic stress-strain data for superelastic nitinol. Of special interest is the fatigue damage that develops during the testing for each method and how functional fatigue effects influence analysis and interpretation of the data.

20.2 Materials and Methods

20.2.1 Specimen Geometry and Fabrication

Smooth specimens were made from seamless drawn pseudoelastic SE-508 nitinol tubing from Nitinol Devices and Components, Inc. (Fremont, CA), with an outer diameter of 5 mm and a 0.6 mm wall. Two pairs of 1 mm diameter holes were laser cut in the ends of the specimens and then carefully honed to ensure precision pin loading in either tension or four point bending. Windows were cut from the specimens with EDM. A flat surface was gently ground on the front face of the test specimens to provide an optically flat surface for the grating replication process. The back surface was similarly ground to make a rectangular cross section 1.25 mm wide by 0.57 mm thick over a gage length of 10 mm. A photograph of the specimen mounted in the load frame is shown in Fig. 20.1.

The specimen was mechanically polished in the shoulder regions to eliminate strain concentrators that could compete with the transition to the smooth section. The specimens were thermomechanically processed with a single heat treatment at 505 °C to have an austenite finish (A_f) temperature of 5 °C, as measured by bend free-recovery.

20.2.2 Dynamic Test System and Data Acquisition

The specimens were dynamically tested with a Bose 3230 ElectroForce test system (Eden Prairie, MN). Global measurements of the applied load were made using a 450 N range load cell and the crosshead transducer was used for displacement sensing. Load control was used for all of the testing. A loading rate of 0.5 N/s was used for both loading and unloading of the specimens.

A thermocouple on the back side of the specimen was used to monitor temperature of the specimen during the experiments. This data was recorded along with load and deflection. All specimens were tested under isothermal conditions in air at 29 °C, ±1 °C for all of the testing.

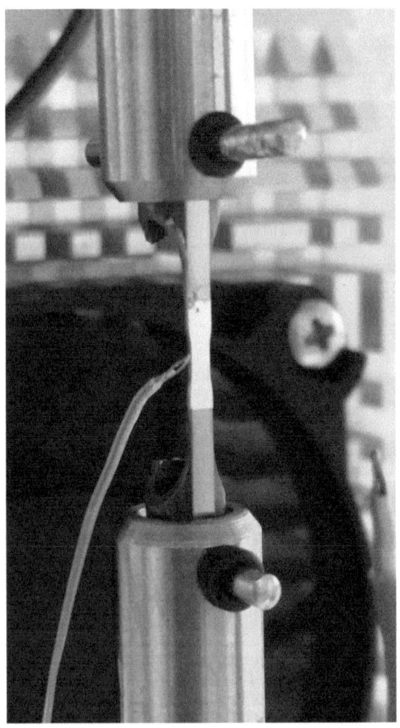

Fig. 20.1 Specimen in the load fixture (*left*)

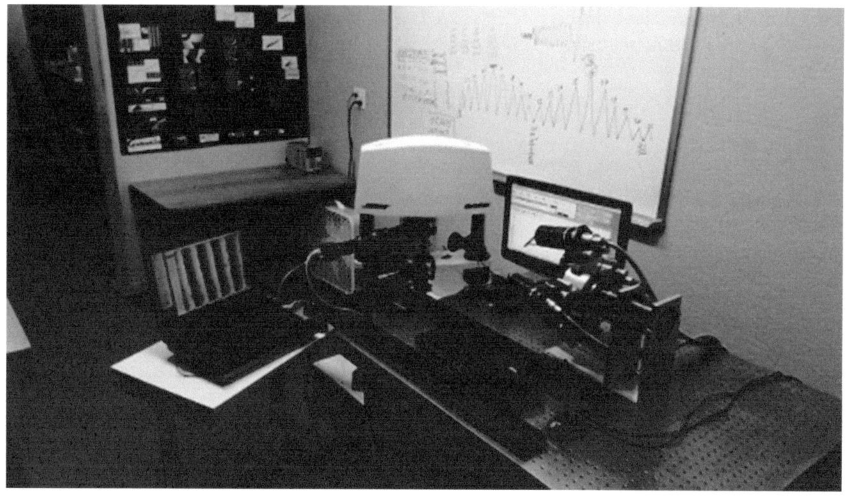

Fig. 20.2 Photograph of the optical table and laboratory setup (*right*)

The experiments employed two personal computers, one each for the Bose test system and phase shifting moiré interferometer system. These are shown Fig. 20.2. The first computer was responsible for controlling the load frame, and the acquisition of all non-moiré data. Load, cross-head deflection, cycle count and thermocouple measurements were digitized to 16 bits and routinely recorded at appropriate intervals during testing for all of the specimens.

A second computer manages the moiré interferometer, phase shifter, high speed camera and the control interface described above. Full field data was available at the maximum resolution of 1360 × 1024 for all frame rates during testing. Figure 20.2 is a photograph of the optical table, interferometer, camera, test frame, environment chamber and computers.

20.2.3 Phase-Shifting Moiré Interferometry

Moiré interferometry provides in-plane displacement data in the form of sinusoidally modulated fringe patterns. The general technique is described in detail in [5]. Phase shifting extends the basic moiré interferometry technique [6] by temporally oversampling the interference patterns and introducing a known phase shift into one of the beam paths. The sampled intensities and the known phase shift are used to create a significantly noise reduced "wrapped" fringe pattern with no loss of spatial resolution.

The interferometer used in these experiments was custom built with an integrated fiber optic beam splitter and phase shifter to provide exceptional stability and configurability [7]. The four-beam interferometer head is mounted on the optical table shown in Fig. 20.2. The horizontal pair of beams was used to produce the U-field fringe patterns and the pair of vertical beams was used for the V-field.

Crossed diffraction gratings were epoxy replicated from 300 lines/mm holographic diffraction grating plates onto the specimens using Micro-measurements (Raliegh, NC) PC-8 epoxy. The final grating thickness was minimized and estimated to be $5 - 20\,\mu\text{m}$.

Digital images were recorded using the Qioptiq T-100 Optem telecentric 10× zoom lens and an Allied Vision Technologies, Canada, Inc. (Burnaby-BC) GC1380H Prosilica GC high speed camera. The specialized equipment has a range of high performance spatial resolution and frame rate combinations from $1360 \times 1024/30$ to $170 \times 128/120$ fps, both at 12 bit gray scale resolution.

A custom interface was written to integrate the cameras capabilities into the phase shifted moiré interferometry [6] software system. The interface integrates control of the camera and fiber phase shifter with two digital control lines from the dynamic test machine. The analog actuator output signal is rectified and used to hardware-trigger the camera, allowing frames to be captured at the min and max loading conditions. A user-controllable digital output is used to allow the fatigue

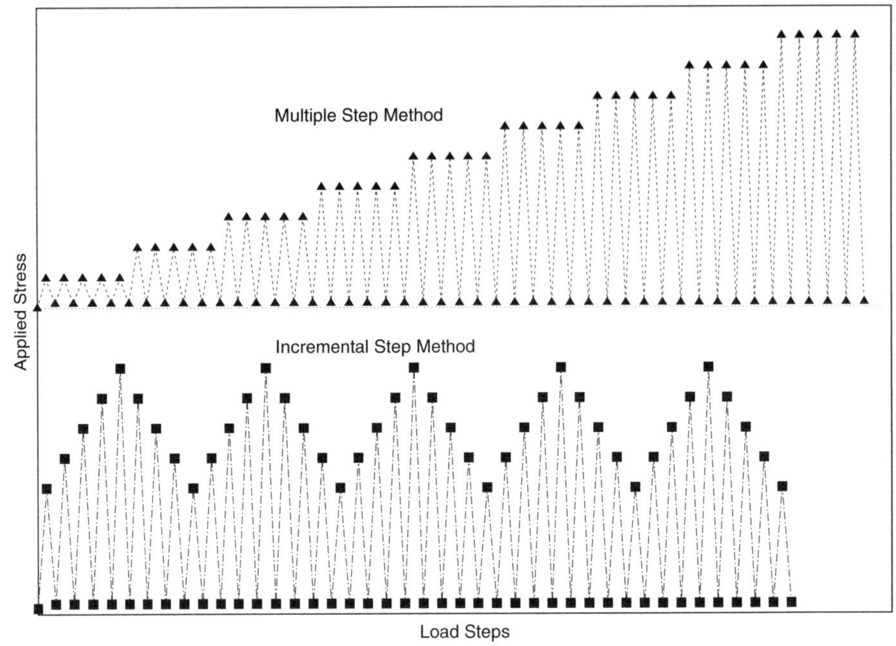

Fig. 20.3 Schematic load history for the multiple and incremental step test methods

testing control software to trigger taking phase shifted moiré interferometry data while the fatigue test machine is paused. Using the block configuration mode within the fatigue test machine software, the machine could fatigue the specimen for a given time, pause the machine, start phase shifting and acquiring image data, and automatically restart cycling after the data was recorded.

To permit fully automatic data collection, a quantitative signal to noise measurement algorithm was implemented to verify the quality of data taken during automatic tests, and to recollect any data smeared by vibrations or other environmental factors. The surface images produced by the wrapping algorithm are digitally correlated to automate the measurement of strain in the presence of moving bimaterial interfaces.

20.2.4 Loading Sequences and Cyclic Test Parameters

Various test methods are available to obtain the cyclic stress-strain behavior for a given material [8]. In all methods, a specimen is cycled between fixed load or displacement conditions until some criteria of stability is met. The cyclic stress-strain curve is determined from the stabilized load or stress plotted as a function of strain. Three test methods are considered in this study: the multiple step, incremental step and single step tests.

The multiple step method uses a single specimen tested at progressively higher load levels. Specimens tested according to this protocol received a minimum of five cycles at load levels from 50 to 300 N in increments of 50 N. Additional cycles were applied as necessary at each level until a stabilized response was obtained. The load history of a multiple step test is shown in the top of Fig. 20.3.

The incremental step method applies a specific sequence or ‘block’ of loading to a single specimen. The block loading sequence is applied again until the stress-strain curve of the specimen has stabilized. The block sequence is an increasing then decreasing waveform of peak loads and unloads of the specimen, generally with five load levels within the block. A typical load history for the incremental method is shown in the lower part of Fig. 20.3. Five blocks of loading are shown in the diagram.

Several combinations of test parameters for the incremental step method were explored and results for two cases are included here. The first specimen was tested with peak loads from 150 to 250 N in 25 N increments and the other from 150 to 275 N in 30 N increments.

The single step method uses individual specimens tested at constant load levels. Results for a single load level at 275 N are included here.

20.3 Results

The first three load levels for the multiple step specimen required only five cycles to stabilize. The stress-strain behavior was linear and there was no appreciable hysteresis. Once above about 250 N, forward and reverse material phase transformation led to localized deformation and damage. Twenty-five additional cycles were applied before reaching a measure of stability.

Figure 20.4 shows the typical stress-strain behavior during cyclic loading for the first 25 cycles at 300 N for the multiple step specimen. Specimen temperature for the first five of these cycles is shown in Fig. 20.4. Temperature spikes as large as 10 °C were routinely observed during the experiments and were associated with forward and reverse phase transformation of the material. Generally, the temperature history was retraced each cycle and had similar characteristics to the hysteresis inherent in the load-deflection data.

The decrease in upper plateau stress for the data in Fig. 20.4 is shown in Fig. 20.5. Also shown in Fig. 20.5 is the inelastic strain obtained at an arbitrary location from analyzing the associated phase shifted moiré interferometry fringe patterns. As the loading cycles progressed, a uniform inelastic strain developed over the gage section, reaching a value of about 0.5 % after 25 cycles. Neither the plateau stress nor inelastic strain appears to have fully stabilized after 25 cycles.

Two specimens were tested using the incremental step method. Full-field strain data was obtained automatically at all peak loads during a block and also with the specimen unloaded at the end of each block. Figure 20.6 are wrapped fringe patterns of axial displacement obtained during a typical incremental step test. The top image shows the inelastic strain after

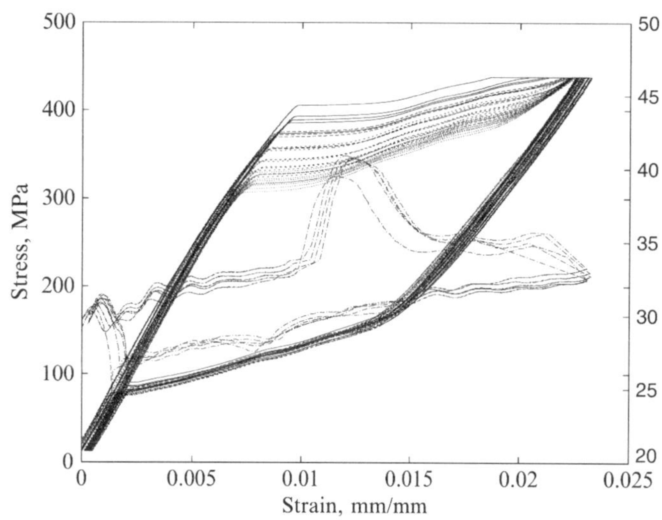

Fig. 20.4 Typical loading and unloading behavior with material phase transformation. Stress versus strain shown for five cycles during a typical multiple step test. Shown on the *right axis* is the specimen temperature

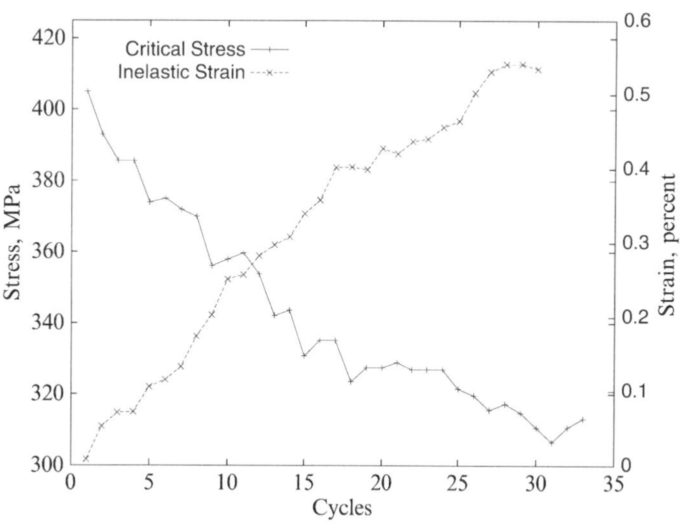

Fig. 20.5 Decrease in plateau stress and increase in local inelastic strain for a typical multiple step test

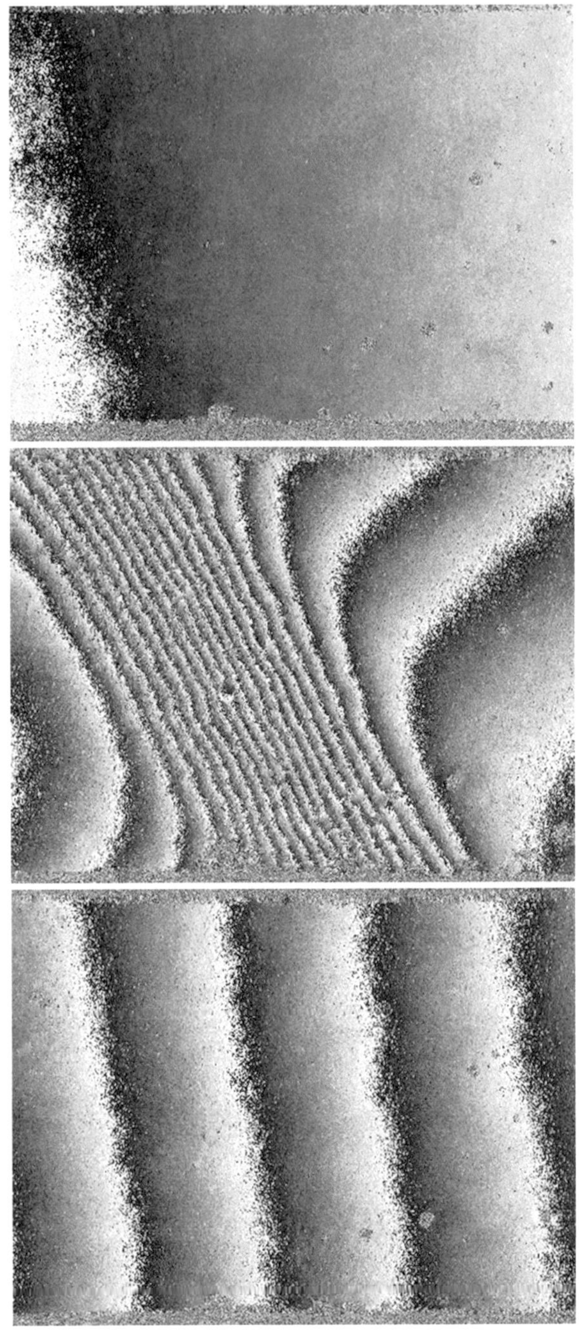

Fig. 20.6 Wrapped fringe patterns of axial displacement obtained during a typical incremental step test. The *top image* shows the inelastic strain after the first block of loading, the *middle image* was captured during loading to show shear band localization and the *bottom image* shows the inelastic strain after five blocks of the loading sequence have been applied

the first block of loading, the middle image was captured during loading to highlight the shear band localization and the bottom image shows the inelastic strain after five blocks of the loading sequence have been applied.

The local strain at any selected point is available during the experiment, or can be retrieved through post-processing. Figure 20.7 is an example of local strain history measurements from an incremental step test. The application of five blocks of loading are shown and local strain measurements are made at the peak loads during each block. Inelastic strain is recorded at the end of each block at load steps 11, 21, 31, 41 and 51 when the specimen has been unloaded to 5 N. This data was taken over a period of about 10 h—the time required to cycle through five blocks of incremental step testing at 0.5 N/s.

A final example of local strain measurement are shown in Fig. 20.8 for a single step specimen. Residual deformation after each of the first 25 loading cycles is shown for a specimen cycled at 275 N.

The stress-strain curves of all four samples after testing are shown in Fig. 20.4.

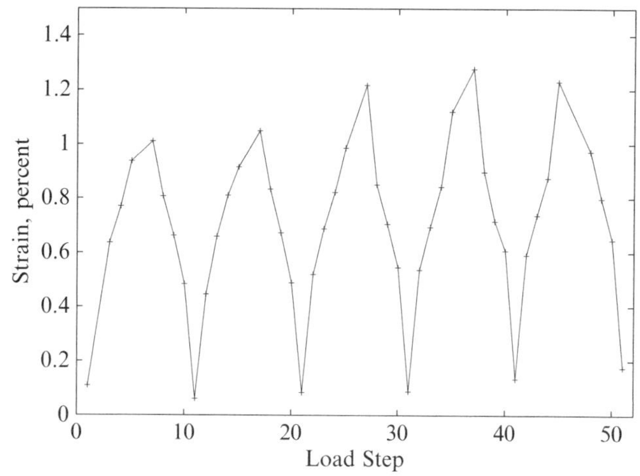

Fig. 20.7 Local strain at a selected location during an incremental step test. The application of five blocks of loading are shown and local strain measurements are made at the peak loads during each block. Inelastic strain is recorded at the end of each block with the specimen unloaded at steps 11, 21, 31, 41 and 51

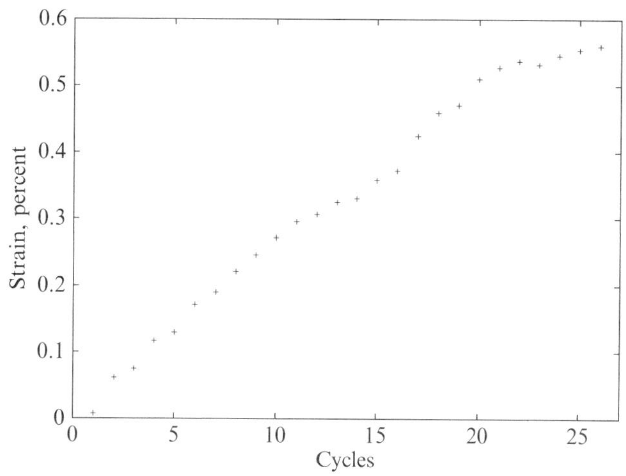

Fig. 20.8 Local strain at a selected point after each load cycle for a typical single step test

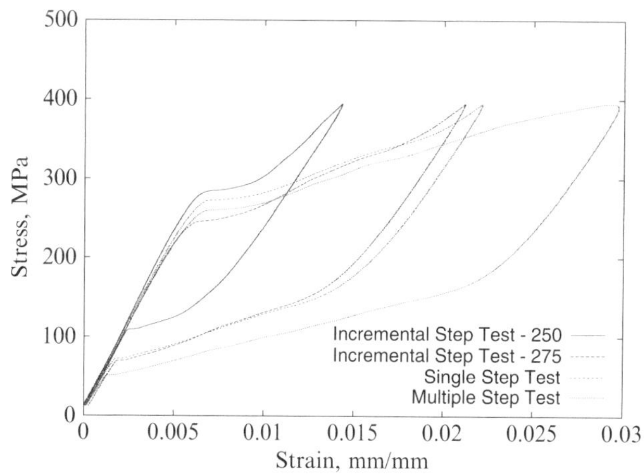

Fig. 20.9 Representative stress-strain curves for the different test methods, including data for two peak load parameters for the incremental step method

20.4 Conclusions

Three methods for obtaining a cyclic stress-strain curve for superelastic nitinol were evaluated. The different methods produced different results. Each approach has advantages and disadvantages as test methods when considered for application to medical device design and analysis.

Strain localization in the form of shear bands was observed in all specimens. Once initiated, shear bands would propagate across the specimens and remain until fully unloaded. Inelastic deformation in the wake of the shear band was uniformly distributed within the gage section of the specimens. The threshold stress for inducing phase transformation decreased throughout loading of the specimens. The rate of decrease varied for the different test methods, as did the maximum transformation strain.

Acknowledgements The authors would like to thank Tim Johnson for a critical review of an early draft of this manuscript.

References

1. Lagoudas, D.C., Entchev P.B.: Modeling of transformation-induced plasticity and its effect on the behavior of porous shape memory alloys. Mech. Mater. **36**, 865–892 (2004)
2. Lubliner J., Auricchio, F.: Generalized plasticity and shape memory alloys. Int. J. Solids Struct. **33**(7), 991–1003 (1996)
3. Perry, K.E., Teiche, A.: Fatigue crack initiation in superelastic nitinol. In: Mitchell, M., Smith, S., Woods, T., Berg, B. (eds.) Fatigue and Fracture Metallic Medical Materials and Devices, pp. 35–52. American Society for Testing and Materials, Conshohocken (2013). ASTM STP 1559
4. Perry, K.E., Labossiere, P.E., Steffler, E.: Measurement of deformation and strain in nitinol. Exp. Mech. **43**(3), 373–380 (2007)
5. Post, D., Han, B., Ifju, P.: High Sensitivity Moiré. Springer, New York (1994)
6. Perry, K.E.: Delamination and damage in composite materials using phase-shifting moiré interferometry. Opt. Lasers Eng. **24**, 467–483 (1996)
7. Deason, V.A., Ward, M.B.: A compact portable difraction moiré inerferometer. In: Laser Interferometry: Quantitative Analysis of Interferograms, vol. 1162, pp. 26–35. SPIE, Bellingham (1989)
8. Manson, S.S., Halford, G.R.: Fatigue and Durability of Structural Materials. ASM International, Materials Park (2006)

Chapter 21
Environmental Protection by the Opto-acoustic and Neutron Emission Seismic Precursors

O. Borla, G. Lacidogna, E. Di Battista, M. Costantino, and A. Carpinteri

Abstract The acoustic (AE) and electromagnetic (EM) emission signals detected during the failure of brittle materials are analogous to the anomalous mechanical and geoelectromagnetic waves observed before major earthquakes. These phenomena reinforce the idea that opto-acoustic emissions can be applied as a forerunning tool for seismic events.

The elastic energy released by micro-cracking eventually yields to form macroscopic fractures, whose mechanical vibrations are converted into electromagnetic oscillations over a wide range of frequencies, from Hz to THz. This excited state of the matter could be the cause of resonance phenomena at the nuclear level producing neutron bursts, in particular during stress-drops or sudden catastrophic failures.

The authors present the results they are obtaining at a gypsum mine located in Northern Italy. The observations revealed a strong correlation between AE/NE events and the closest and most intense earthquakes. Thanks to the position of the monitoring station (100 m under the ground level), the acoustic and electromagnetic noise from human activities is greatly reduced, as well as the neutron background. An integration of AE/EME/NE data with CO_2 and Radon variations, that are considered as additional seismic precursors, is planned.

Keywords Acoustic emission • Electromagnetic emission • Neutron emission • CO_2 and Radon variations • Seismic activity

21.1 Introduction

Monitoring and detection of the different forms of energy emitted during the failure of natural and artificial brittle materials allow an accurate interpretation of damage in the field of Fracture Mechanics. These phenomena have been mainly measured based on the signals captured by the acoustic emission (AE) measurement systems [1–10], or on the detection of the electromagnetic (EM) charge [11–18].

Nowadays, the AE technique is well-known in the scientific community and applied for monitoring purpose. In addition, based on the analogy between AE and seismic activity, AE associated with microcracks are monitored and power-law frequency vs. magnitude statistics are observed. With frequencies up to GigaHertz, and therefore cracks between the micron and the millimetre scale, pressure waves can generate electromagnetic waves of the same frequency, which turn out to be even a more advanced seismic precursor (up to a few days before).

The EM signals are usually related to those materials in which the fracture propagation occurs suddenly and it is accompanied by stress drops in the stress-strain curve. A number of laboratory studies revealed the existence of EM signals during fracture experiments carried out on a wide range of materials [11]. Studies on electrical properties of rocks have demonstrated that a correlation exists between electrical resistivity and cumulative damage [19–21].

Very recently, it has been also argued that the elastic energy due to micro-crack production eventually yields to forming macroscopic fractures, whose mechanical vibrations are converted into electromagnetic oscillations over a wide range of frequencies, from few Hz to MHz, and even up to "microwaves". This excited state of the matter could be a cause of subsequent resonance phenomena of nuclei able to produce neutron bursts in the presence of stress-drops or sudden catastrophic fractures [22–25]. When pressure waves show frequencies between Giga- and TeraHertz, and then with cracks below the micron scale, we are witnessing a phenomenon partially unexpected: phonons resonate with the crystal lattices and, through a complex cascade of events (acceleration of electrons, bremsstrahlung gamma radiation, photo-fission, etc.), may produce nuclear fission reactions [24–28].

O. Borla (✉) • G. Lacidogna • E. Di Battista • M. Costantino • A. Carpinteri
Department of Structural, Geotechnical and Building Engineering, Politecnico di Torino, Corso Duca degli Abruzzi 24, Torino 10129, Italy
e-mail: oscar.borla@polito.it

A.M. Beese et al. (eds.), *Fracture, Fatigue, Failure and Damage Evolution, Volume 8*, Conference Proceedings of the Society for Experimental Mechanics Series, DOI 10.1007/978-3-319-21611-9_21

As regards the neutron emissions (NE), original experimental tests were performed by Carpinteri et al. on brittle rock specimens [29–36]. Different kinds of compression tests under monotonic, cyclic and ultrasonic mechanical loading have been carried out fully confirming the hypothesis of piezonuclear reactions, giving rise to neutron emissions up to three orders of magnitude higher than the background level at the time of catastrophic failure.

These phenomena have important implications also at the Earth's crust scale. Recent neutron emission detections by the authors [37] and other Russian researchers [38–41] have led to consider the Earth's crust as a relevant source of neutron flux variations.

In this work the authors, after summarizing the main laboratory experimental tests on gypsum samples, describe the preliminary results acquired at a gypsum mine situated in Northern Italy (Murisengo, Alessandria) and related to the evaluation of acoustic and nuclear phenomena. The monitoring system, based on the simultaneous acquisition of the various physical quantities, control the structural stability of the mine carrying out, at the same time, the environment monitoring for the seismic risk evaluation.

Taking into account the relationship between AE, EME, NE and seismic activity, it will be possible to set up a sort of territorial database station that could be at the base of a warning network.

This warning system could combine the signals from other stations to prevent the effects of seismic events and to identify the earthquakes' epicentres. Similar networks are being utilized all over the World in countries like Mexico, Taiwan, Turkey, Romania and Japan [42]. However, these already established seismic monitoring systems are usually based on the kinematic quantities of the ground motion: displacement, velocity, and acceleration. The last, in particular, is proportional to the inertial forces transmitted by the ground shaking to the masses. On the other hand, the use of AE, EME, NE will represent a huge step forward, not only for their monitoring capabilities during the earthquake, but also for their forecasting potentialities before the event. Neutrons therefore appear to be as the most advanced earthquake precursor (up to 2 weeks before) [37–41]. An integration of AE/EME/NE data with CO_2 and Radon variations, that are considered as additional seismic precursors [43], is currently planned and the ad-hoc instrumentation is under installation and testing.

21.2 Material and Methods

21.2.1 Acoustic Piezoelectric Transducer

The AE activity emerging from the compressed specimens was detected by a piezoelectric (PZT) transducer glued on the external surface, resonant at 78 kHz, which is able to convert the high frequency surface motions due to the acoustic wave into electric signals (the AE signal). The transducer sensitivity in the low-frequency range was measured by placing it on shaker excited by frequencies in the range 0–10 kHz (white noise). The result of this calibration at low frequencies was 1.2 μV/(mms^{-2}). Resonant sensors are more sensitive than broadband sensors, which are characterized by a flat frequency response in their working range, and then they can be successfully used in monitoring of large-sized structures [6, 7].

21.2.2 Dedicated Loop and Telescopic Antennas

EM signals were monitored using a device, calibrated according to metrological requirements, constituted by three winding loops with different number of turns that can be positioned around the monitored specimen. The working principle is based on the induction Faraday's law. It states that the electromagnetic force (voltage) in a closed circuit (loop) is proportional to the change of the magnetic flux in the windings section. In fact, the three coaxial coils with an increasing number of turns are capable to perform the measurement from very low (Hertz) to high (MHz) frequencies in the magnetic field. The first coil, constituted by 5 turns, works in a frequency range from 300 kHz to 4 MHz. The other two coils constituted by 125 turns and 500 turns, work in the frequency range from 0 to 20 kHz, and from 0 to 1 kHz, respectively. Each turn is realized by a 0.2 mm copper wire, mounted on two coaxial PVC tubes embedded in a two components resin, in order to allow a large range of measurements.

In addition, a telescopic antenna was tested in view of a permanent use at the gypsum mine. The employed antenna, having a maximum length of 125 cm, was coupled with an Agilent DSO1052B oscilloscope (50 MHz, 2 channels) that allows appropriate monitoring of EM signals with frequencies up to tens of MHz.

21.2.3 Radon-Monitor: Ramon 2.2

The Ramon 2.2 Radon monitor is an automatic, electronic Radon dosimeter. The device is designed to detect alpha particles from Radon decay and it is not sensitive to other types of ionizing radiation such as gamma- and beta- radiation. The detector is equipped with a special silicon semiconductor sensor surrounded by a light-proof plastic housing called "metering cell". Radon gas from ambient air diffuses into the metering cell through small openings that are covered by a filter keeping dust outside. The Ramon 2.2 complies with the general requirements for the competence of calibration and testing laboratories according to DIN EN ISO/IEC 17025.

The Ramon 2.2 can operate in two different measurement modes that give results for the averages of Radon levels on a long-term and on a short-term basis. The Radon monitor collects data in measurement cycles of 1 h and the results for newly calculated averages of Radon levels are therefore automatically updated. When the long-term reading is selected, the Radon monitor indicates the average Radon level over the entire measurement period since the last reset of the memory. When the short-term reading is selected, the Ramon 2.2 Radon monitor indicates the average of the Radon level that the instrument detected in the last 7 days.

21.2.4 Carbon Dioxide Meter CO210

The CO210 Carbon Dioxide Meter measures and datalogs CO_2 level, air temperature, humidity, date and time. With visible and audible alarms, this is an ideal instrument for air quality diagnosis. This meter is shipped fully tested and calibrated. The meter can record readings of CO_2, temperature and humidity for long time environment monitoring. The memory capacity is 15,999 points. The Windows based pc software provided is used to set-up the datalogger, download data and to view the data in graphical or text format.

21.2.5 ^{3}He Proportional Counter

Since neutrons are electrically neutral particles, they cannot directly produce ionization in a detector, and therefore cannot be directly detected. This means that neutron detectors must rely upon a conversion process where an incident neutron interacts with a nucleus to produce a secondary charged particle. These charged particles are then detected, and from them the neutrons presence is deduced. For an accurate neutron evaluation, a ^{3}He radiation monitor was used.

During the experimental trial the neutron field monitoring was carried out in "continuous mode". The AT1117M (ATOMTEX, Minsk, Republic of Belarus) neutron device is a multifunctional portable instrument with a digital readout consisting of a processing unit (PU) with an internal Geiger-Müller tube and external smart probes (BDKN-03 type). This type of device provides a high sensitivity and wide measuring ranges (neutron energy range 0.025 eV–14 MeV), with a fast response to radiation field change ideal for environmental monitoring purpose.

21.3 Experimental Results

21.3.1 Laboratory Tests on Cylindrical Gypsum Specimens

Preliminary laboratory compression tests on some gypsum specimens were conducted to verify the behaviour of gypsum under mechanical loading and to assess the validity and efficiency of the dedicated antennas in view to a permanent installation for in-situ monitoring.

Gypsum samples of different size and shape taken from Murisengo mine, were used. For these tests a standard servo-hydraulic press with a maximum capacity of 1000 kN, equipped with control electronics, was employed. This machine makes it possible to carry out tests in either load or displacement control. The tests were performed in piston travel displacement control by setting, for all the tested specimens, a velocity of 0.001 mm/s during compression.

Specific tests were previously conducted to assess the EM environmental noise. In particular, the EM probe was used to detect the EM background for about 12 h before to start the compression tests. By means FFT (Fast Fourier Transform) analysis the main contribution of EM environmental noise is calculated with signals peaked at 50 Hz, the typical frequency of domestic power supply.

Fig. 21.1 The electromagnetic coaxial coils device (loop antenna) and AE piezoelectric transducer positioned around the monitored gypsum specimen (*left*). The telescopic antenna during the mechanical loading test (*right*)

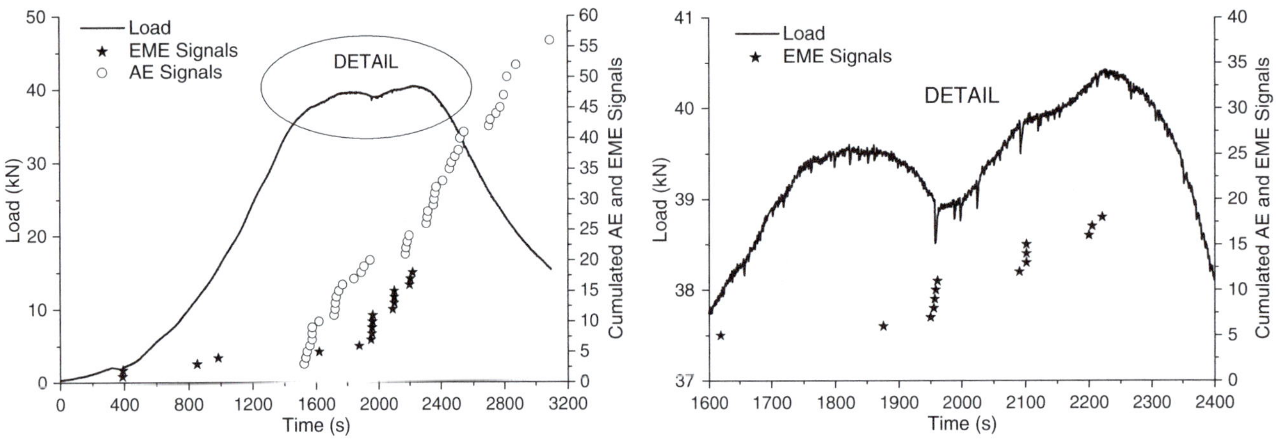

Fig. 21.2 Load vs. time curve, AE (*circles*) and EM (*stars*) signals cumulated number (*left*). Detail of load vs. time curve and EM cumulated number in the vicinity of the peaks load (*right*)

In addition, to better monitoring the fracture precursors, the AE signals were detected by applying to the sample surface the piezoelectric transducer described above (Fig. 21.1 left).

In Fig. 21.2 (left), the load vs time and the cumulated AE and EM signals for one of the tested gypsum specimen are represented. Similar results were obtained for the other specimens. This sample, that is characterized by an evident ductile behaviour, reached a maximum load of about 40 kN. In particular, starting from the first peak load, a significant increase in AE rate is observed. The AE signals achieved the maximum rate in proximity of the peak loads, while in the post peak phase, at the end of specimen softening behaviour, the AE rate diminished.

On the other hand, EM emissions were detected for all the duration of the experiment, with a sensible growth rate just after the second peak load. This behaviour, both for AE and EM signals clearly represents a precursor of specimen failure.

Moreover, as shown in detail in Fig. 21.2 (right), it is observed that gypsum is characterized by weak and frequent stress drops, around which EM signals have accumulated. This EM distribution reinforces the idea that in these particular phases the accumulated energy is suddenly released, due to micro-cracks formation, and the mechanical vibrations are converted into electromagnetic oscillations. A wide range of EM spectrum with respects to the one related to environmental noise was observed. In particular frequencies up to 500 kKz with the loop antenna and up to 7 MHz with the telescopic antenna were estimated.

21.4 In-Situ Environmental Monitoring: San Pietro Prato Nuovo Gypsum Mine

From June 24, 2013 a dedicated in-situ monitoring at the San Pietro Prato Nuovo gypsum quarry located in Murisengo (AL), Italy is started and it is still in progress. The San Pietro Prato Nuovo quarry in Murisengo, is currently structured in five levels of underground development and from which high quality gypsum is extracted every day. Gypsum is a soft sulfate mineral composed of calcium sulphate dihydrate, $CaSO_4{\cdot}2H_2O$. It can be used as a fertilizer, is the main constituent in many forms of plaster and is widely mined. The structural stability in each level is assured by an archway-pillar system (Fig. 21.3 left) which unloads over the underlying floor of average thickness of 4 m. Through accurate topographic surveys, it was possible to ensure a good coaxiality of the pillars between the different levels. In this way dangerous loads eccentricity were avoided. At the fifth level the average pressure to which the pillars are usually subjected is of about 6 MPa.

Since June 2013 the quarry is subjected to a multiparameter monitoring, by the AE technique and the detection of the environmental neutron field fluctuations, in order to assess the structural stability and, at the same time, to evaluate the seismic risk of the surrounding area. Moreover, from the end of January 2015, the Radon and CO_2 monitoring equipments and the telescopic antenna tested in laboratory were installed. Currently, the devices are acquiring the first experimental data and in the coming months it will be possible to evaluate the correlations between these parameters and the seismic activity of the monitored area. In Fig. 21.3 (right) the in-situ experimental set-up is reported.

The dedicated "USAM" AE acquisition system consists of 6 PZT transducers, calibrated over a wide range of frequency comprised between 50 and 800 kHz, 6 units of data storage provided of triggers and a central unit for the data synchronization.

The AE signals received by all the transducers are analysed by means of a threshold detection device that counts the signal bursts exceeding a certain electric tension (measured in volts (V)). Throughout the monitoring period, the threshold level for the detection of the input signals coming from the PZT transducers was kept at 100 μV. Based on the authors' experience, in fact, this level is the most significant for the detection of AE signals from damage processes in non-metallic materials. In addition, also a neutron field evaluation was carried out in "continuous mode" by means of the ATOMTEX ^{3}He neutron radiation monitor. The comparison between the earthquakes [44] occurred in the immediate vicinity of the monitored area during the experimental campaign was carried out. In Fig. 21.4 the occurrence of quakes (with magnitude higher or equal to 2.5 degree in the Richter scale) correlated with acoustic and neutron emissions in the period from April 1st, 2014 to February 19th, 2015 is reported. From the analysis of the experimental data it was observed that the most significant AE and NE increments happen respectively about one day and one week before earthquakes. These increments reached values of one or more order of magnitude higher than the ordinary background. In particular bursts of neutron radiation of at least a couple of hours long were detected.

Furthermore, an analysis of the AE frequencies in the 3 days preceding the earthquake, the day of the seismic event and in the next 3 days was performed. From this survey it was experimentally verified that the recorded average frequencies were respectively of about (38 ± 10) kHz, (52 ± 13) kHz and (63 ± 14) kHz (Fig. 21.5).

These results show that the lower frequencies monitored before the earthquake are mainly due to the formation of fractures up to the scale of meter that occur from large distances in the ground, and anticipate the quake shaking. On the other hand the monitored frequencies just during the seismic event and in the following days tend to increase. This is due to the fact that, from a situation of maximum damage, perceived by the fractured ground, the effects caused by the earthquake in the pillar are

Fig. 21.3 San Pietro Prato Nuovo gypsum quarry (*left*). The six AE devices and the ATOMTEX ^{3}He neutron radiation monitor applied to the selected pillar (*right*)

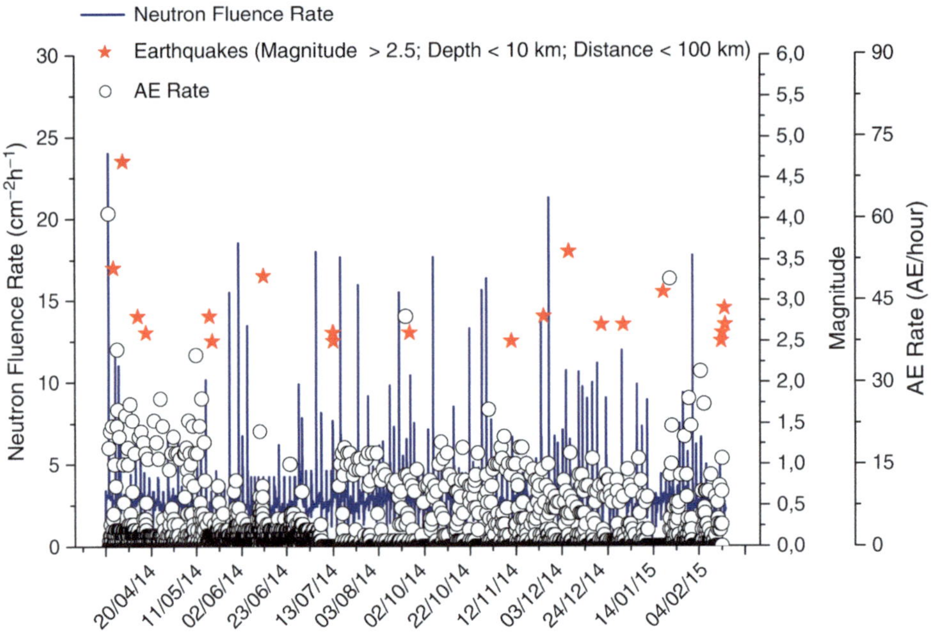

Fig. 21.4 Neutron ambient fluence rate and AE hourly rate compared to local seismic activity

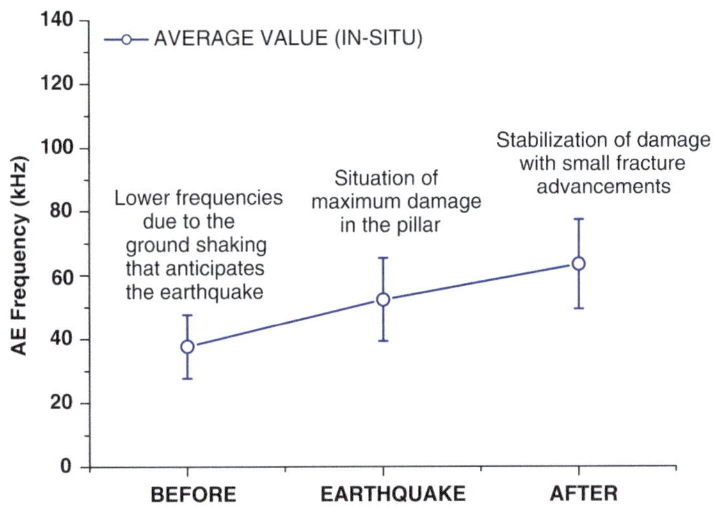

Fig. 21.5 Average AE frequencies monitored in the days around the seismic events

subsequently identified. These involve the formation of small cracks that tend to stabilize with small fracture advancements which are associated to higher AE frequencies.

This further experimental evidence strengthens the idea that integrating neutron fluctuations with acoustic and electromagnetic emissions—and also considering gas radon and CO_2 emission that are considered as additional reliable seismic precursors [43]—it will be possible to set up a sort of monitoring systems that could perform a warning environmental monitoring.

Moreover, an accurate localization of the AE sources on the monitored pillar was obtained using the USAM AE acquisition system composed of 6 PZT transducers, which had permitted to measure the arrival times of acoustic signals. Applying the source location method [6], the AE sources are determined. Considering that the six sensors are positioned on a plane, the cracks location is distributed along the x and y axes. From Fig. 21.6, it is possible to evaluate the main propagation direction of the microcracking during the monitoring period. In particular, many AE localized points are concentrated between the sensor S2–S4, S3–S5 and S3–S6. Therefore, three preferential crack propagation paths can be observed.

Finally, just as an example, the CO_2 and Radon mine background acquired in a period of about 20 days long are reported (Fig. 21.7 right, left).

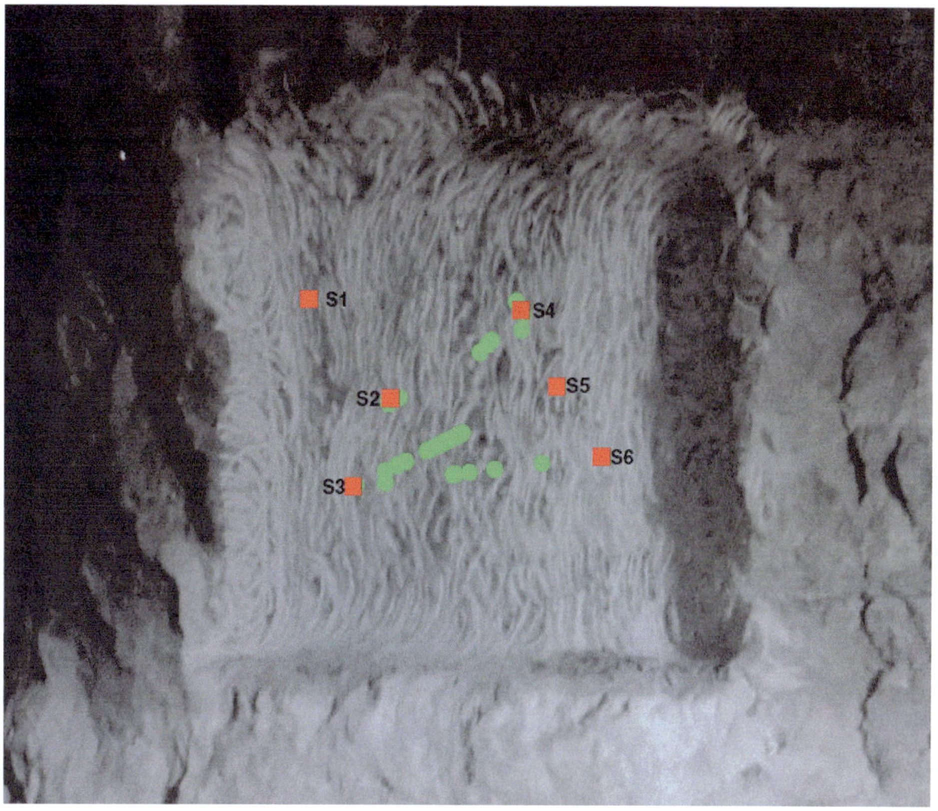

Fig. 21.6 PZT USAM sensors (*square*) and AE source localization with preferential crack propagation (*circle*)

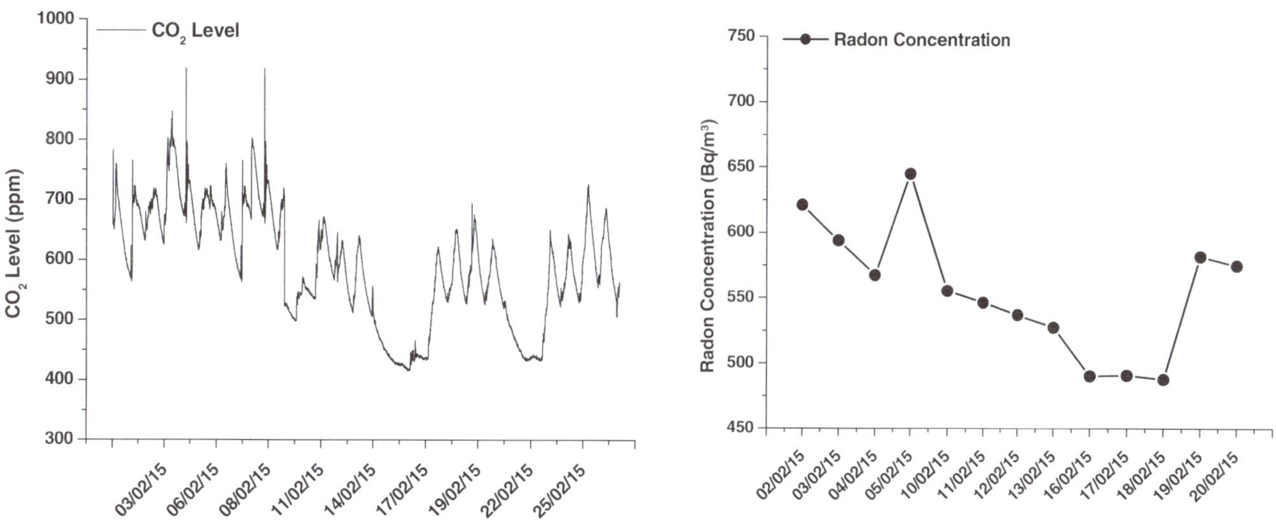

Fig. 21.7 CO_2 (*left*) and radon (*right*) mine average background

21.5 Conclusions

Dedicated instrumentations were used to detect Acoustic, Electromagnetic and Neutron emissions generated during micro-cracking. The experimental data strengthen the hypothesis that the released energy due to micro-crack production can yield to forming macroscopic fractures, whose mechanical vibrations are converted into electromagnetic oscillations over a wide range of frequencies, from few Hz to MHz, and even up to microwaves. This excited state of the matter could be a cause of subsequent resonance phenomena of nuclei able to produce neutron bursts in the presence of stress-drops or sudden catastrophic fractures.

It is also known that the EM signals detected during failure of materials are analogous to the anomalous radiation of geoelectromagnetic waves observed before major earthquakes, reinforcing the idea that the EM effect can be applied as a forecasting tool for seismic events.

On the other hand, recent neutron emission detections have led to consider also the Earth's crust as a relevant source of neutron flux variations.

Preliminary results, acquired at a gypsum mine related to the evaluation of acoustic, and nuclear phenomena are reported. The experimental data acquired emphasize the close correlation between acoustic, electromagnetic, nuclear emissions and seismic activity. Currently the mine is also monitored by means of Radon and CO_2 emissions that are considered as additional reliable seismic precursors. The monitoring campaign is still in progress and in the coming months it will be possible to evaluate the correlations between these parameters and the seismic activity of the monitored area.

By integrating all these signals, it will be possible to set up a sort of territorial database network that combine AE, EM and neutron sensors for the prediction and diagnosis of earthquakes. These sensors could be applied at certain depths in the soil, along the most important faults, or very close to the most seismic areas to prevent well in advance the effects of seismic events and to identify the epicentre of an earthquake.

Acknowledgement The authors are grateful to Dr. Sandro Gennaro from San Pietro PratoNuovo gypsum mine for his assistance during the in-situ environmental monitoring.

References

1. Mogi, K.: Study of elastic shocks caused by the fracture of heterogeneous materials and its relation to earthquake phenomena. Bull. Earthquake Res. Inst. **40**, 125–173 (1962)
2. Lockner, D.A., Byerlee, J.D., Kuksenko, V., Ponomarev, A., Sidorin, A.: Quasi static fault growth and shear fracture energy in granite. Nature **350**, 39–42 (1991)
3. Shcherbakov, R., Turcotte, D.L.: Damage and self-similarity in fracture. Theor. Appl. Fract. Mech. **39**, 245–258 (2003)
4. Ohtsu, M.: The history and development of acoustic emission in concrete engineering. Mag. Concr. Res. **48**, 321–330 (1996)
5. Carpinteri, A., Lacidogna, G., Pugno, N.: Richter's laws at the laboratory scale interpreted by acoustic emission. Mag. Concr. Res. **58**, 619–625 (2006)
6. Carpinteri, A., Lacidogna, G., Niccolini, G.: Critical behaviour in concrete structures and damage localization by acoustic emission. Key Eng. Mater. **312**, 305–310 (2006)
7. Carpinteri, A., Lacidogna, G., Pugno, N.: Structural damage diagnosis and life-time assessment by acoustic emission monitoring. Eng. Fract. Mech. **74**, 273–289 (2007)
8. Carpinteri, A., Lacidogna, G., Niccolini, G.: Fractal analysis of damage detected in concrete structural elements under loading. Chaos Soliton. Fract. **42**, 2047–2056 (2009)
9. Carpinteri, A., Lacidogna, G., Puzzi, S.: From criticality to final collapse: evolution of the b-value from 1.5 to 1.0. Chaos Soliton. Fract. **41**, 843–853 (2009)
10. Carpinteri, A., Lacidogna, G., Niccolini, G., Puzzi, S.: Morphological fractal dimension versus power-law exponent in the scaling of damaged media. Int. J. Damage Mech. **18**, 259–282 (2009)
11. Miroshnichenko, M., Kuksenko, V.: Study of electromagnetic pulses in initiation of cracks in solid dielectrics. Sov. Phys. Solid State **22**, 895–896 (1980)
12. Warwick, J.W., Stoker, C., Meyer, T.R.: Radio emission associated with rock fracture: possible application to the great Chilean earthquake of May 22, 1960. J. Geophys. Res. **87**, 2851–2859 (1982)
13. O'Keefe, S.G., Thiel, D.V.: A mechanism for the production of electromagnetic radiation during fracture of brittle materials. Phys. Earth Planet. Inter. **89**, 127–135 (1995)
14. Scott, D.F., Williams, T. J. and Knoll, S.J.: Investigation of electromagnetic Emissions in a deep underground mine. In: Proceedings of the 23rd International Conference on Ground Control in Mining, Morgantown, pp. 125–132. 3–5 August 2004
15. Frid, V., Rabinovitch, A., Bahat, D.: Fracture induced electromagnetic radiation. J. Phys. D Appl. Phys. **36**, 1620–1628 (2003)
16. Rabinovitch, A., Frid, V., Bahat, D.: Surface oscillations. A possible source of fracture induced electromagnetic oscillations. Tectonophysics **431**, 15–21 (2007)
17. Lacidogna, G., Carpinteri, A., Manuello, A., Durin, G., Schiavi, A., Niccolini, G., Agosto, A.: Acoustic and electromagnetic emissions as precursor phenomena in failure processes. Strain **47**(2), 144–152 (2011)
18. Carpinteri, A., Lacidogna, G., Manuello, A., Niccolini, A., Schiavi, A., Agosto, A.: Mechanical and electromagnetic emissions related to stress-induced cracks. Exp. Tech. **36**, 53–64 (2012)
19. Russell, J.E., Hoskins E.R.: Correlation of electrical resistivity of dry rock with cumulative damage. In: Proceedings of the 11th U.S. Symposium on Rock Mechanics (USRMS), Berkeley, 16–19 June 1969
20. Brace, W.F., Orange, A.S.: Electrical resistivity changes in saturated rocks during fracture and frictional sliding. J. Geophys. Res. **73**(4), 1433–1445 (1968)
21. Lacidogna, G., Carpinteri, A., Manuello, A., Niccolini, G., Agosto, A., Borla, O.: Acoustic emission and electrical properties of quasi-brittle materials under compression. In: Proceedings of SEM Annual Conference & Exposition on Experimental and Applied Mechanics, Uncasville, 13–16 June 2011, Paper No. 112

22. Carpinteri, A.: Chapter 1: TeraHertz phonons and piezonuclear reactions from nano-scale mechanical instabilities. In: Carpinteri, A., Lacidogna, G., Bertetto, A.M. (eds.) Acoustic, electromagnetic, neutron emissions from fracture and earthquakes. Springer, Heidelberg (2015)
23. Hagelstein, P.L., Chaudhary, I.U.: Anomalies in fracture experiments and energy exchange between vibrations and nuclei. Meccanica **50**, 1189–1203 (2015). doi:10.1007/s11012-014-9988-8
24. Widom, A., Swain, J., Srivastava, Y.N.: Neutron production from the fracture of piezoelectric rocks J. Phys. G: Nucl. Part. Phys. 40, 015006 (8pp) (2013)
25. Widom, A., Swain, J., Srivastava, Y.N.: Photo-disintegration of the iron nucleus in fractured magnetite rocks with magnetostriction. Meccanica **50**, 1205–1216 (2014). doi:10.1007/s11012-014-0007-x
26. Bridgman, P.W.: The breakdown of atoms at high pressures. Phys. Rev. **29**, 188–191 (1927)
27. Batzel, R.E., Seaborg, G.T.: Fission of medium weight elements. Phys. Rev. **82**, 607–615 (1951)
28. Fulmer, C.B., et al.: Evidence for photofission of iron. Phys. Rev. Lett. **19**, 522–523 (1967)
29. Carpinteri, A., Cardone, F., Lacidogna, G.: Piezonuclear neutrons from brittle fracture: early results of mechanical compression tests. Strain **45**, 332–339 (2009)
30. Cardone, F., Carpinteri, A., Lacidogna, G.: Piezonuclear neutrons from fracturing of inert solids. Phys. Lett. A **373**, 4158–4163 (2009)
31. Carpinteri, A., Cardone, F., Lacidogna, G.: Energy emissions from failure phenomena: mechanical, electromagnetic, nuclear. Exp. Mech. **50**, 1235–1243 (2010)
32. Carpinteri, A., Borla, O., Lacidogna, G., Manuello, A.: Neutron emissions in brittle rocks during compression tests: Monotonic vs. cyclic loading. Phys. Mesomech. **13**, 268–274 (2010)
33. Carpinteri, A., Lacidogna, G., Manuello, A., Borla, O.: Evidence of piezonuclear reactions: From geological and tectonic transformations to neutron detection and measurements. In: Proceedings of SEM Annual Conference & Exposition on Experimental and Applied Mechanics, Indianapolis, 7–10 June 2010, Paper No. 458
34. Carpinteri, A., Lacidogna, G., Manuello, A., Borla, O.: Energy emissions from brittle fracture: neutron measurements and geological evidences of piezonuclear reactions. Strength Fract. Complex. **7**, 13–31 (2011)
35. Carpinteri, A., Chiodoni, A., Manuello, A., Sandrone, R.: Compositional and microchemical evidence of piezonuclear fission reactions in rock specimens subjected to compression tests. Strain **47**(2), 267–281 (2011)
36. Carpinteri, A., Manuello, A.: Geomechanical and geochemical evidence of piezonuclear fission reactions in the Earth's crust. Strain **47**(2), 282–292 (2011)
37. Borla, O., Lacidogna, G., Di Battista, E., Niccolini, G., Carpinteri, A.: Multiparameter approach for seismic risk evaluation through environmental monitoring. Atti del 21 Congresso Nazionale di Meccanica Teorica ed Applicata (AIMETA), Torino, 2013, CD-ROM
38. Volodichev, N.N., Kuzhevskij, B.M., Nechaev, O. Yu., Panasyuk M., Podorolsky M.I.: Lunar periodicity of the neutron radiation burst and seismic activity on the Earth. In: Proceedings of the 26th International Cosmic Ray Conference, Salt Lake City, 17–25 August 1999
39. Kuzhevskij, M., Nechaev, O.Y., Sigaeva, E.A.: Distribution of neutrons near the Earth's surface. Nat. Hazards Earth Syst. Sci. **3**, 255–262 (2003)
40. Kuzhevskij, M., Nechaev, O.Y., Sigaeva, E.A., Zakharov, V.A.: Neutron flux variations near the Earth's crust. A possible tectonic activity detection. Nat. Hazards Earth Syst. Sci. **3**, 637–645 (2003)
41. Antonova, V.P., Volodichev, N.N., Kryukov, S.V., Chubenko, A.P., Shchepetov, A.L.: Results of detecting thermal neutrons at Tien Shan high altitude station. Geomagn. Aeron. **49**, 761–767 (2009)
42. Allen, R.: "Seconds before the big one" Seismology. ScientificAmerican.com, 54–59 (2011)
43. Padron, E., et al.: Changes on diffuse CO_2 emission and relation to seismic activity in and around El Hierro, Canary Islands. Pure Appl. Geophys. **165**, 95–114 (2008)
44. ISIDe Working Group: (INGV, 2015), Italian Seismological Instrumental and parametric database. http://iside.rm.ingv.it. Accessed February 2015

Chapter 22
Neutron Emissions from Hydrodynamic Cavitation

A. Manuello, R. Malvano, O. Borla, A. Palumbo, and A. Carpinteri

Abstract During the last few years, some investigators reported interesting results regarding neutron emissions from ultra-sonic cavitation in liquids and solids. In the present paper, the described experiments were conducted in order to evaluate neutron emissions from liquids subjected to hydrodynamic cavitation by an hydraulic circuit prototype. In particular, different aqueous iron salt solutions were tested in order to correlate neutron emissions and evolution of chemical element concentrations after different operating hours. The experiments were conducted using an hydraulic circuit fine-tuned by the authors. The pilot plant, which also includes the power supply and the electronic control of the recirculation pump, was realized entirely by plastic material (with the exception of the centrifugal pump and the hydraulic cavitator). The pump will be equipped with a system of six stages and with a maximum flow rate of 6 m^3/h. The maximum working pressure is equal to 10 bar. The evidence obtained during the tests returned an appreciable neutron emission, about 30 % greater than the background level. A significant decrement in Fe concentration was detected at the end of the test, whereas a considerable amount of aluminum—previously absent—was found on the internal walls of the pipe.

Keywords Hydrodynamic Cavitation • Turbulence • Neutron Emissions • Chemical Concentration

22.1 Introduction

Hydrodynamic cavitation has been increasingly used as a substitute of conventional acoustic (or ultrasonic) cavitation due to its efficiency in homogenizing, or mixing and breaking down, suspended particles in colloidal liquid compounds, such as paint mixtures or milk. A new kind of hydrodynamic horn is used in the present research in order to detect neutron emissions and chemical variations in the different Fe salt solutions. In this case a converging–diverging nozzle is used during the experiments.

From a general point of view, the use of cavitation for chemical processes is well known. The good mixing effect is attributed to the production of bubbles driven in liquids by pressure variations. Under the influence of pressure variations, gas bubbles in the medium undergo highly nonlinear amplitude and volume oscillations. In the expansion phase, liquid vapor diffuses into the bubble due to evaporation at gas–liquid interface. During the subsequent compression phase, pressure inside the bubbles increases and vapor starts to condense. However, depending on the nature of the pressure variations, the collapse can be so quick that the bubble wall velocity reaches or even exceeds the velocity of sound in the medium. At this time, the vapor present in the central core of the bubble has not sufficient time to escape. This "trapped" vapor is subjected to extreme conditions of temperature and pressure reaching the adiabatic collapse of the bubble. These conditions of temperature and pressure are adequate to cause the cleavage of vapour molecules. In this context, chemical reactions are induced in the medium bulk when the bubble contents mix with the surrounding liquid [1].

Several papers appeared in the last decades, reporting successful applications of hydrodynamic cavitation to physical and chemical processing. These include hydrolysis of fatty oils [2], polymerization and depolymerization of aqueous

A. Manuello (✉) • O. Borla • A. Carpinteri
Department of Structural, Geotechnical and Building Engineering, Politecnico di Torino, Corso Duca degli Abruzzi 24, Torino 10129, Italy
e-mail: amedeo.manuellobertetto@polito.it

R. Malvano
Department of Mechanical and Aerospace Engineering, Politecnico di Torino, Corso Duca degli Abruzzi 24, Torino 10129, Italy

A. Palumbo
Private Research, Torino, Italy

A.M. Beese et al. (eds.), *Fracture, Fatigue, Failure and Damage Evolution, Volume 8*, Conference Proceedings of the Society for Experimental Mechanics Series, DOI 10.1007/978-3-319-21611-9_22

polymeric solutions [3], microbial cell disruption [4–7], wastewater treatment [8–10], oxidation of arylalkanes [11,12], water disinfection [12–15], synthesis of nano-structured catalysts [16,17]. A state-of-the-art review of literature published in the area of hydrodynamic cavitation reactors is given by Shah et al. [18] and Gogate and Pandit [19].

Due to the various effects on fluid power systems, cavitation is normally to be avoided as much as possible in many cases. When actions for preventing cavitation are considered, it is essential to recognise the existence of cavitation and the location of cavitation inception point. The problem of cavitation detection can be solved directly only by verifying the existence of cavities and bubbles. Direct detection is performed by observing visually the population of developed bubbles in flow passages. However, fluid power components encompass usually complicated constructions and cavitation can occur in various locations where the access for visualisation instruments is limited. In addition, the very high speed of the cavitation phenomenon makes the task difficult. Under the light of the experiment proposed in this paper, the authors investigated the possibility to check the cavitation activation in correlation to the monitoring of neutron and alpha particle emissions during the experiments.

As stated before, when cavities are carried to higher-pressure regions, they implode violently and high-pressure shock waves can occur contributing to give a suitable environmental condition to induce also in the solution the anomalous nuclear reactions already encountered in solids (piezonuclear reactions) [20–24].

The hydrodynamic cavitation performed in the present experiments is due to the variation in pressure of the flux that occurs when the speed of the fluid itself changes due to a geometric variation in the duct. For instance, the flux in exit from an orifice increases its velocity when pressure decreases. If the pressure decreases until the value of the vapour pressure, a number of bubbles is generated as a result of the partial vaporization of the liquid. When pressure increases again, the collapse of the bubbles occurs. The hydrodynamic cavitation, in fact, permits to obtain a very high energy density in a local portion of the volume, as well as high pressure and temperature of the fluid.

The hydrodynamic cavitation reactor used in this work is a novel reactor that uses a converging–diverging nozzle for creating pressure variations in the flow necessary for driving bubble motion. This kind of reactor is very suitable to control several parameters such as the bubble nucleation rate. During the experiments, different solutions were used. In particular, pure water and water with iron salts were analysed in experiments with different time durations. During these tests, neutron emissions were detected by thermodynamic dosimeters and by a He^3 proportional counter. Alpha particle emissions were also analysed by a 6150AD-k probe, and the chemical analysis of the solutions before and during the experiments were performed by the inductively coupled plasma mass spectrometry (ICP-MS).

22.2 Experimental Set Up and Measurement Devices

The hydraulic circuit was design and developed to include the cavitation reactor in order to experiment the cavitation in pure water and in aqueous solutions with iron salts. The system provides a constant water flow by means of a centrifugal pump. In order to inspect the water before and after the cavitation, a plastic tank was placed upstream of the pump. Due to the high level of energy expected during piezonuclear reactions, each material of each component has been carefully selected basing on its chemical stability and in order to avoid any kind of contamination. The water flows into the centrifugal pump passing trough a hole in the tank by gravity; then the water is entered in the circuit to be cavitated by the reactor, see Fig. 22.1.

The ducts are made of high density polyethylene (according to PE100 UNI EN 12201—EN 12201). The section of the ducts is circular with an inner diameter of 30 mm. The pressure of the water is measured just before the reactor by means of an analogical pressure gauge (fs = 16 bar). The overall length of the circuit it is about 7.1 m.

Each test was executed twice, by introducing into the hydraulic circuit both the pure water and aqueous solutions of iron nitrate. The concentration of iron nitrate in the pure water was 2.3 mg/L. Both the liquids were provided by the Analytical Chemistry Department of the University of Turin. Each experiment lasted several hours and the pressure was maintained at about 9 atm.

The reactor is a steel hydraulic component with a overall length of 17 cm, see Fig. 22.2. It is made up of two sections of different pipes. Considering the flow direction, the first pipe, characterized by a larger circular section (33 mm, external diameter; 27.3 mm, inner one), is about 80 mm long. The second pipe has got a smaller circular section (26.5 mm, external diameter; 10 mm, inner one) and is about 80 mm long. An intermediate connection between the two pipes is provided in order to gradually reduce the section: it consists in a short segment (10 mm long) having the shape of a truncated cone with its generatrices inclined of 45°. The reactor features an external input source so that a desired gas or liquid can be introduced in the flow just after the first segment. The sparger comprises of a small tube. After many preliminary test it was shown that best results in terms of nucleation rate occurs if no gas or liquid is introduced through the sparger. We fixed a first point in the experimental set-up: the tap of the sparger was kept closed.

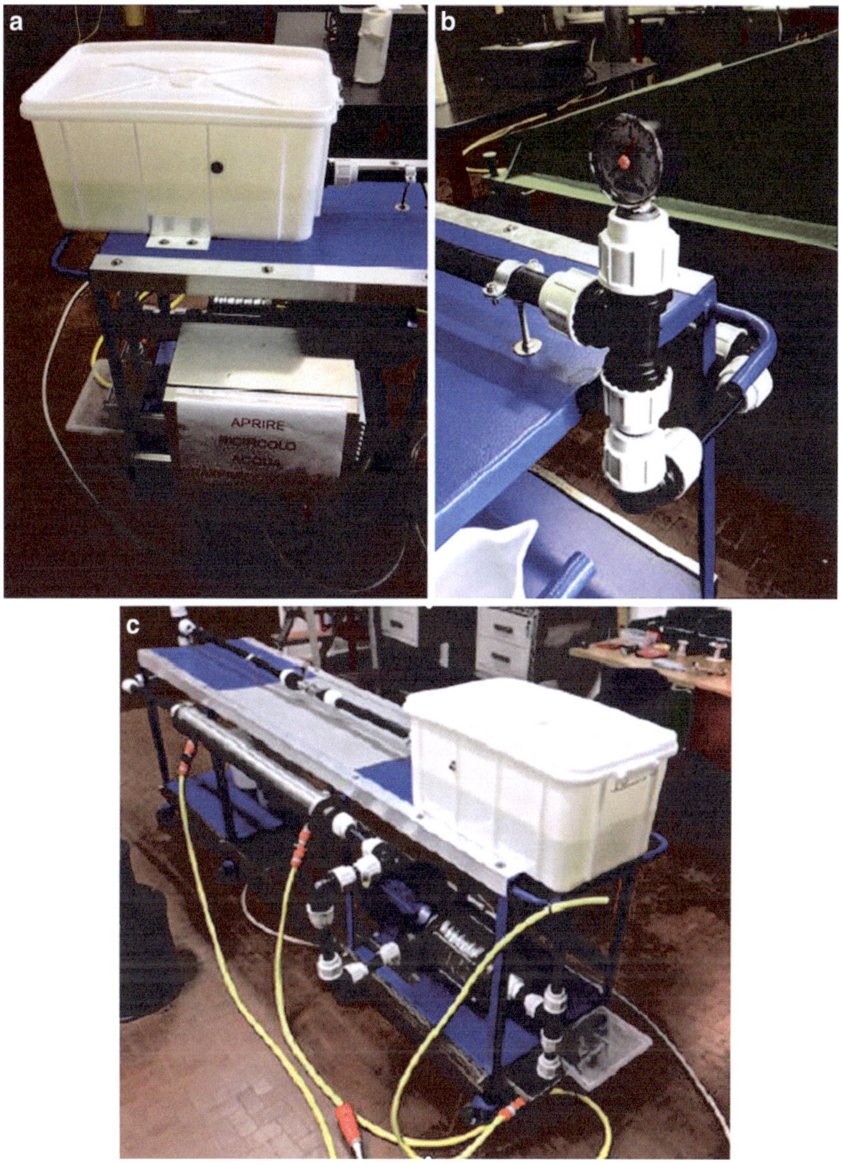

Fig. 22.1 The tank and the pump equipped with an inverter installed in the laboratory (**a**). The analogical pressure gauge as it appears installed in the circuit (**b**). The complete circuit with a coaxial water cooling system in order to maintain the temperature constant (**c**)

Fig. 22.2 Hydrodynamic cavitation reactor installed in the circuit

The terminal section of the sparger (external diameter of 5.9 mm) starts from the beginning of the second pipe of the reactor and its length is 11.4 mm. Its external shape (cylinder), together with the internal profile of the second pipe and with the two connecting elements, provide the main turbulence to the flow. The shape of the two elements that provide

Fig. 22.3 Alpha particle and neutron emission detection devices close to the reactor during the experiment. (**a**) Lateral view and (**b**) Frontal view

the connection between the sparger and the second pipe has been studied. The flow swirls with a helical twist along the connection surface and, due to this further diversion, the cavitation takes place (nucleation). A third turbulence source is provided to the flow by the presence of six peculiar half-spheres called "mushroom heads". Several conjectures were formulated about the effects these half-spheres have on the flow and a numerical analysis has been developed to this purpose. It is opinion of the author that this third turbulence source could make some of the bubbles to explode (see Fig. 22.2).

For what concerns the neutron emission measurements, since neutrons are electrically neutral particles, they cannot directly produce ionization in a detector, and therefore cannot be directly detected. This means that neutron detectors must rely upon a conversion process accounting the interaction between an incident neutron and a nucleus, which produces a secondary charged particle. Such charged particle is then detected and the neutron's presence is revealed from it. For an accurate neutron evaluation a He^3 proportional counter was employed. The detector used in the tests is a He^3 type (Xeram, France) with pre-amplification, amplification, and discrimination electronics directly connected to the detector tube. The detector is supplied with a high voltage power (about 1.3 kV) via NIM (Nuclear Instrument Module). The logic output producing the TTL (transistor–transistor logic) pulses is connected to a NIM counter. The logic output of the detector is enabled for analog signals exceeding 300 mV. This discrimination threshold is a consequence of the sensitivity of the He^3 detector to the gamma rays ensuing neutron emission in ordinary nuclear processes. This value has been determined by measuring the analog signal of the detector by means of a Co-60 gamma source. The detector is also calibrated at the factory for the measurement of thermal neutrons; its sensitivity is 65 cps/$n_{thermal}$ ($\pm$10 % declared by the factory), i.e., the flux of thermal neutrons was one thermal neutron/s cm^2, corresponding to a count rate of 65 cps.

For the alpha particle emission, a 6150AD-k probe a sealed proportional counter was used, which does not require refilling or flushing from external gas reservoirs. The probe is sensitive to alpha, beta, and gamma radiation. An electronic switch allows for the operating mode "alpha" to detect alpha radiation only, such that in this mode the radiation recognition is very sensitive because the background level is much lower. A removable discriminator plate (stainless steel, 1 mm) distinguishes between beta and gamma radiation detection. An adjustable handle can be locked to the most convenient orientation. During the experiments the 6150AD-k probe was used in the operating mode alpha to monitor the background level before and after the switching on of the cell (see Fig. 22.3). Finally, inductively coupled plasma mass spectrometry (ICP-MS) is a type of mass spectrometry which is capable of detecting metals and several non-metals at concentrations as low as one part in 1015 (part per quadrillion, ppq) on non-interfered low-background isotopes. This is achieved by ionizingthe sample with inductively coupled plasma and then using a mass spectrometer to separate and quantify those ions.

22.3 Neutron Emission Measurements

In Fig. 22.4(a–f) results coming from the neutron and alpha particle emissions in the cases of four different tests are reported.

In Fig. 22.4a the neutron emissions acquired by thermodynamic dosimeters are shown for the first experiment characterized by a duration of 32 h. In this case the equivalent neutron dose presented a maximum value two times greater

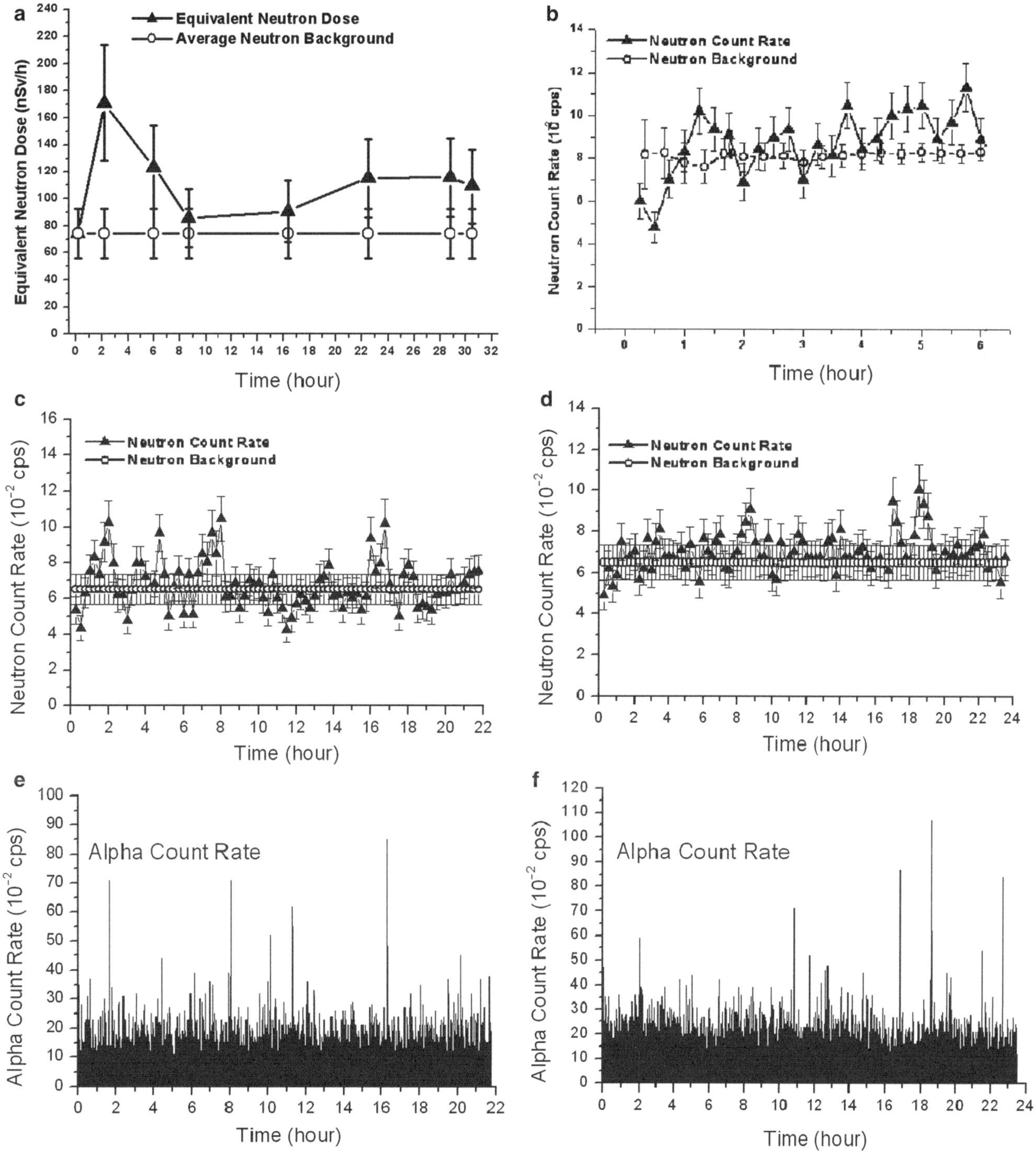

Fig. 22.4 Neutron emission results for test durations of 32 (**a**), 6 (**b**), 22 (**c**) and 24 (**d**) hours. The alpha particle emissions for the last two test are reported (**e**, **f**). This kind of emissions can be compared to the neutron emissions acquired for the same tests

than the background level. In Fig. 22.4b very similar results are shown in the case in which the neutron emissions is obtained by the He^3 proportional counter. The time duration of the experiments is equal to 6 h. In this particular case, after 2 h from the beginning of the experiment a neutron emission greater than the background level has been recognized. At the end of the experiments an emission of about two times the background level is observed.

In Fig. 22.4c, d neutron emissions behaviour have been reported for two tests with a total duration of 22 and 24 h. In these two cases different impulsive emissions with an amplitude of about two times the background level may be observed. In the

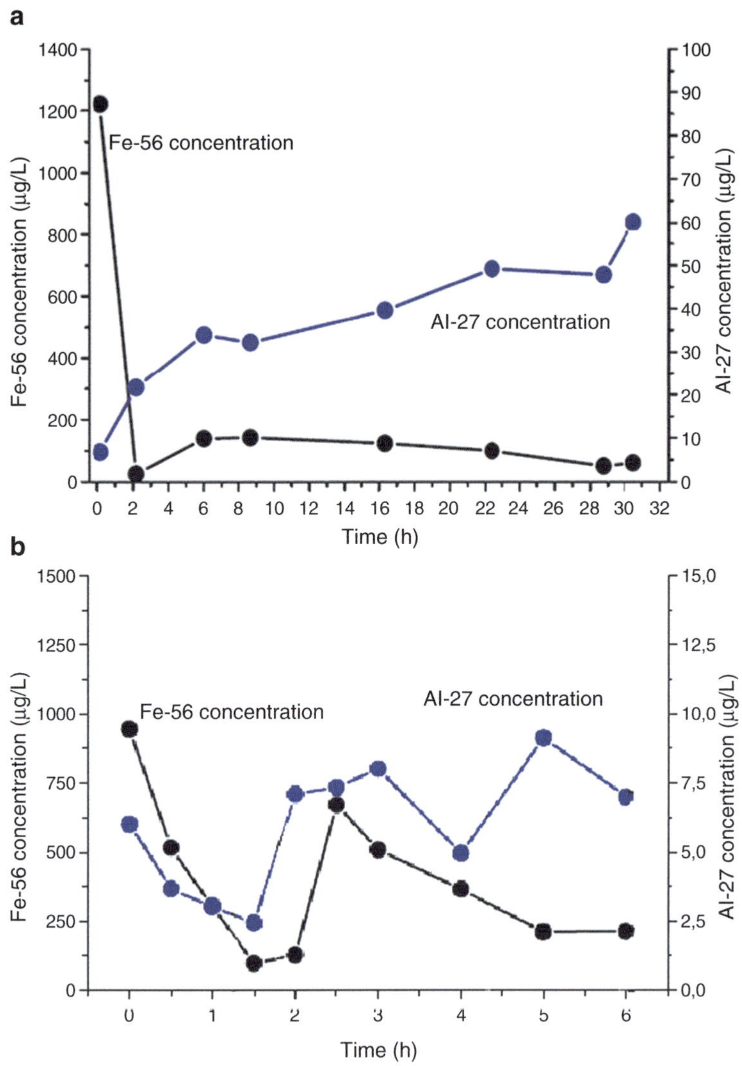

Fig. 22.5 (ICP-MS) analysis performed on solution samples taken during the tests in the case of two experiments with a duration of 32 h (**a**) and 6 h (**b**) respectively

case of Fig. 22.4c these emissions can be recognised after 2, 4, 8 and 16 h from the beginning of the experiment. In the case of Fig. 22.4d the neutron emissions can be recognised after 4, 8, 16 and 18 h from the beginning of the experiment. According to these evidence it is also interesting to make a comparison between neutron emissions and alpha particle emissions reported in Fig. 22.4e, f for these two experiments. It easy to observe that the most important event in the alpha particle rate can be correlated to the neutron emissions acquired during the same test. In particular, the comparison of the two kind of emissions for the same test: Fig. 22.4c–f denounced an overlapping between the two kind of detections. This last data correlation seems to be very important under the light of piezonuclear reaction hypothesis recently proposed in other kind of experiments [20–24].

22.4 Chemical Analysis of the Cavitated Solutions: Preliminar Remarks

In Fig. 22.5a, b (ICP-MS), the analyses are reported for the tests with a duration of 32 and 6 h, respectively. The corresponding neutron emissions evidence for these experiments are reported in Fig. 22.4a, b. In the test reported in Fig. 22.5a, the liquid solution samples are taken and analysed every 2 h. In the second case, in order to obtain additional information about the first stage, the samples are taken for the analysis every 30 min.

It is evident that the results coming from the (ICP-MS) technique describe Fe decrement just at the beginning of each test. At the same time, Al increased up to six times the initial value (see Fig. 22.5a, b). It is also important to observe that the concentration of iron nitrate in pure water changes totally in correspondence to neutron emissions. Furthermore, neutron emission has been highlighted by the two different devices adopted: bubble detectors and He^3 proportional counter. In the second test, analysed by The (ICP-MS) technique, a cyclic behaviour can be observed (Fig. 22.5b). According to this evidence, a sliding increase is detected from the fourth hour till the end of the test. In this case, the overall average rate of the deviation between the two neutron emission measurements is about +29 % (Fig. 22.5b).

22.5 Conclusions

Neutron emissions up to two times higher than the background level were observed during the operating time of an hydraulic prototype circuit. In particular, during two of these experiments, the hydrodynamic effect of a converging–diverging nozzle horn seems to be correlated to neutron and alpha particle emissions.

By the ICP-MS analysis performed on the solutions used in the test, evident variations in Fe and Al concentrations were be observed. The last evidence appears to be analogous to the results obtained by Carpinteri et al. from fracture, fatigue, and ultrasonic tests on rocks [20–24]. Further tests will be conducted in order to evaluate more in details these chemical variations. In particular, additional tests will be carried out in order to detect also other elements in addition to Fe and Al. In particular, some devices will be implemented to exclude the deposition of Fe on the internal circuit surface and to correlate directly the Fe decrement to the increment in lighter elements, such Al, under the light of the piezonuclear assumption.

Acknowledgements The authors gratefully acknowledge the economical support of Metalwork srl (Brescia-Italy). Special thanks are due to Prof. C. Baiocchi and Dr. G. Mariella for the solution preparations and the ICP-MS analysed before and during the experimental campaign. Mr. F. Alasia is gratefully acknowledged for his support in the hydraulic circuit construction.

References

1. Yasui, K., Tuziuti, T., Sivakumar, M., Iida, Y.: Theoretical study of single-bubble sonochemistry. J. Chem. Phys. **122**(224706), 1–12 (2005)
2. Joshi, J.B., Pandit, A.B.: Hydrolysis of fatty oils: effect of cavitation. Chem. Eng. Sci. **48**, 3440–3442 (1993)
3. Chivate, M.M., Pandit, A.B.: Effect of sonic and hydrodynamic cavitation on aqueous polymeric solutions. Indian Chem. Eng. **35**, 52–57 (1993)
4. Save, S.S., Joshi, J.B., Pandit, A.B.: Microbial cell disruption—role of cavitation. Chem. Eng. J. **55**(3), B67–B72 (1994)
5. Save, S.S., Joshi, J.B., Pandit, A.B.: Microbial cell disruption in hydrodynamic cavitation. Chem. Eng. Res. Des. **75**(Part C), 41–49 (1997)
6. Balasundaram, B., Pandit, A.B.: Selective release of invertase by hydrodynamic cavitation. Biochem. Eng. J. **8**, 251–256 (2001)
7. Balasundaram, B., Pandit, A.B.: Significance of location of enzymes on their release during microbial cell disruption. Biotechnol. Bioeng. **75**, 607–614 (2001)
8. Kalumuck, K.M., Chahine, G.L.: The use of cavitating jets to oxidize organic compounds in water. J. Fluids Eng. **122**, 464–470 (2000)
9. Sivakumar, M., Pandit, A.B.: Wastewater treatment: a novel energy efficient hydrodynamic cavitational technique. Ultrason. Sonochem. **9**, 123–131 (2002)
10. Ambulgekar, G.V., Samant, S.D., Pandit, A.B.: Oxidation of alkylarenes using aqueous potassium permanganate under cavitation: comparison of acoustic and hydrodynamic techniques. Ultrason. Sonochem. **11**, 191–196 (2004)
11. Ambulgekar, G.V., Samant, S.D., Pandit, A.B.: Oxidation of alkylarenes using aqueous potassium permanganate under cavitation: comparison of acoustic and hydrodynamic techniques. Ultrason. Sonochem. **12**, 85–90 (2005)
12. Botha, C.J., Buckley, C.A.: Disinfection of potable water: the role of hydrodynamic cavitation. Water Supply **13**(2), 219–229 (1995). IWSA International Specialized Conference on Disinfection of Potable Water, 1994
13. Jyoti, K.K., Pandit, A.B.: Hybrid cavitation methods for water disinfection: simultaneous use of chemicals with cavitation. Ultrason. Sonochem. **10**, 255–264 (2003)
14. Jyoti, K.K., Pandit, A.B.: Effect of cavitation on chemical disinfection efficiency. Water Res. **38**, 2249–2258 (2004)
15. Jyoti, K.K., Pandit, A.B.: Ozone and cavitation for water disinfection. Biochem. Eng. J. **18**, 9–19 (2004)
16. Moser, W.R., Marshik, B.J., Kingsley, J., Lemberger, M., Willette, R., Chan, A., Sunstrom, J.E., Boye, A.: The synthesis and characterization of solid-state materials produced by high shear-hydrodynamic cavitation. J. Mater. Res. **10**, 2322–2335 (1995)
17. Find, J., Emerson, S.C., Krausz, I.M., Moser, W.R.: Hydrodynamic cavitation as a tool to control macro-, micro-, and nano-properties of inorganic materials. J. Mater. Res. **16**, 3503–3513 (2001)
18. Shah, Y.T., Pandit, A.B., Moholkar, V.S.: Cavitation reaction engineering. Kluwer Academic/Plenum Publishers, New York (1999)
19. Gogate, P.R., Pandit, A.B.: Hydrodynamic cavitation reactors: a state of the art review. Rev. Chem. Eng. **17**, 1–85 (2001)
20. Carpinteri, A., Cardone, F., Lacidogna, G.: Piezonuclear neutrons from brittle fracture: Early results of mechanical compression tests. Strain **45**, 332–339. (2009). Atti dell'Accademia delle Scienze di Torino. 33, 27–42, (2009)

21. Cardone, F., Carpinteri, A., Lacidogna, G.: Piezonuclear neutrons from fracturing of inert solids. Phys. Lett. A **373**, 4158–4163 (2009)
22. Carpinteri, A., Cardone, F., Lacidogna, G.: Energy emissions from failure phenomena: mechanical, electromagnetic, nuclear. Exp. Mech. **50**, 1235–1243 (2010)
23. Carpinteri, A., Lacidogna, G., Manuello, A., Borla, O.: Piezonuclear fission reactions: evidences from microchemical analysis, neutron emission, and geological transformation. Rock Mech. Rock. Eng. **45**, 445–459 (2012)
24. Carpinteri, A., Lacidogna, G., Manuello, A., Borla, O.: Piezonuclear fission reactions from earthquakes and brittle rocks failure: evidence of neutron emission and nonradioactive product elements. Exp. Mech. **53**, 345–365 (2013)

Chapter 23
From Dark Matter to Brittle Fracture

P.C.F. Di Stefano, C. Bouard, S. Ciliberto, S. Deschanel, O. Ramos, S. Santucci, A. Tantot, L. Vanel, and N. Zaïm

Abstract Prompted by the intriguing results obtained by some of the rare-event searches looking for the dark matter that may make up the bulk of the matter in the Universe, we have studied brittle fracture as a background in scintillation detectors. Under conditions of ambient temperature and pressure, we have demonstrated a correlation between fracture, acoustic emission, and emission of light in several common scintillators. We present early results from an improved setup. When commissioned, it will provide additional channels to study these phenomena, in controllable atmospheres.

Keywords Dark matter • Mechanoluminescence • Scintillator • Brittle Fracture

Our interest in fractoluminescence has its origins in astroparticle physics and particle physics. Since the astronomical observations of F. Zwicky in the 1930s [1], confirmed at many different scales and with many different techniques, astronomers have known that most of the matter in the universe only appears through its gravitational interactions. This is referred to as the mystery of dark matter. Over the years, various attempts have been made to solve this problem. One well-motivated class of candidates is Weakly Interacting Massive Particles (WIMPs) [2]. These particles have the advantage that they appear, independently of astronomy, in many of the extensions to the standard model of particle physics. Evidencing the WIMPs that may be around us would therefore solve a longstanding astrophysical problem, and also open a whole new region of particle physics. The direct detection method is to build a detector and wait for the WIMPs to interact in them. This seemingly simple program is unfortunately a huge experimental challenge because, first, the energies that WIMPs will deposit in a detector are quite small, with a typical scale of the order of keV to tens of keV. Second, the rate of WIMP interactions is tiny, less than one per month per kilogram of detector, which is tiny compared to typical ambient levels of radioactivity, for instance that of a human body which is of hundreds of radioactive decays per second and per kilogram. Experiments looking for WIMPs therefore take draconian steps to eliminate all forms of background, including careful screening of all materials to ensure radiopurity, specialized shielding against ambient radioactivity, working in deep underground labs to escape cosmic radiation [3], and developing detectors with some form of particle identification to separate the rare signal from the abundant backgrounds. In the late 1990s, the CRESST dark matter search deployed sapphire crystals cooled to $\sim$10 mK. The detection principle was that particles interacting in them would create a measurable rise in temperature; the attained energy thresholds were less than 1 keV [4]. In the first phase, an unexpectedly high background was encountered. After troubleshooting, this background was identified as originating in small cracks propagating in the crystals because of unintentional indentation at the mounting points, and was eliminated by modifying the contacts between the crystal and its environment. The analysis of these data yielded a very rich data set of precisely calibrated fracture energies down to a values equivalent to a few hundred bonds. Results included demonstrating the Gutenberg-Richter law of fracture energy distribution holds at energies thirty six orders of magnitude below those for a large earthquake, and evidencing Omori's law at similarly small scales [5].

P.C.F. Di Stefano (✉) • C. Bouard • S. Deschanel • A. Tantot • L. Vanel • N. Zaïm
Department of Physics, Queen's University, Kingston, ON, Canada K7L 3N6
e-mail: philippe.distefano@gmail.com

O. Ramos
Institut Lumière Matière, UMR5306 Université Lyon 1-CNRS, Université de Lyon, Villeurbanne Cedex F-69622, France

S. Ciliberto • S. Santucci
Laboratoire de Physique, Université de Lyon, Ecole Normale Supérieure de Lyon,
Centre National de la Recherche Scientifique UMR 5672, 46 Allée d'Italie, Lyon Cedex 07 F-69364, France

S. Deschanel
MATEIS, UMR 5510 CNRS and INSA-Lyon, Université de Lyon, Villeurbanne F-69621, France

A.M. Beese et al. (eds.), *Fracture, Fatigue, Failure and Damage Evolution, Volume 8*, Conference Proceedings of the Society for Experimental Mechanics Series, DOI 10.1007/978-3-319-21611-9_23

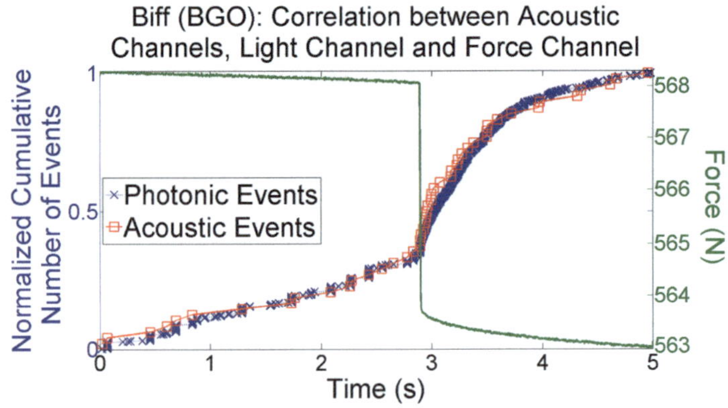

Fig. 23.1 Correlation between fracture, acoustic emission, and emission of light during the rupture of a DCDC sample of scintillator BGO (from Ref. [SOUND])

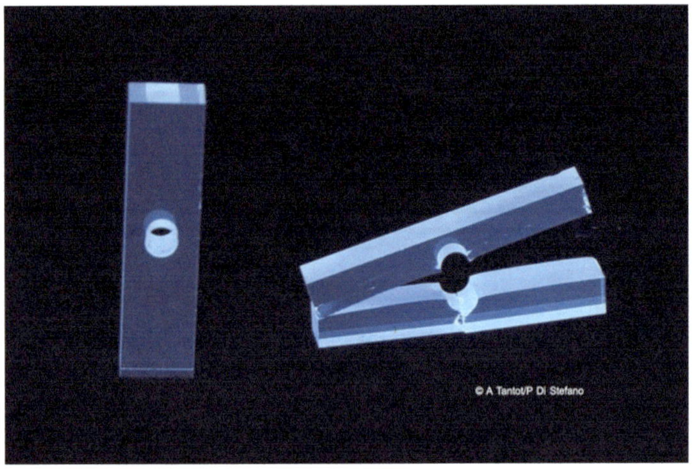

Fig. 23.2 Intact and broken DCDC samples of scintillator BGO exhibiting luminescence under X-ray illumination. Sample dimensions are $20 \times 5 \times 3$ mm^3. Broken sample shows main fracture along long axis, and a secondary fracture that occurred after the main one when force was kept after main fracture, along one of the short axes

More recently, several dark matter searches have observed intriguing events incompatible with known backgrounds [6–9], in tension with the null results of other experiments [10–16]. We have therefore decided to study fracture as a potential background in rare-event searches, in particular as a source of light in the experiments deploying scintillating detectors. Such devices convert the energy deposited by an interacting particle into light that can be detected by photomultipliers (PMTs) or cryogenic detectors for instance. Detectors of this type are widespread in particle and nuclear physics [17]. For our purpose, after various tests, we converged on scintillator samples compressed in the double-cleavage drilled compression (DCDC) geometry. The samples were $20 \times 5 \times 3$ mm^3 rectangular prisms, polished to optical quality, with a 1 mm diameter circular hole drilled perpendicularly in the middle of the 20×5 mm^2 face. Under compression along the long axis, cracks formed reproducibly on either side of the hole in the plane parallel to the 20×3 mm^2 face. Applied force and acoustic emission (AE) were monitored; the emitted light was also recorded using a PMT. All channels were recorded simultaneously, at 1 Gsample/s for the light channel and at 10 Msample/s for the others. Our results evidenced a clear correlation between the rupture of various common scintillators (Bi4Ge3O12 aka BGO, ZnWO4 and CdWO4), AE and light emission, in various mechanical configurations at room temperature and ambient pressure as shown in Fig. 23.1 taken from [18]. We have also demonstrated that the spectrum of emitted light is compatible with that of the final stage of the scintillation process, itself linked to transitions in the Bi3+ ion. Because in other cases emission of charged and neutral particles from fracture surfaces has been observed [19], we hypothesise that the mechanism is as follows: as the fracture advances, fractoemitted particles from one surface interact in the other as they do in a scintillator, emitting light. Other mechanisms may be possible, including electric arcs forming, and causing the BGO to luminesce, or some direct stimulation of the Bi3+ ions. Indeed, DCDC BGO samples excited by UV light will luminesce as shown in Fig. 23.2. Using the precise energy calibration of the light channel possible with radioactive sources, we were also able to set a lower limit on the fraction of elastic energy converted to light. We find a value of 3×10^{-5}, limited by saturation of the light detection channel.

To better understand the mechanisms involved, and avoid saturation of the light detector, we are commissioning an improved setup in the form of a vacuum chamber that can be evacuated or filled with various gases [20]. Other improvements include a distance sensor to monitor the actual compression of the sample, an infrared (IR) camera to track the progression of the fracture in the DCDC sample, filters on the PMT to block IR light and also to reduce saturation from the fracture itself, and a new streaming DAQ that allows resolution of individual photons. An IR picture of a sample being broken in the dark with the PMT running is shown in Fig. 23.3. Offline image processing allows to track the progression of the crack. Lastly, the DAQ controlling the actuator which imposes the force on the sample has been upgraded to allow regulation at a constant

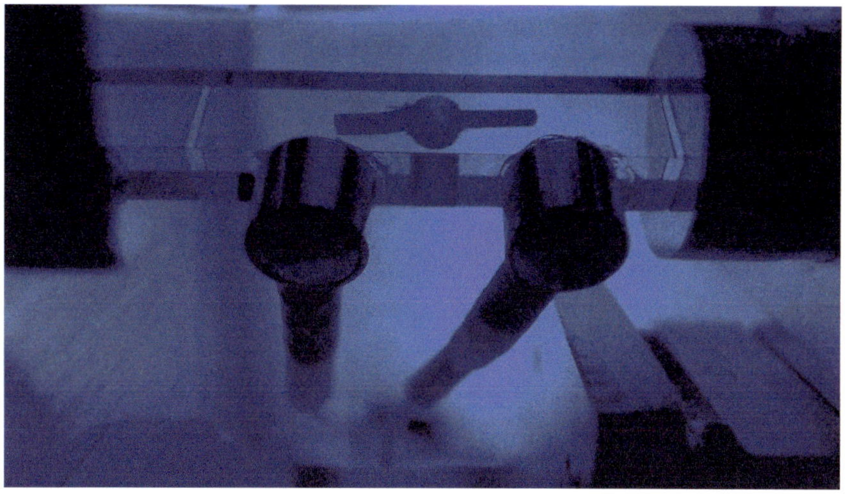

Fig. 23.3 Propagation of crack in DCDC sample as visualized with camera under IR lighting. IR lighting is chosen so as not to interfere with the PMT which is sensitive to visible wavelengths. Offline data analysis allows to track propagation of crack tip. Sample is squeezed between fixed backstop and movable pushrod. The two cylindrical objects in foreground are AE sensors

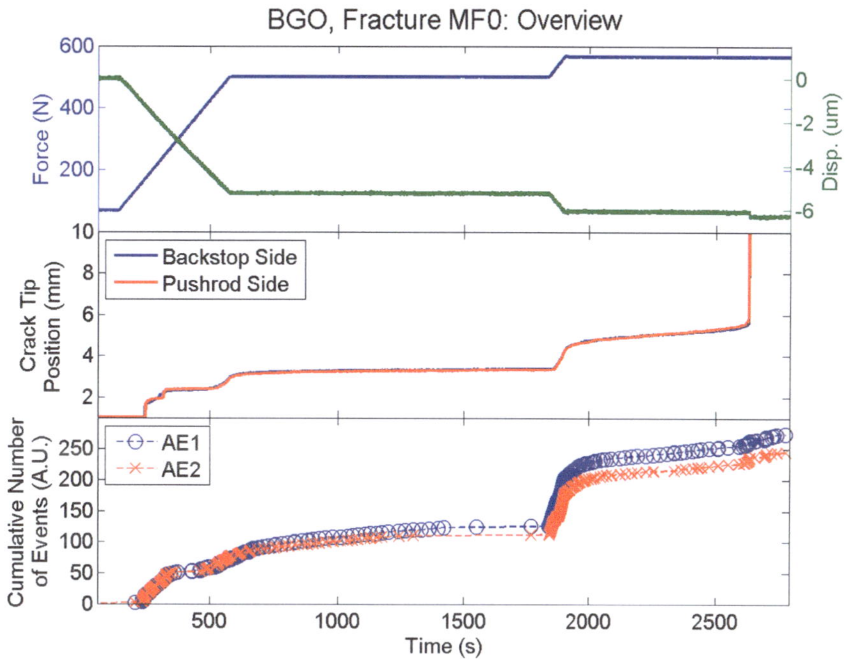

Fig. 23.4 Partial preliminary results from new setup, in vacuum at ambient temperature. Force is ramped up to two different values, allowing the crack to propagate freely the rest of the time. Propagation of the crack is slow, until critical length of fracture is reached, whereupon fracture accelerates. Main failure of sample occurs around 2600 s from start. A strong correlation between the acoustic emission and the fracture propagation is observed

force, enabling studies of sub-critical fracture for instance. Figure 23.4 shows an example from a sample broken at ambient temperature in vacuum. Over the course of nearly an hour, the force was first set at 500 N while the fracture was allowed to propagate freely and slowly, then ramped up to 565 N, and the fracture propagated freely on its own again, slowly at first before eventually accelerating as expected in the DCDC geometry, and destroying the sample. We observe an excellent correlation between the fracture propagation and the AE. The light channel is not presented for this data set.

In conclusion, we are commissioning an improved setup that will allow us to study brittle fracture in scintillators under various atmospheres with multiple channels: compression distance, crack length, force, acoustic emission and light emission. We expect that this work will be beneficial to development of detectors for rare-event searches, and may lead to insight into fracture mechanisms.

References

1. Zwicky, F.: Die rotverschiebung von extragalaktischen nebeln. Helv. Phys. Acta **6**, 110 (1933)
2. Schnee, R.W.: Physics of the Large and Small. In Csaki, C., Dodelson, S. (eds.) Proceedings of the 2009 Theoretical Advanced Study Institute in Elementary Particle Physics, p. 629. World Scientific, Singapore (2010)

3. Smith, N.J.T.: The SNOLAB deep underground facility. Eur. Phys. J. Plus **127**, 108 (2012)
4. Angloher, G., et al.: Limits on WIMP dark matter using sapphire cryogenic detectors. Astropart. Phys. **18**, 43–55 (2002)
5. Åström, J., et al.: Fracture processes observed with a cryogenic detector. Phys. Lett. A **356**, 262–266 (2006)
6. Agnese, R., et al.: Silicon detector dark matter results from the final exposure of CDMS II. Phys. Rev. Lett. **111**, 251301 (2013). CDMS Collaboration
7. Aalseth, C.E., et al.: CoGeNT: a search for low-mass dark matter using p-type point contact germanium detectors. Physical Review D **88**, 012002 (2013). CoGeNT Collaboration
8. Angloher, G., et al.: Results from 730 kg days of the CRESST-II dark matter search. Eur. Phys. J. C **22**, 1971 (2012). CRESST-II Collaboration
9. Bernabei, R., et al.: First results from DAMA/LIBRA and the combined results with DAMA/NaI. Eur. Phys. J. C **56**, 333–355 (2008)
10. Aprile, A., et al.: Dark matter results from 225 live days of XENON100 data. Phys. Rev. Lett. **109**, 181301 (2012)
11. Ahmed, Z., et al.: Dark matter search results from the CDMS II experiment. Science **327**, 1619–1621 (2010). CDMS II Collaboration
12. Agnese, R., et al.: Search for low-mass weakly interacting massive particles with SuperCDMS. Phys. Rev. Lett. **112**, 241302 (2014). SuperCDMS Collaboration
13. Agnese, R., et al.: Search for low-mass weakly interacting massive particles using voltage-assisted calorimetric ionization detection in the SuperCDMS experiment. Phys. Rev. Lett. **112**, 041302 (2014). SuperCDMS Collaboration
14. Armengaud, E., et al.: Search for low-mass WIMPs with EDELWEISS-II heat-and-ionization detectors. Phys. Rev. D. **86**, 051701(R) (2012). EDELWEISS Collaboration
15. Akerib, D.S., et al.: First results from the LUX dark matter experiment at the sanford underground research facility. Phys. Rev. Lett. **112**, 091303 (2014). LUX Collaboration
16. Archambault, S., et al.: Constraints on low-mass WIMP interactions on 19F from PICASSO. Phys. Lett. B **711**, 153–161 (2012). PICASSO Collaboration
17. Knoll, G.F.: Radiation detection and measurement, 3rd edn. Wiley, New York (2000)
18. Tantot, A., et al.: Sound and light from fractures in scintillators. Phys. Rev. Lett. **111**, 154301 (2013)
19. Langford, S.C., et al.: Simultaneous measurements of the electron and photon emission accompanying fracture of single-crystal MgO. J. App. Phys. **62**, 1437 (1987)
20. Tantot, A.: in preparation

Chapter 24
Compositional Variations in Palladium Electrodes Exposed to Electrolysis

A. Carpinteri, O. Borla, A. Goi, S. Guastella, A. Manuello, R. Sesana, and D. Veneziano

Abstract Literature presents several cases of nuclear anomalies occurring in condensed matter, during fracture of solids, cavitation of liquids, and electrolysis experiments.

Previous papers by the authors have recently shown that, on the surface of the electrodes exposed to electrolysis visible cracks and compositional changes are strictly related to nuclear particle emissions. In particular, a mechanical interpretation of the phenomenon was provided accounting to the hydrogen embrittlement effects. Piezonuclear reactions were considered responsible for the neutron and alpha particle emissions detected during the electrolysis. Such effects are thoroughly studied in a new experimental campaign, where three pure palladium (100 % Pd) cathodes coupled with Ni anodes are used for electrolysis, separately exposed to processes of different duration: 2.5 h, 5 h and 10 h, respectively. In this paper, the authors intend to show the new results concerning the changes on the surface of the electrodes in terms of composition and presence of cracks after the electrolytic process. Measures of heat generation as well as of neutron emission will be reported.

Keywords Hydrogen embrittlement • Cold fusion • Electrolysis • Piezonuclear reactions • Neutron emission • Energy balance

24.1 Introduction

Several evidences of anomalous nuclear reactions occurring in condensed matter were observed by different authors [1–34]. These experiments are characterized by extra heat generation, neutron emission, and alpha particle detection. Some of these studies, using electrolytic devices, reported also significant evidences of compositional variation after the microcraking of the electrodes. At the same time, recent experiments provided evidence of piezonuclear reactions occurring in condensed matter during fracture of solids, cavitation of liquids, and electrolysis experiments [35–40]. These experiments were characterized by significant neutron and alpha particle emissions, together with appreciable variations in the chemical composition. Based on these experimental evidences, a mechanical reason for the so-called Cold Nuclear Fusion was recently proposed by the same authors [41–43]. The hydrogen embrittlement due to H atoms produced by the electrolysis itself seems to play an essential role for the observed microcracking in the electrode host metals (Pd, Ni, Fe, etc.). Consequently, the hypothesis recently proposed is that piezonuclear fission reactions may occur in correspondence to the microcrack formation [41–45].

Classical experiments, belonging to the so called "Cold Fusion" research, are characterized by a generated heat several times greater than the input energy. In some cases, neutron emission rate, during electrolysis, was measured to be about three times the natural background level [6]. In 1998, Mizuno presented the results of the measurements conducted by means of neutron detectors and compositional analysis techniques related to different electrolytic experiments. According to Preparata, "despite the great amount of experimental results observed by a large number of scientists, a unified interpretation and theory of these phenomena has not been accepted and their comprehension still remains unsolved" [6–9, 26, 27].

A. Carpinteri (✉) • O. Borla • A. Manuello • D. Veneziano
Department of Structural, Geotechnical and Building Engineering, Politecnico di Torino, Corso Duca degli Abruzzi 24, Torino 10129, Italy
e-mail: alberto.carpinteri@polito.it; amedeo.manuellobertetto@polito.it

S. Guastella
Department of Applied Science and Technology, Geotechnical and Building Engineering, Politecnico di Torino, Corso Duca degli Abruzzi 24, Torino 10129, Italy

R. Sesana
Department of Mechanical and Aerospace Engineering, Politecnico di Torino, Corso Duca degli Abruzzi 24, Torino 10129, Italy

A. Goi
Private Research, Torino, Italy

A.M. Beese et al. (eds.), *Fracture, Fatigue, Failure and Damage Evolution, Volume 8*, Conference Proceedings of the Society for Experimental Mechanics Series, DOI 10.1007/978-3-319-21611-9_24

On the other hand, as shown by most of the articles devoted to Cold Nuclear Fusion, one of the principal features is the appearance of microcracks on electrode surfaces after the tests [26, 27]. Such evidence might be directly correlated to hydrogen embrittlement of the material composing the metal electrodes (Pd, Ni, Fe, Ti, etc.). This phenomenon, well-known in Metallurgy and Fracture Mechanics, characterizes metals during forming or finishing operations [44]. In the present study, the host metal matrix (Pd and Ni) is subjected to mechanical damaging and fracturing due to hydrogen atoms penetrating into the atomic lattice and forcing it, during the gas loading. Hydrogen effects are largely studied especially in metal alloys where the hydrogen absorption is particularly high. The hydrogen atoms generate an internal stress that lowers the fracture resistance of the metal, so that brittle crack growth can occur with a hydrogen partial pressure below 1 atm [44, 45]. Some experimental evidences show that neutron emissions may be strictly correlated to fracture of non-radioactive or inert materials. From this point of view, anomalous nuclear emissions and heat generation had been verified during fracture in fissile materials [2–4] and in deuterated solids [5, 8, 30].

In order to confirm the earlier results obtained by Co–Cr and Ni–Fe electrodes [41, 42], and by Pd and Ni electrodes [43], electrolytic tests have been conducted using 100 % Pd at the cathode with different operating time intervals. As a result, relevant compositional changes and traces of elements previously absent have been observed on the Pd and Ni electrodes after the experiments and significant neutron emissions were observed during the test. In addition to the previous experimental campaign, also the amount of escaped gas during electrolysis was measured together with the evaluation of the extra heat generation during the tests in order to obtain an energy balance.

24.2 Experimental Set Up

Over the last ten years, specific experiments have been conducted on an electrolytic reactor (owners: Mr. A. Goi et al.). The aim was to investigate whether the anomalous heat generation may be correlated to a new type of nuclear reactions during electrolysis phenomena. The reactor was built in order to be appropriately filled with a salt solution of water and Potassium Carbonate (K_2CO_3). The electrolytic phenomenon was obtained using two metal electrodes immersed in the aqueous solution for three different tests of 2.5, 5 and 10 h (Fig. 24.1). The solution container, named also reaction chamber in the following, is a cylinder-shaped element of 100 mm diameter, 150 mm high, and 5 mm thick. For the reaction chamber Inox AISI 316 L steel was used. The two metallic electrodes were connected to a source of direct current: a Ni–Fe-based electrode as the positive pole (anode), and a Pd-based electrode as the negative pole (cathode).

The reaction chamber base consists of a ceramic plate preventing the direct contact between liquid solution and Teflon lid. Two threaded holes host the electrodes, which are screwed to the bottom of the chamber successively filled with the solution. A valve at the top of the cell allows the vapor to escape from the reactor and condense in an external collector. Externally, two circular Inox steel flanges, fastened by means of four threaded ties, hold the Teflon layers. The inferior steel flange of

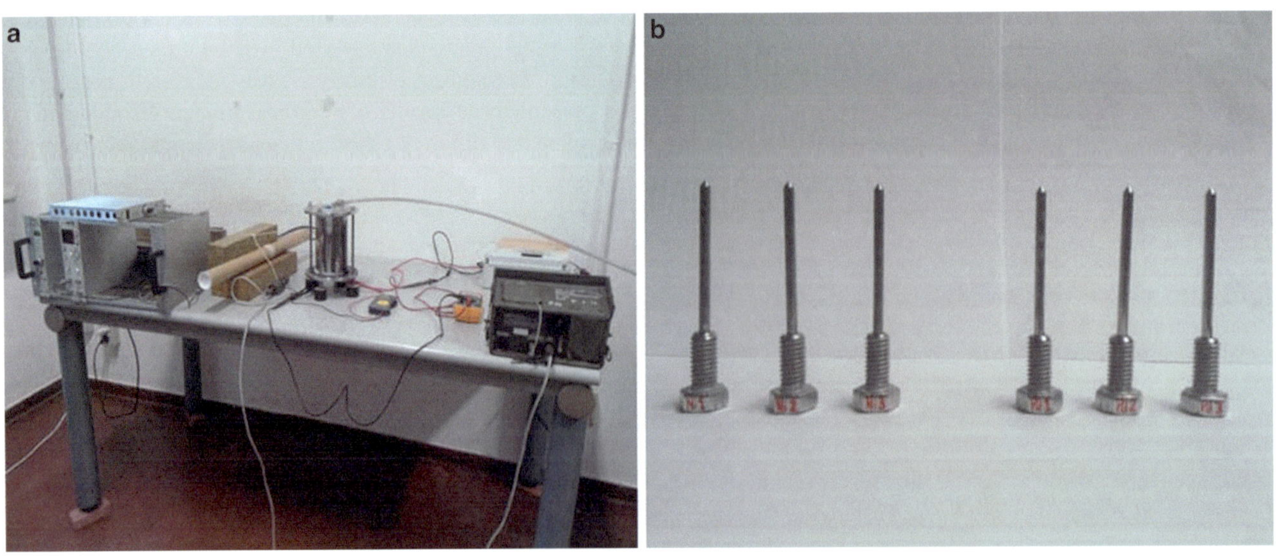

Fig. 24.1 Experimental set-up (**a**) Pd and Ni electrodes (**b**)

the reactor is connected to four supports isolated from the ground by means of rubber-based material. As mentioned before, a direct current passes through the anode and the cathode electrodes, provided by a power circuit connected to the power grid through an electric socket. The components of the circuit are an isolating transformer, an electronic variable transformer (Variac), and a diode bridge linked in series.

Electric current and voltage probes were positioned in different parts of the circuit. The voltage measurements were performed by a differential voltage probe of 100 MHz with a maximum rated voltage of 1400 V. The current was measured by a Fluke I 310S probe with a maximum rated current of 30 A. Current intensity and voltage measurements were also taken by means of a multimeter positioned at the input line. From the turning on to the switching off of the electrolytic cell, current and voltage were found to vary in a range from 3 to 5 A and from 20 to 120 V, respectively.

During testing an infrared thermo tracer IR Tech Timage Radiamatic XT was placed in front of the specimens. A dedicated software, Timage Radiamatic, allowed both acquisition and data processing. The equipment acquires a surface thermal contour of surfaces in front of the camera. Data are acquired as temperature data related to pixels and plotted by a colour map. Data processing procedure requires the definition of a ROI (Region of Interest) on the thermal map. The region can have any shape and dimension and it is possible to extract maximum, minimum, average and whatever temperature in the ROI. To optimize thermocamera measurement, reactor surface has been black painted, to minimize reflected radiation and maximize emitted radiation. The width of the area is of 20 pixels that is where the reactor surface can be approximated as parallel to the camera lens. In this ROI maximum, minimum and average temperature have been acquired with relation to time development. For what concerns temperature distribution in the reacting solution, the thermal map acquired by means of the thermo-camera shows convection streams during reaction taking place during reaction. Anyway the average temperature measured in the selected ROI shows a constant trend as long as maximum and minimum temperature trends in the same ROI. This allows us to assume the average temperature in the ROI equivalent to the average temperature in the solution.

For an accurate neutron emission evaluation, a He^3 proportional counter was employed (Xeram, France) with pre-amplification, amplification, and discrimination electronics directly connected to the detector tube. The detector is supplied by a high voltage power (about 1.3 kV) via NIM (Nuclear Instrument Module). The logic output producing the TTL (transistor–transistor logic) pulses is connected to a NIM counter. The logic output of the detector is enabled for analog signals exceeding 300 mV (see Fig. 24.1a). This discrimination threshold is a consequence of the sensitivity of the He^3 detector to the gamma rays ensuing neutron emission in ordinary nuclear processes. This value has been determined by measuring the analog signal of the detector by means of a Co-60 gamma source. The detector is also calibrated at the factory for the measurement of thermal neutrons; its sensitivity is 65 cps/$n_{thermal}$ ($\pm$10 % declared by the factory), i.e., the flux of thermal neutrons is one thermal neutron/s cm^2, corresponding to a count rate of 65 cps. Finally, before and after the experiments, Energy Dispersive X-ray Spectroscopy has been performed in order to recognise possible direct evidence of piezonuclear reactions that can take place during the electrolysis. The elemental analyses were performed by a ZEISS Auriga field emission scanning electron microscope (FESEM) equipped with an Oxford INCA energy-dispersive X-ray detector (EDX) with a resolution of 124 eV @ MnKa. The energy used for the analyses was 18 KeV.

24.3 Neutron Emission Measurements

Neutron emission measurements performed during the experimental activity are represented in Figs. 24.2, 24.3 and 24.4. The measurements performed by the He^3 detector were conducted for different phases of 2.5, 5 and 10 h. The background level was measured for different time intervals before and after switching on the reaction chamber. These measurements reported an average neutron background of about $(6.00 \pm 2.45) \times 10^{-2}$ cps. Furthermore, when the reaction chamber is active, it is possible to observe that, after a time interval of about 80 min, neutron emissions greater than the background level may be detected in the first test (Fig. 24.2a). This evidence may be observed also in the cumulated curve of CPS respect to the cumulated average background level (Fig. 24.2b). During the experiment of 5 h After 260 min from the beginning of the measurements, it is possible to observe a neutron emission level of about 5 times greater than the background level (Fig. 24.3a). A small difference between the cumulated curves could be observed also in this case (Fig. 24.3b). Similar results were observed for the experiments of 10 h, after 200 min and up to 400 neutron emissions up to more than ten times the background were measured (Fig. 24.4a). This evidence may be observed also in the cumulated curve of CPS after 200 min from the beginning of the test (Fig. 24.4b).

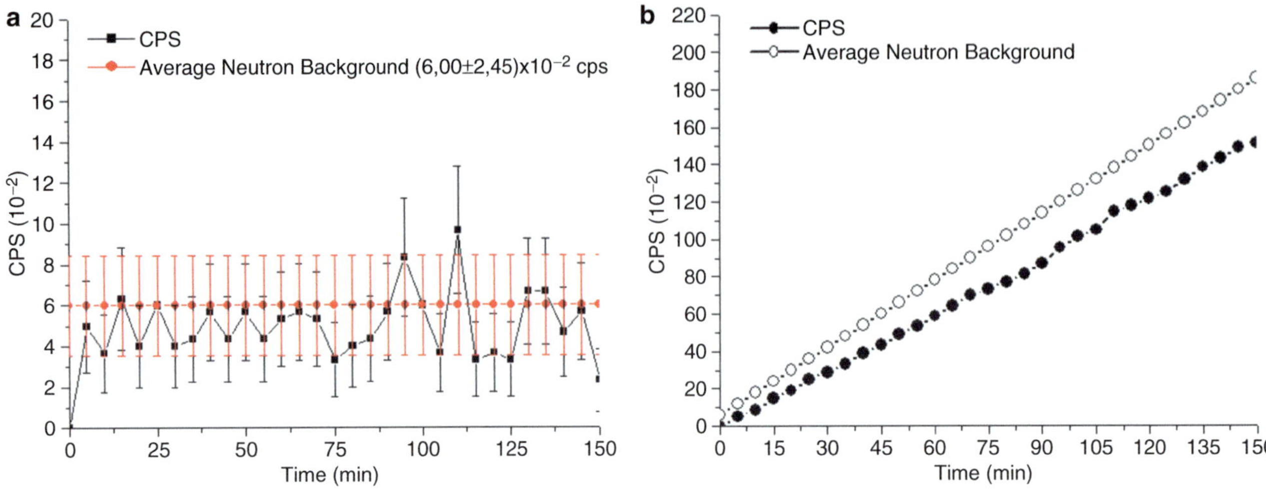

Fig. 24.2 Neutron emission measurements for the test with a time duration of 2.5 h (**a**). Cumulated curve of CPS respect to the cumulated average background level (**b**)

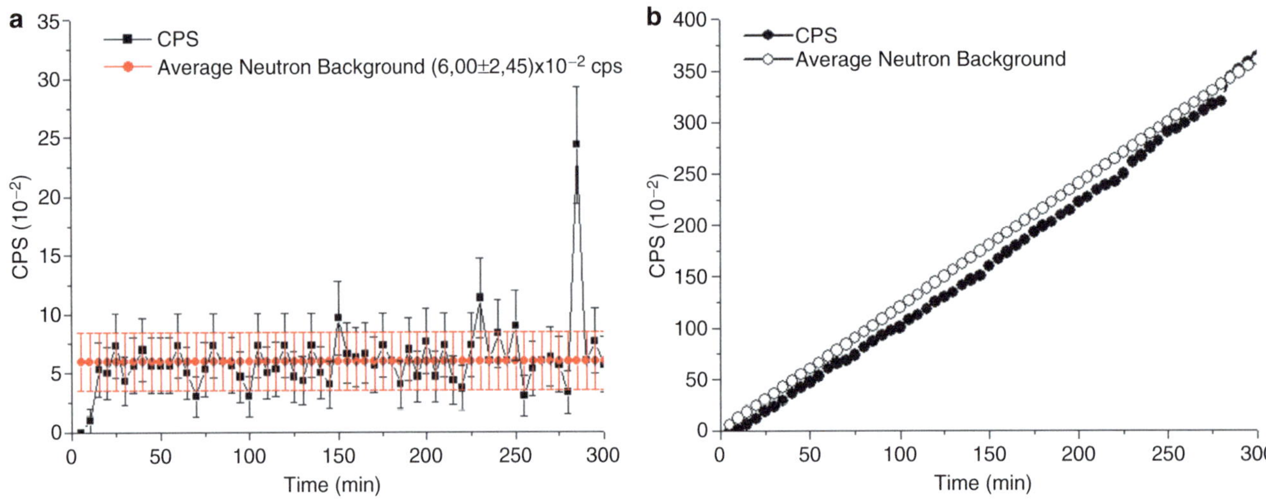

Fig. 24.3 Neutron emission measurements for the test with a time duration of 5 h (**a**). cumulated curve of CPS compared to the cumulated average background level (**b**)

24.4 Compositional Analysis of the PD Electrode

In the present section, the chemical compositions before and after the experiment will be taken into account (Fig.24.5a).

In particular, under the light of what can be deduced from the neutron emissions measurements and according to the hydrogen embrittlement hypothesis suggested by Carpinteri et al. [41–43], the presence of microcracks and macrocracks on the electrode surface (Fig. 24.5b) is accounted in the mechanical interpretation of the phenomena. These evidences are particularly strong in the case of the Pd electrode, where a macroscopic fracture took place during the test. The fracture presented a width of about 40 μm observable at naked eyes (see Fig. 24.5b).

Considering the average decrement of Pd (−28.9 %) after 10 h, reported in Table 24.1, according to the piezonuclear hypothesis a first fission reaction can be assumed:

$$\mathrm{Pd}_{46}^{106} \rightarrow \mathrm{Zn}_{30}^{65} + \mathrm{Si}_{14}^{28} + 10 \text{ neutrons}, \tag{24.1}$$

according to reaction (24.1), a Pd decrement of about 2.07 % can be counterbalanced by the Zn and Si increments at the end of the experiment. reported in Table 24.1. By this reaction, in fact, increments of 1.46 % and 0.61 can be assumed for Zn

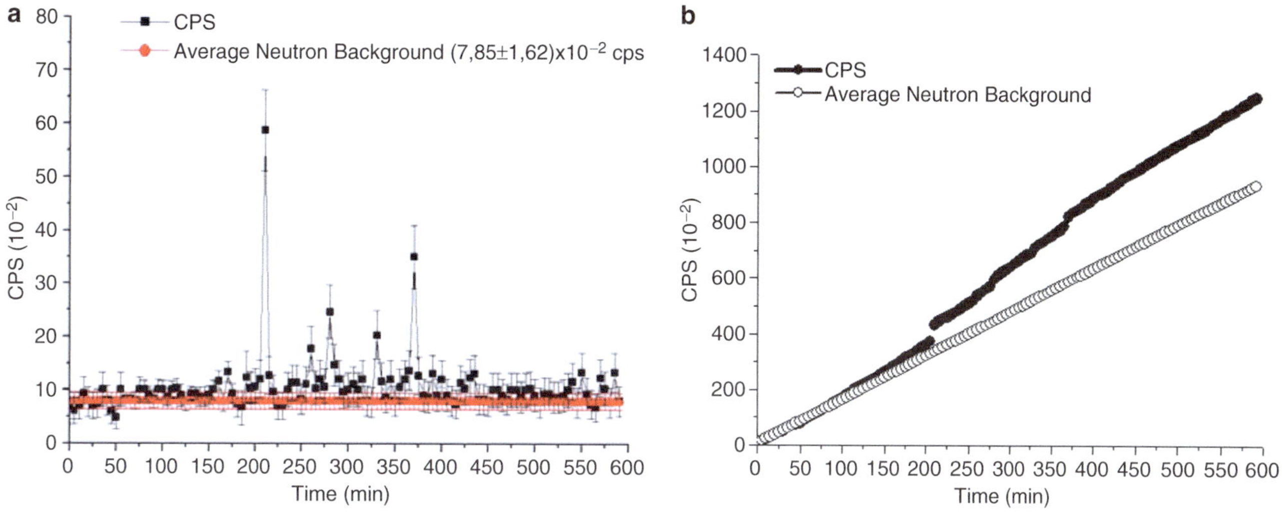

Fig. 24.4 Neutron emission measurements for the test with a time duration of 10 h (**a**); cumulated curve of CPS compared to the cumulated average background level (**b**)

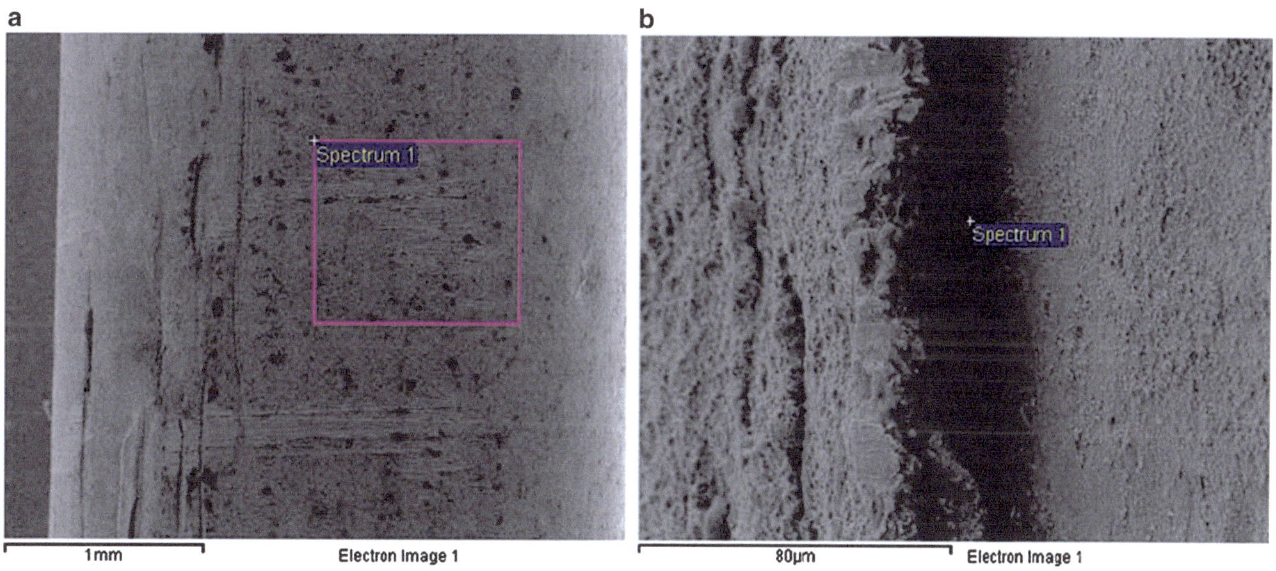

Fig. 24.5 Pd electrode; energy dispersive X-ray spectroscopy image before (**a**) and after (**b**) the test (10 h)

and Si, respectively. A quantity of 26.8 % of Pd remains to be counterbalanced. For these reasons, a second reaction can be assumed in analogy to previous experiments [43]:

$$Pd_{46}^{106} \rightarrow Fe_{26}^{56} + Ca_{20}^{40} + 10 \text{ neutrons}. \quad (24.2)$$

Considering reaction (24.2), Ca and Fe respectively would increase by the following quantities: 10 % and 14 %. These variations may be accompanied by a neutron emission corresponding to the remaining 3 % of the mass concentration. The whole iron increment, according to reaction (24.2), could be entirely considered as the starting element for the production of other elements. Hence, a third hypothesis can be proposed involving Fe as starting element and O as the product, together with alpha and neutron emissions:

$$Fe_{26}^{56} \rightarrow 3O_{8}^{16} + He_{2}^{4} + 4 \text{ neutrons}, \quad (24.3)$$

According to reaction (24.3), the iron depletion produces 11.9 % of oxygen with alpha particles (He) and neutron emissions. The total measured increment in oxygen after the experimental test is equal to 20.2 % (see Table 24.1). This

Table 24.1 Element concentrations before, after 2.5, 5 and 10 h of electrolysis (Pd electrode)

	Mean concentration (mass %)			
Element	0 h	2.5 h	5 h	10 h
Pd	99.85 ± 0.58	92.66 ± 3.42	88.26 ± 2.24	70.91 ± 4.63
O	0.15 ± 0.58	6.20 ± 2.89	9.04 ± 2.11	21.22 ± 3.64
K	–	0.41 ± 0.53	–	3.33 ± 0.86
Zn	–	–	–	1.46 ± 0.74
Fe	–	0.61 ± 0.61	2.67 ± 0.35	1.28 ± 0.45
Si	–	0.06 ± 0.17	–	0.61 ± 0.25
Mg	–	–	0.06 ± 0.2	0.42 ± 0.34
Na	–	0.06 ± 0, 22	–	0.29 ± 0.37
Cu	–	–	–	0.25 ± 0.67
Ni	–	–	–	0.11 ± 0.28
Ca	–	–	–	0.06 ± 0.23
Co	–	–	–	0.06 ± 0.24

Table 24.2 Element concentrations before, after 2.5, 5 and 10 h of electrolysis (Ni electrode)

	Mean concentration (mass %)			
Element	0 h	2.5 h	5 h	10 h
Ni	91.44 ± 0.86	76.57 ± 1.41	71.42 ± 2.43	72.66 ± 4.31
O	4.02 ± 0.70	18.52 ± 1.36	22.88 ± 2.08	16.97 ± 2.78
Pd	–	–	–	4.35 ± 1.12
Ti	3.43 ± 0.20	2.95 ±0.16	2.77 ± 0.26	3.04 ± 0.25
Al	0.44 ± 0.13	0.89 ± 0.27	1.69 ± 0.23	1.57 ±0.36
Si	0.67 ± 0.29	0.26 ± 0.27	0.50 ± 0.19	0.52 ± 0.08
Fe	–	0.03 ± 0.12	0.10 ± 0.21	0.31 ± 0.27
K	–	0.72 ±0.16	0.64 ± 0.32	0.56 ± 0.35

quantity seems to be only partially explained by reaction (24.3). The remaining O concentration could be explained considering other reactions involving Ca (product in reaction 24.2) as the starting element:

$$\mathrm{Ca}_{20}^{40} \rightarrow \mathrm{O}_{8}^{16} + \mathrm{Mg}_{12}^{24}, \tag{24.4}$$

$$\mathrm{Ca}_{20}^{40} \rightarrow 2\mathrm{O}_{8}^{16} + 4\mathrm{H}_{1}^{1} + 4 \text{ neutrons}. \tag{24.5}$$

From reaction (24.4), we can consider a decrease in Ca concentration of 0.7 % and the formation of 0.42 % of Mg and 0.28 % of O. On the other hand, from reaction (24.5), we obtain a decrease in Ca corresponding to the increase of about 8 % in O with H production and neutron emissions. Considering the O increments coming from reactions (24.3), (24.4), and (24.5), totally equal to 19.9 %, and the experimental evidence reporting a total measured O concentration of 20.2 %, we may assume that O seems to be almost perfectly justified by the proposed reactions. At the same time, the Zn, Mg and Si increments observed after the experiment can be explained by the results of reactions (24.1) and (24.4). These evidences seem to be very interesting also considering that they represent a quantitative confirmation for the results previously published by Carpinteri et al. [43].

24.5 Nickel Electrode Composition Analysis

Analogously let us consider the Nickel electrode. Table 24.2 summarizes the concentration variations after the electrolysis. Nickel diminishes by 18.8 %, whereas the most apparent positive variation is that of oxygen (+12.9 %).

On the basis of the piezonuclear reaction conjecture, we could assume the oxygen average variation as a nuclear effect caused by the following hypothesis:

$$\mathrm{Ni}_{28}^{59} \rightarrow 3\mathrm{O}_{8}^{16} + 2\mathrm{He}_{2}^{4} + 3 \text{ neutrons}. \tag{24.6}$$

$$\mathrm{Ni}_{28}^{58} \rightarrow 2\mathrm{Al}_{13}^{27} + \mathrm{He}_{2}^{4} \tag{24.7}$$

Reactions (24.6) and (24.7), imply emissions of neutrons and alpha particles, which will provide a great support to the hypotheses based on piezonuclear reactions. The main idea underlying the hypothesis is that an average decrease of 18.8 % in Nickel underwent a reaction producing at least 14.5 % of oxygen together with alpha and neutron emissions (3 %). Secondly, a further average depletion of 1.5 % in Ni would have produced about 1.3 % of Al accompanied by alpha particle emissions. According to reactions (24.6) and (24.7) the following balance: Ni (−18.8 %) = O (+14.5 %) + Al (+1.5 %) + neutrons and alpha particles may be considered. The amount of O coming from this balance (+14.5 %) and the amount detected after 10 h (+13 %) are very similar.

24.6 Heat Generation and Energy Balance Evaluation

The aim of the following considerations is to provide a thermodynamic background on the multiple phenomena involved in the energy equilibrium of the operating electrolytic cell. In particular, we shall define the physical quantities and the contributions required for an energy balance considering the input and output contributions. Thus, it is critical to define some hypotheses on the considered system. The input energy E_{in} is related to the electric power exchanged between the two electrodes and can be quantified by means of electric power consumption. The corresponding electric power ϕ_{in} can be calculated as the average power absorbed by the system and is given by direct electric measurements:

$$E_{in} = \frac{\int_{ti}^{tf} E_{in}(t)dt}{t_f - t_i}, \tag{24.8}$$

where t_f and t_i are the final and initial time instants respectively of the testing session. The instant input energy $E_{in}(t)$ is:

$$E_{in}(t) = V_{in} \times I(t), \tag{24.9}$$

where $V_{in}(t)$ is the measured voltage between the electrodes and I(t) is the electric current intensity measured by means of the Amperometer placed before the cell, neglecting the circuit dissipations.

The main terms of energy transformation during testing are: (i) vaporization, and (ii) heat convection exchange. They are both "outgoing" power terms. The first term E_v can be computed by means of measuring the condensed water volume, as a pipe drives the exceeding vapour out of the reservoir, and its temperature while flowing in the pipe. The second term E_H can be quantified by means of thermodynamic heat exchange equations.

The energy equilibrium equation, involving the main energy terms, has the following formulation in steady state conditions:

$$E_{in} + E_X = E_v + E_H \tag{24.10}$$

Where E_{in} represents the term due to the power of the electric circuit, measured at the connection between the circuit and the electrodes just before the cell; E_X represents the unknown energy term correlated to piezonuclear reactions; E_v and E_h represent the terms due to vaporization and convection, respectively.

Other terms could be considered (electrolytic transformations, turbulent flow, etc.) but they are considered quantitatively negligible in a first approximation. According to (24.10), in a ten hour testing the input term counts by the electric measurers, neglecting the circuit dissipations, is 1349 kJ, the vaporization term counts 3953 kJ while the heat convection term is equal to 1460 kJ. The unbalance between the two terms of the equilibrium equation is about 4064 kJ, representing in a first approximation the estimation of the term E_X. The difference between output and input energies is about three times the input energy.

24.7 Conclusions

Neutron emissions up to one order of magnitude higher than the background level were observed during the operating time of an electrolytic cell. These evidences are very close to the results coming from previous experiments.

By the EDX analysis performed on the two electrodes, significant compositional variations were recorded. In general, the decrements in Pd at the first electrode seem to be almost perfectly counterbalanced by the increments in lighter elements like oxygen, see reactions (24.1)–(24.5). As far as the Ni electrode is concerned, the Ni decrement is almost perfectly counterbalanced by the O and Al increments according to reactions (24.6) and (24.7).

The unbalance between the two terms of the thermal equilibrium equation is about 4064 kJ, representing in a first approximation the estimation of the energy production term. The amount E_x has been quantified as three times the input energy. Chemical variations, energy balance, and neutron emissions may be accounted for direct and indirect evidence of piezonuclear fission reactions correlated to microcrack formation and propagation as well as to hydrogen embrittlement. The so-called Cold Nuclear Fusion, interpreted under the light of hydrogen embrittlement, may be explained by piezonuclear fission reactions occurring in the host metal, rather than by nuclear fusion reactions of hydrogen isotopes forced by the atomic lattice.

Acknowledgement Special thanks for his collaboration in the realization of the tests are due to Mr. M. Yon.

References

1. Borghi, D.C., Giori, D.C., Dall'Olio, A.: Experimental evidence on the emission of neutrons from cold hydrogen plasma. Proceedings of the International Workshop on Few-body Problems in Low-energy Physics, Alma-Ata, Kazakhstan, pp. 147–154 (1992); Unpublished Communication (1957); Comunicacao n. 25 do CENUFPE, Recife Brazil (1971)
2. Diebner, K.: Fusionsprozesse mit Hilfe konvergenter Stosswellen—einige aeltere und neuere Versuche und Ueberlegungen. Kerntechnik. **3**, 89–93 (1962)
3. Kaliski, S.: Bi-conical system of concentric explosive compression of D-T. J. Tech. Phys. **19**, 283–289 (1978)
4. Winterberg, F.: Autocatalytic fusion–fission implosions. Atomenergie-Kerntechnik. **44**, 146 (1984)
5. Derjaguin, B.V., et al.: Titanium fracture yields neutrons? Nature **34**, 492 (1989)
6. Fleischmann, M., Pons, S., Hawkins, M.: Electrochemically induced nuclear fusion of deuterium. J. Electroanal. Chem. **261**, 301 (1989)
7. Bockris, J.O'M., Lin, G.H., Kainthla, R.C., Packham, N.J.C., Velev, O.: Does tritium form at electrodes by nuclear reactions? In: The First Annual Conference on Cold Fusion. National Cold Fusion Institute, University of Utah Research Park, Salt Lake City (1990)
8. Preparata, G.: Some theories of cold fusion: a review. Fusion Technol. **20**, 82 (1991)
9. Preparata, G.: A new look at solid-state fractures, particle emissions and "cold" nuclear fusion. Il Nuovo Cimento. **104A**, 1259–1263 (1991)
10. Mills, R.L., Kneizys, P.: Excess heat production by the electrolysis of an aqueous potassium carbonate electrolyte and the implications for cold fusion. Fusion Technol. **20**, 65 (1991)
11. Notoya, R., Enyo, M.: Excess Heat Production during Electrolysis of H_2O on Ni, Au, Ag and Sn Electrodes in Alkaline Media, Proc. Third International Conference on Cold Fusion. Universal Academy Press, Tokyo (1992)
12. Miles, M.H., Hollins, R.A., Bush, B.F., Lagowski, J.J., Miles, R.E.: Correlation of excess power and helium production during D_2O and H_2O electrolysis using palladium cathodes. J. Electroanal. Chem. **346**, 99–117 (1993)
13. Bush, R.T., Eagleton, R.D.: Calorimetric studies for several light water electrolytic cells with nickel fibrex cathodes and electrolytes with alkali salts of potassium, rubidium, and cesium. In: Fourth International Conference on Cold Fusion. Lahaina, Maui. Electric Power Research Institute 3412 Hillview Ave., Palo Alto. (1993)
14. Fleischmann, M., Pons, S., Preparata, G.: Possible theories of cold fusion. Nuovo Cimento. Soc. Ital. Fis. A. **107**, 143 (1994)
15. Szpak, S., Mosier-Boss, P.A., Smith, J.J.: Deuterium uptake during Pd-D codeposition. J. Electroanal. Chem. **379**, 121 (1994)
16. Sundaresan, R., Bockris, J.O.M.: Anomalous reactions during arcing between carbon rods in water. Fusion Technol. **26**, 261 (1994)
17. Arata, Y., Zhang, Y.: Achievement of solid-state plasma fusion ("cold-fusion"). Proc. Jpn Acad. **71B**, 304–309 (1995)
18. Ohmori, T., Mizuno, T., Enyo, M.: Isotopic distributions of heavy metal elements produced during the light water electrolysis on gold electrodes. J. New Energy. **1**(3), 90 (1996)
19. Monti, R.A.: Low energy nuclear reactions: experimental evidence for the alpha extended model of the atom. J. New Energy. **1**(3), 131 (1996)
20. Monti, R.A.: Nuclear transmutation processes of lead, silver, thorium, uranium. In: The Seventh International Conference on Cold Fusion. ENECO Inc. Vancouver (1998)
21. Ohmori, T., Mizuno, T.: Strong excess energy evolution, new element production, and electromagnetic wave and/or neutron emission in light water electrolysis with a tungsten cathode. Infinite Energy. **20**, 14–17 (1998)
22. Mizuno, T.: Nuclear Transmutation: The Reality of Cold Fusion. Infinite Energy Press, Concord (1998)
23. Little, S.R., Puthoff, H.E., Little, M.E.: Search for Excess Heat from a Pt Electrode Discharge in K_2CO_3-H_2O and K_2CO_3-D_2O Electrolytes (1998)
24. Ohmori, T., Mizuno, T.: Nuclear transmutation reaction caused by light water electrolysis on tungsten cathode under incandescent conditions. Infinite Energy. **5**(27), 34 (1999)
25. Ransford, H.E.: Non-stellar nucleosynthesis: transition metal production by DC plasma-discharge electrolysis using carbon electrodes in a non-metallic cell. Infinite Energy. **4**(23), 16 (1999)
26. Storms, E.: Excess power production from platinum cathodes using the Pons-Fleischmann effect. In: 8th International Conference on Cold Fusion. Lerici (La Spezia). Italian Physical Society, Bologna. pp. 55–61 (2000)
27. Storms, E.: Science of Low Energy Nuclear Reaction: a Comprehensive Compilation of Evidence and Explanations about Cold Fusion. World Scientific, Singapore (2007)
28. Mizuno, T., et al.: Production of heat during plasma electrolysis. Jpn. J. Appl. Phys. **39**, 6055–6061 (2000)
29. Warner, J., Dash, J., Frantz. S.: Electrolysis of D2O with titanium cathodes: enhancement of excess heat and further evidence of possible transmutation. In: The Ninth International Conference on Cold Fusion. Tsinghua University, Beijing, p. 404 (2002)

30. Fujii, M.F., et al.: Neutron emission from fracture of piezoelectric materials in deuterium atmosphere. Jpn. J. Appl. Phys. **41**, 2115–2119 (2002)
31. Mosier-Boss, P.A., et al.: Use of CR-39 in Pd/D co-deposition experiments. Eur. Phys. J. Appl. Phys. **40**, 293–303 (2007)
32. Swartz, M.: Three physical regions of anomalous activity in deuterated palladium. Infinite Energy **14**, 19–31 (2008)
33. Mosier-Boss, P.A., et al.: Comparison of Pd/D co-deposition and DT neutron generated triple tracks observed in CR-39 detectors. Eur. Phys. J. Appl. Phys. **51**(2), 20901–20911 (2010)
34. Kanarev, M., Mizuno, T.: Cold fusion by plasma electrolysis of water. New Energy Technol. **1**, 5–10 (2002)
35. Cardone, F., Mignani, R.: Energy and Geometry. World Scientific, Singapore (2004). Chapter 10
36. Carpinteri, A., Cardone, F., Lacidogna, G.: Piezonuclear neutrons from brittle fracture: early results of mechanical compression tests. Strain. 45, 332–339 (2009). Atti dell' Accademia delle Scienze di Torino. 33, 27–42 (2009)
37. Cardone, F., Carpinteri, A., Lacidogna, G.: Piezonuclear neutrons from fracturing of inert solids. Phys. Lett. A. **373**, 4158–4163 (2009)
38. Carpinteri, A., Cardone, F., Lacidogna, G.: Energy emissions from failure phenomena: mechanical, electromagnetic, nuclear. Exp. Mech. **50**, 1235–1243 (2010)
39. Carpinteri, A., Lacidogna, G., Manuello, A., Borla, O.: Piezonuclear fission reactions: evidences from microchemical analysis, neutron emission, and geological transformation. Rock. Mech. Rock. Eng. **45**, 445–459 (2012)
40. Carpinteri, A., Lacidogna, G., Manuello, A., Borla, O.: Piezonuclear fission reactions from earthquakes and brittle rocks failure: evidence of neutron emission and nonradioactive product elements. Exp. Mech. **53**, 345–365 (2013)
41. Carpinteri, A., Borla, O., Goi, A., Manuello, A., Veneziano, D.: Mechanical conjectures explaining cold nuclear fusion. Proceedings of the Conference & Exposition on Experimental and Applied Mechanics (SEM), Lombard, CD-ROM, p. 481 (2013)
42. Veneziano, D., Borla, O., Goi, A., Manuello, A., Carpinteri A.: Mechanical conjectures based on hydrogen embrittlement explaining cold nuclear fusion. Proceedings of the 21° Congresso Nazionale di Meccanica Teorica ed Applicata (AIMETA), Torino, CD-ROM (2013)
43. Carpinteri, A., Borla, O., Goi, A., Guastella, S., Manuello, A., Veneziano, D.: Hydrogen embrittlement and cold fusion effects in palladium during electrolysis experiments. In: Conference & Exposition on Experimental and Applied Mechanics (SEM), Greenville, vol. 6, pp. 37–47 (2014)
44. Milne, I., Ritchie, R.O., Karihaloo, B.: Comprehensive Structural Integrity: Fracture of Materials from Nano to Macro, vol. 6, pp. 31–33. Elsevier, Amsterdam (2003)
45. Liebowitz, H.: Fracture an Advanced Treatise. Academic, New York (1971)

Chapter 25
Strain-Rate-Dependent Yield Criteria for Composite Laminates

Joseph D. Schaefer and Isaac M. Daniel

Abstract The strain-rate-dependent failure of a fiber-reinforced toughened-matrix composite (IM7/8552) was experimentally characterized over the range of quasi-static (10^{-4}) to dynamic ($10^3\ s^{-1}$) strain rates using off-axis lamina and angle-ply laminate specimens. A progressive failure paradigm was proposed to describe the matrix-dominated transition from linear elastic to non-linear material behavior, and the Northwestern Failure Theory was adapted to develop a set of yield criteria for predicting the matrix-dominated yielding of composites using the lamina-based transverse tension ($F_{2t}{}^{y}$), transverse compression ($F_{2c}{}^{y}$), and shear ($F_6{}^{y}$) yield strengths. A verification and validation protocol was employed to evaluate the applicability of the new failure-mode-based yield criteria. Starting with the lamina, the proposed criteria were validated to accurately predict matrix-dominated yielding. Angle-ply laminates were investigated to isolate the matrix-dominated laminate behavior, and the predictions were found to be in superior agreement with the experimental results compared to the classical failure theories in all cases using simply determined average lamina yield properties.

Keywords Composites • Failure • Strain rate • Northwestern failure theory • Verification and validation

25.1 Introduction

Structural composite failure is designed to be a controlled process comprised of scale-relevant phenomena. Proper implementation of pre and post-failure computational models for the various (and interacting) matrix and fiber-dominated mechanisms must be completed with the understanding of the direct model outputs to be correlated to experiment in order for proper validation. Further, effective model validation is only achieved when the approach is proven to accurately predict failure for a specific failure type at multiple levels of the component hierarchy [1].

While predicting the ultimate failure of composite lamina remains a robust field of research, it has become increasingly clear that the translation of the classical approaches to predicting failure in composite laminates leaves much to be desired [2]. The inability of the classical theories to accurately predict ultimate lamina failure is of critical interest to the computational analyses that must utilize the theory as the basis for predicting embedded ply failure in complex composite laminates. Any investigation of progressive ply failure in composite laminates must first accurately predict in-situ layer failure initiation so that the damage propagation models may be effectively implemented.

In light of this 'progressive failure' framework, it is critical to determine not only when a lamina or laminate fails, but when it *begins* to fail. In the current work, matrix yielding is defined as the initiation of the failure process in composite laminates. Macroscopically, the yield point is the point at which the stress-strain behavior breaks linearity; the yield point for an off-axis 75° lamina specimen tested in compression is shown in Fig. 25.1. The yield point represents the point at which stress and strain may be accurately determined utilizing a linear elastic theoretical basis to predict failure for non-linear materials. Beyond the yield point, the stress predicted by linear elastic failure models no longer correlates to the actual material strain. Importantly, this discrepancy may lead to significant and propagating error when such approaches are used to model the progressive failure of composite laminates; thus, evaluation and quantification of model uncertainty becomes all but impossible.

J.D. Schaefer
The Boeing Company, St. Louis, MO, USA
e-mail: joseph.d.schaefer@boeing.com

I.M. Daniel (✉)
Northwestern University, Evanston, IL, USA
e-mail: imdaniel@northwestern.edu

A.M. Beese et al. (eds.), *Fracture, Fatigue, Failure and Damage Evolution, Volume 8*, Conference Proceedings of the Society for Experimental Mechanics Series, DOI 10.1007/978-3-319-21611-9_25

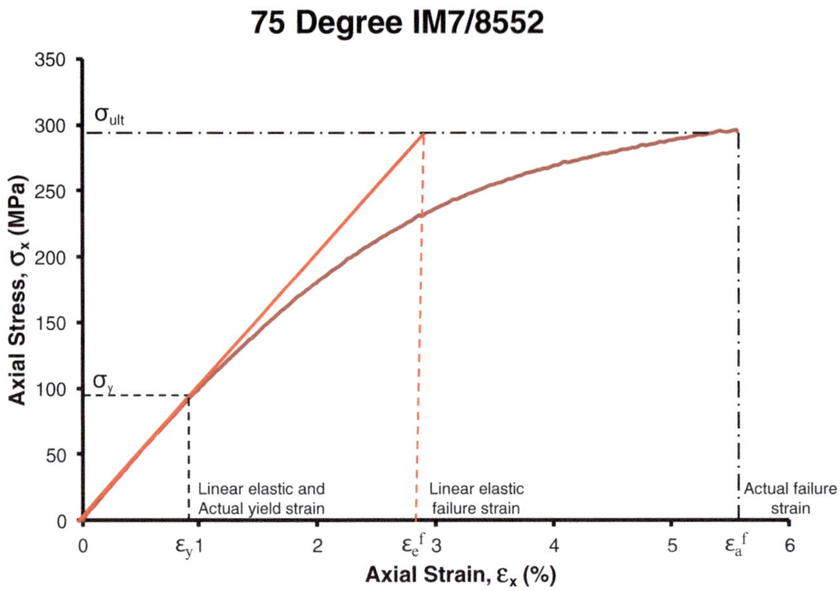

Fig. 25.1 Failure framework for lamina yielding and ultimate failure

For the first time, the Northwestern Yield Criteria are introduced as the theoretical basis for predicting matrix-dominated yield behavior in composites. Sub-component validation is addressed through investigation of the matrix-dominated failure modes in composite lamina. The investigation then extends to the component case of angle-ply laminates wherein thermally-induced residual stresses are present. The Northwestern Yield Criteria are presented as strain-rate-dependent criteria for determining the yielding of composite laminates, both lamina and angle-ply, over applicable ranges of loading rate. The work serves to provide an accurate representation of the elastic behavior of matrix-dominated layups to potentially enhance the analysis strategies of computational solid mechanics approaches.

25.2 Lamina Failure

Hexcel IM7/8552 was selected for this work. IM7/8552 is a popular research medium for the aerospace field [3–10] and has been qualified by NCAMP [11]. Hexply® 8552 prepreg is an amine cured, toughened resin system and was provided on a 48 in. wide roll with the fibers aligned along the rolling direction. The epoxy resin has been toughened with a thermoplastic polymer that provides the material with increased damage resistance and increased strain to failure; thus, considerable non-linear response results under loading for certain ply layup configurations.

Developed by Daniel et al. [12], the Northwestern University (NU) Theory is a failure mode based failure theory derived from the micromechanical resin behavior of composites yet dependent on macroscopic material properties. As noted by the authors, the NU criteria are suitable for interfiber and interlaminar failure under transverse normal, and parallel to the fibers shear loading on the 1–2 or 1–3 planes. They are especially applicable to highly anisotropic composites with failure occurring due to a low interaction of modes. It was previously found that the NU theory performs quite well at predicting the ultimate failure of both unidirectional and fabric composites [12–16].

25.3 Northwestern University Yield Criteria

The Northwestern Failure Theory (NU Theory) for composites was recast as a set of yield criteria based on transverse tensile yield, transverse compression yield, in-plane shear yield, transverse modulus, and in-plane shear modulus (F_{2t}^{y}, F_{2c}^{y}, F_{6}^{y}, E_2, and G_{12}). The underlying NU Theory formulation remains the same [12]; however, the noted yield properties were used as the failure basis instead of the previously presented ultimate lamina failure values:

Table 25.1 Lamina axial yield stresses at quasi-static, intermediate, and dynamic strain rates

$[\theta]_{54}$	$10^{-4}s^{-1}$ (MPa)	$1\ s^{-1}$ (MPa)	300–1000 s^{-1} (MPa)
90	105	146	175
75	100	146	159
60	100	135	147
45	96	128	140
30	94	123	140
15	133	180	198

Table 25.2 Strain-rate-dependence of lamina properties for NU criteria

Property	$10^{-4}\ s^{-1}$	$1\ s^{-1}$	300–1000 s^{-1}
Transverse modulus, E_2(GPa)	9	10.6	11.2
Shear modulus, G_{12} (GPa)	5.6	6.23	6.8
Transverse tensile yield, F^y_{2t} (MPa)	41	51	(66)
Transverse compressive yield, F^y_{2c} (MPa)	105	146	173
Shear yield, F^y_6, (MPa)	36	50	(61)

$$\left(\frac{\sigma_2^*}{F_{2c}^y}\right)^2+\left(\frac{E_2}{G_{12}}\right)^2\left(\frac{\tau_6^*}{F_{2c}^y}\right)^2=1 \quad \text{Compression dominated yield} \tag{25.1}$$

$$\left(\frac{\tau_6^*}{F_6^y}\right)^2+2\left(\frac{G_{12}}{E_2}\right)\left(\frac{\sigma_2^*}{F_6^y}\right)=1 \quad \text{Shear dominated yield} \tag{25.2}$$

$$\left(\frac{\sigma_2^*}{F_{2t}^y}\right)+\frac{1}{4}\left(\frac{E_2}{G_{12}}\right)^2\left(\frac{\tau_6^*}{F_{2t}^y}\right)^2=1 \quad \text{Tension dominated yield} \tag{25.3}$$

where σ_2* and τ_6* are functions of strain rate.

25.4 Strain-Rate-Dependent Yielding of Lamina

Further complicating the analysis of composite materials is the tendency for the matrix-dominated properties to be influenced by strain rate [13]. Werner et al. [15] recently performed extensive characterization of matrix property-dependence on strain rate and developed constitutive relations predictive of biaxial stress states; however, carbon fibers have been noted to have no such direct dependence. The influence of strain rate on composite materials has become a considerably active area for application [14, 16–20].

In the current work, the experimental data was previously obtained by Schaefer et al. [16]. From this work, the yield points for off-axis lamina specimens ($\theta = 15$–$90°$) tested in uniaxial compression were determined based on Fig. 25.1 for each tested strain rate. The off-axis fiber orientation to the loading axis induces a state of biaxial compressive and in-plane shear stress in the specimen. The average values are provided in Table 25.1.

The five lamina properties used in the NU Yield Criteria ($F_{2t}{}^y$, $F_{2c}{}^y$, $F_6{}^y$, E_2, and G_{12}) are shown in Table 25.2 for each tested strain rate.

The table shows that the properties increase with increasing strain rate. As previously noted [16], the values for shear strength and transverse tensile strength at the high strain rate were obtained by the data trend, which was established by plotting the individual test values for $F_{2c}{}^y$, $F_{2t}{}^y$, and $F_6{}^y$ against the log of strain rate. The result is shown in Fig. 25.2.

The data was fitted with a linear relationship with respect to the log of strain rate in the following form:

$$F^y(\dot{\varepsilon}) = F^y(\dot{\varepsilon_o})\left[m\log_{10}\left(\frac{\dot{\varepsilon}}{\dot{\varepsilon}_0}\right)+1\right] \tag{25.4}$$

where F^y is the yield stress ($F_{2c}{}^y$, $F_{2t}{}^y$, $F_6{}^y$), m is 0.092, and $\dot{\varepsilon_o}$ is reference strain rate of $10^{-4}\ s^{-1}$.

Fig. 25.2 Strain rate dependence for F^y_{2c}, F^y_{2t}, and F^y_6

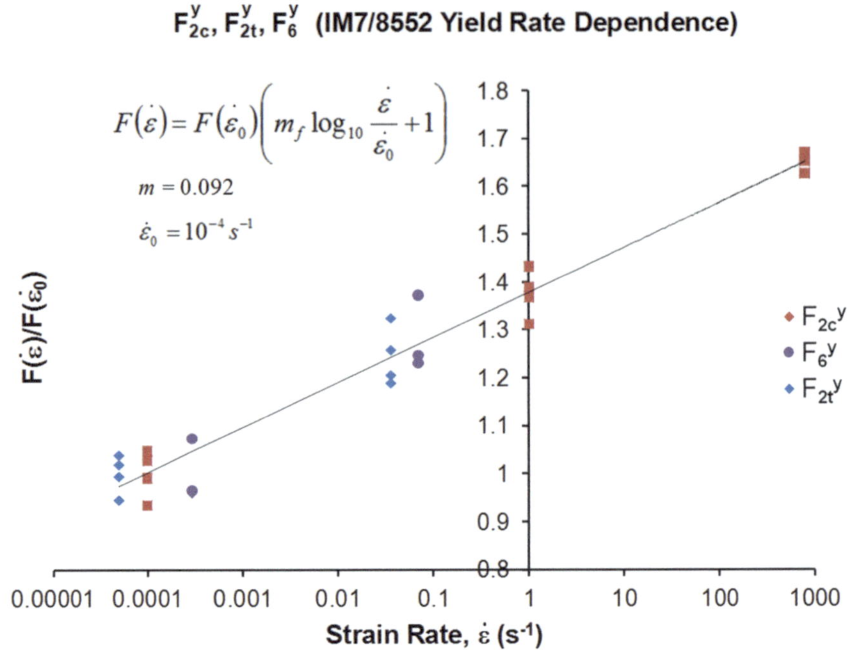

The yield property strain rate dependence of 0.092 was found to be higher than that of the unidirectional lamina ultimate failure ($m = 0.055$) [16]. In terms of material properties, the matrix yield strength is considerably lower than the brittle fibers; thus, the yield strength strain-rate-dependence for the unidirectional lamina is expected to coincide closely to that of the matrix for matrix-dominated layups.

From (25.4), the stresses were normalized to the reference strain rate, $\dot{\varepsilon_o}$, according to the relation:

$$f^y(\dot{\varepsilon}) = \left[m\log_{10}\left(\frac{\dot{\varepsilon}}{\dot{\varepsilon}_0}\right) + 1 \right] \tag{25.5}$$

The values for σ_2 and τ_6 were then determined via the conversion:

$$\dot{\sigma_i} = \sigma_i f^y(\dot{\varepsilon})^{-1}$$

$$\sigma_i = \sigma_2, \tau_6$$

The off-axis uniaxial yield stresses were transformed using the standard transformation relations and plotted on the $\sigma_2 - \tau_6$ plane (Fig. 25.3). The NU Yield Criteria and classic failure theories are plotted for comparison.

The NU Yield Criteria fit the data in superior agreement to the classic approaches (recast as yield criteria). The Sun criteria provide a similarly accurate prediction for yielding; however, a fitting parameter independent of material properties is required to enhance the fit. The values for $F_{2c}{}^y$, $F_{2t}{}^y$, $F_6{}^y$, E_2, and G_{12} were adjusted to those given by the intermediate rate experiments. This was also done for the dynamic testing; however, the general trend of the data was used to extrapolate the expected value for $F_{2t}{}^y$ and $F_6{}^y$ due to experimental limitations. The results are shown in Fig. 25.4. The intermediate and dynamic envelopes again provide a good fit to the plotted yield points in the $\sigma_2 - \tau_6$ plane.

The strain rate dependence relation (25.4) was used with $m = 0.092$ to produce a master data set by which the NU Criteria could be compared, as shown in Fig. 25.5. The master normalized lamina ultimate failure data from [16] and the yield analysis from the current investigation were integrated to create a master 'design envelope', shown in Fig. 25.6.

The combined master yield and ultimate lamina failure envelope creates a unique design plot. The plot provides a potential designer or structural engineer with ample flexibility for considering the strain rate effects on the failure as well as the yielding for a particular lamina orientation. The NU Yield Criteria fit the data for each of the tested strain rates (10^{-4}, 1, 10^3 s^{-1}) well, and provide an additionally useful tool for design. Considering that the same testing done to determine F_{2c}, F_{2t}, and F_6 may be used to also obtain $F_{2c}{}^y$, $F_{2t}{}^y$, and $F_6{}^y$, this material knowledge comes at no additional experimental cost. Now, only 12 characterization tests are required to obtain the yield and failure envelopes for a composite lamina over seven decades of strain rate. Thus, the NU Failure Theory and the NU Yield Criteria may be simply used for determining these key material properties.

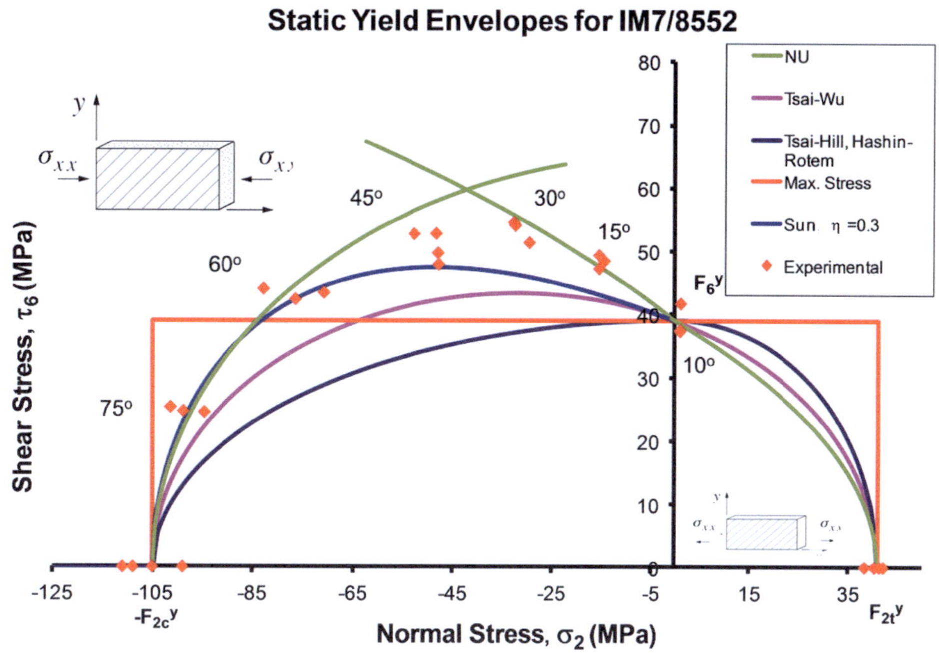

Fig. 25.3 Comparison of static yield envelopes for IM7/8552

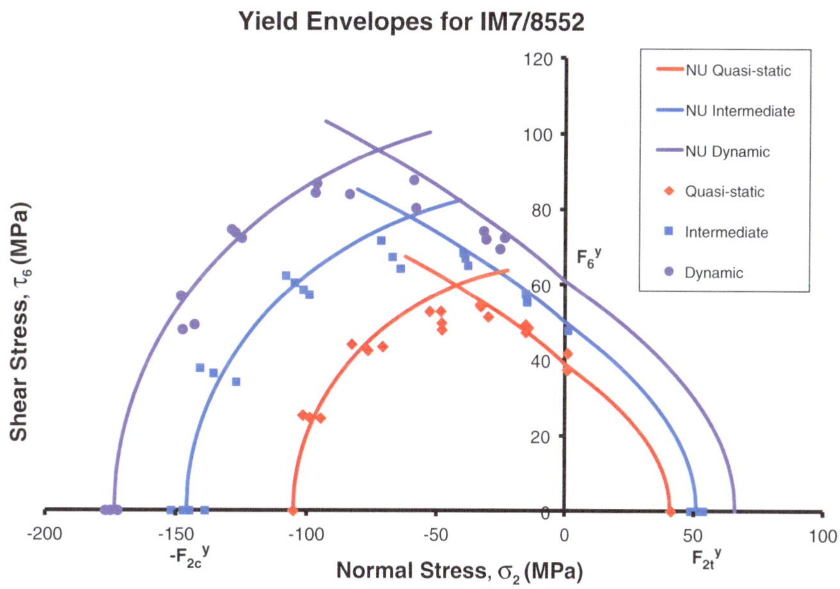

Fig. 25.4 NU yield criteria envelopes for quasi-static, intermediate, and dynamic strain rates

25.5 Yielding in Angle-Ply Laminates

To validate the NU Yield Criteria, it is required to isolate the underlying theoretical assumptions in more hierarchically complex scenarios for evaluation. In this light, angle-ply laminates, $[\pm\theta^o]_{14s}$ (where $\theta^o = 15°, 30°, 45°, 60°, 75°, 90°$), were tested under uniaxial compression. By alternating the orientation of the layer after each ply during layup, the effect of interlaminar stresses is kept to a minimum [21]. The $\pm 90°$ laminate is simply a unidirectional plate tested in the direction transverse to the fibers. Cuntze [22] relates that the matrix-dominated 'inter-fiber' failure modes present in such laminates are those least-well modeled by the leading theories in the recent worldwide failure exercises (WWFE), and Jadhav et al. [23] noted the potential for significant strain-rate-dependence for such modes in angle-ply laminates of a similar material system. To determine the thermally-induced residual cure stresses, the temperature difference (ΔT) between the curing stress-free temperature and ambient temperature was determined to be -150 K per [24]. Additionally, the embedded ply stresses were derived to include the thermally-induced residual stresses (σ_{1res}, σ_{2res}, τ_{6res}):

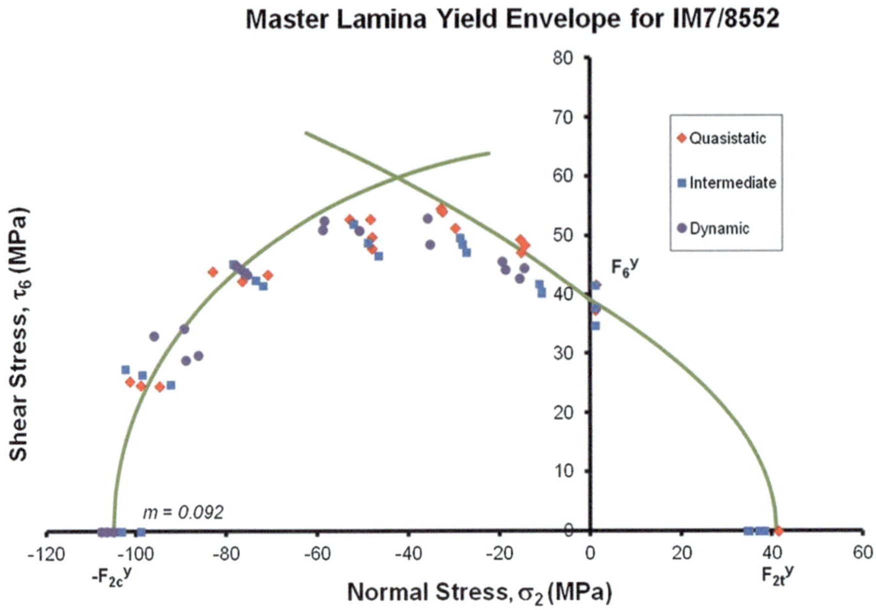

Fig. 25.5 Master lamina yield envelope

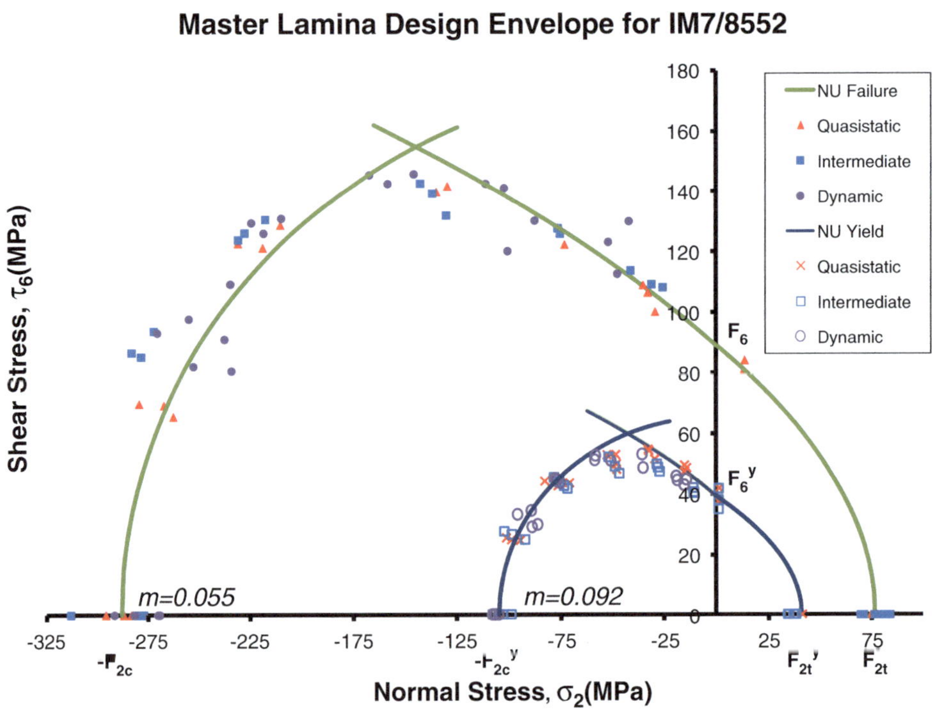

Fig. 25.6 Master NU lamina design envelope

$$\sigma_1 = \sigma_x \left[Q_{11} \left(\frac{m^2}{Q_{xx} - \frac{Q_{xy}^2}{Q_{yy}}} - \frac{n^2 Q_{yx}}{Q_{yy}Q_{xx} - Q_{xy}^2} \right) + Q_{12} \left(\frac{n^2}{Q_{xx} - \frac{Q_{xy}^2}{Q_{yy}}} - \frac{m^2 Q_{yx}}{Q_{yy}Q_{xx} - Q_{xy}^2} \right) \right] + \sigma_{1res} \quad (25.6)$$

$$\sigma_2 = \sigma_x \left[Q_{21} \left(\frac{m^2}{Q_{xx} - \frac{Q_{xy}^2}{Q_{yy}}} - \frac{n^2 Q_{yx}}{Q_{yy}Q_{xx} - Q_{xy}^2} \right) + Q_{22} \left(\frac{n^2}{Q_{xx} - \frac{Q_{xy}^2}{Q_{yy}}} - \frac{m^2 Q_{yx}}{Q_{yy}Q_{xx} - Q_{xy}^2} \right) \right] + \sigma_{2res} \quad (25.7)$$

Table 25.3 Angle-ply laminate embedded ply thermally induced residual stresses

$\pm\theta$	σ_{1res} (MPa)	σ_{2res} (MPa)	$\tau\sigma_{6res}$ (MPa)
15	−7.2	7.2	−12.4
30	−25.9	25.9	−15
45	−38.6	38.6	0
60	−25.9	25.9	15
75	−7.2	7.2	12.4

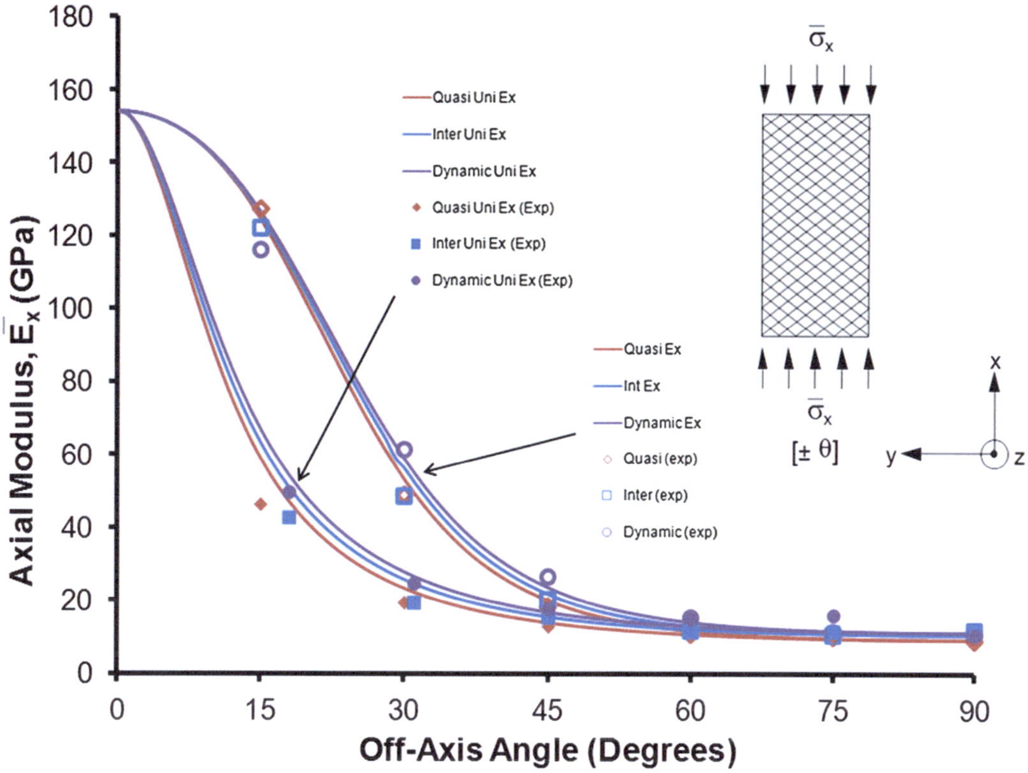

Fig. 25.7 Axial moduli for unidirectional lamina and angle-ply laminates

$$\tau_6 = -2mnQ_{66}\sigma_x \left[\frac{1}{Q_{xx} - \frac{Q_{xy}^2}{Q_{yy}}} + \frac{Q_{yx}}{Q_{yy}Q_{xx} - Q_{xy}^2} \right] + \tau_{6res} \tag{25.8}$$

where $m = cos(\theta)$, $n = sin(\theta)$, and σ_x is the axial stress.

The embedded ply thermal residual stresses are provided in Table 25.3.

The axial angle-ply laminate moduli were determined using lamination mechanics for symmetric balanced laminates [25]. The axial Young's modulus is related to the in-plane laminate stiffness, [A], and thickness, h, as follows:

$$\overline{E_x} = \frac{1}{h}\left[A_{xx} - \frac{A_{xy}^2}{A_{yy}} \right] \tag{25.9}$$

which reduces to a relation in terms of the transformed lamina stiffnesses, [Q]:

$$\overline{E_x} = \left[Q_{xx} - \frac{Q_{xy}^2}{Q_{yy}} \right] \tag{25.10}$$

The average axial angle-ply laminate moduli were plotted against axial strain; the average off-axis lamina axial moduli are included for comparison in Fig. 25.7.

The predicted values were in good agreement with the experimentally obtained data. In the figure, a clear correlation exists between the fiber orientation and composite moduli, as the fibers more effectively carry the load when they are closely aligned with the loading axis. A transition from the 'fiber-dominated' to the 'matrix-dominated' moduli is seen to occur between 45 and 50°. Angle-ply stress-strain behavior was recorded using the same approach as that for the lamina at three strain rates [24]. The results are shown in Fig. 25.8.

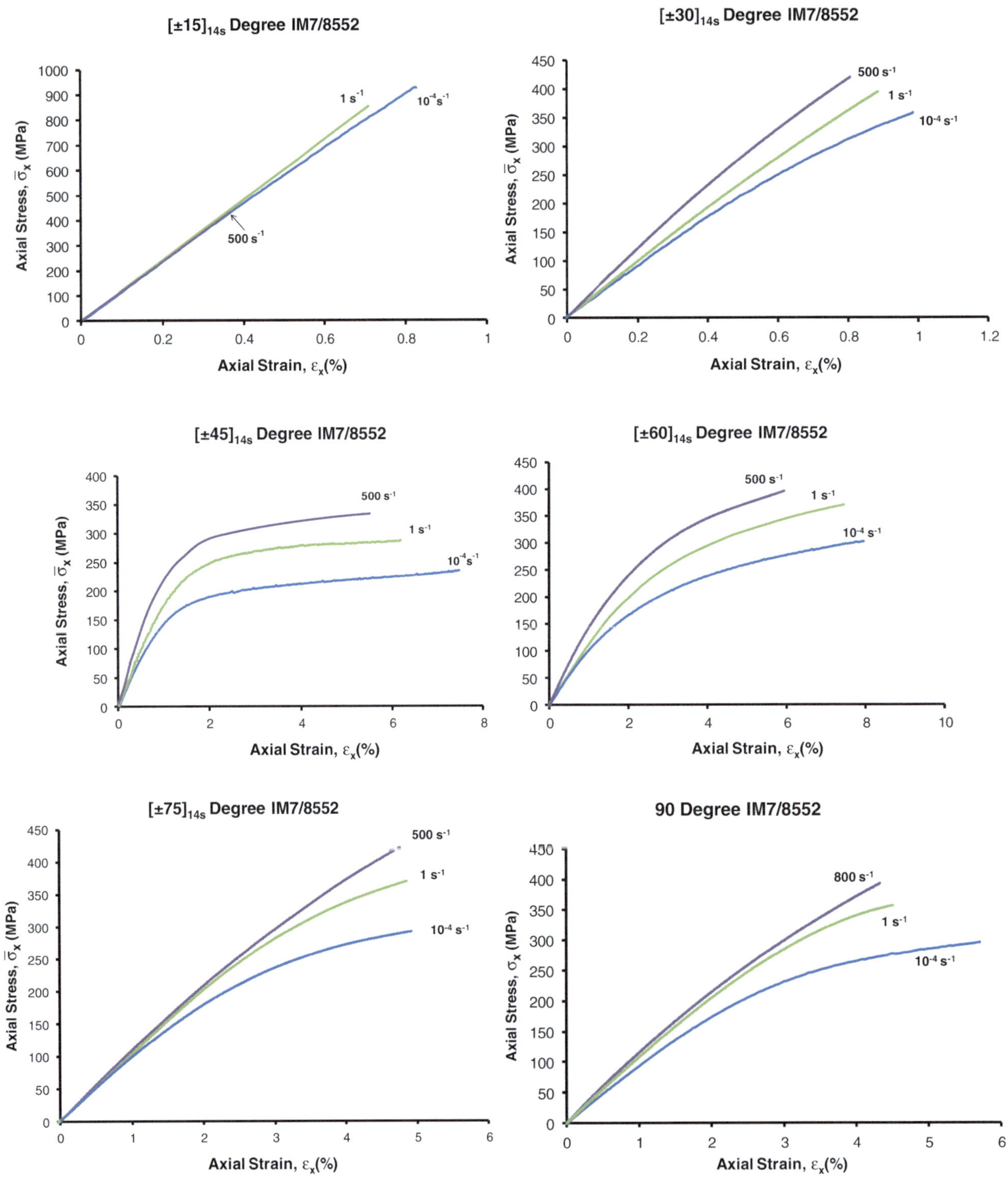

Fig. 25.8 Angle-ply axial stress behavior at quasi-static, intermediate and dynamic strain rates

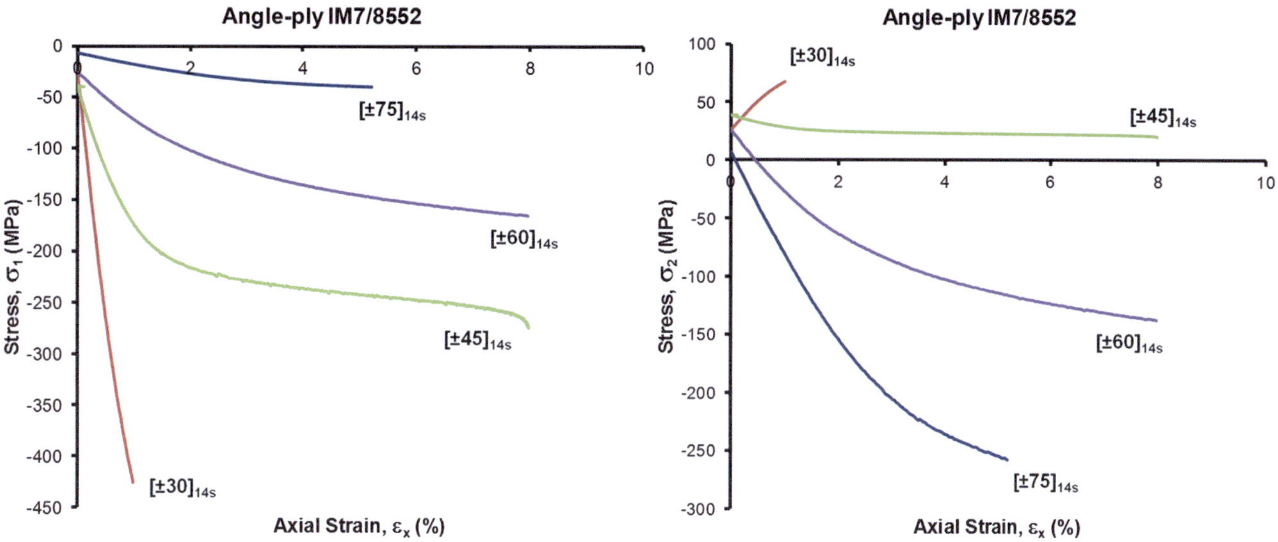

Fig. 25.9 Influence of thermally induced residual stress in angle-ply laminate

Table 25.4 Yield points of angle-ply compression samples at quasi-static, intermediate, and dynamic strain rates

$[\pm\theta]_{14s}$	$10^{-4}s^{-1}$ (MPa)	$1\ s^{-1}$ (MPa)	$500\ s^{-1}$ (MPa)
90	105	146	175
75	97	133	160
60	46	75	113
45	40	60	94
30	120	130	182
15	540	700	–

The angle-ply maximum stress increases with increasing strain rate while the maximum strain decreases; this is considered a 'stiffening' and 'strengthening' effect. The σ_1 and σ_2 stress-strain behavior for the $[\pm30]_{14s}$, $[\pm45]_{14s}$, $[\pm60]_{14s}$, and $[\pm75]_{14s}$ laminates is compared in Fig. 25.9.

The transverse stress provides an interesting perspective on the transition from tensile to compressive loading experienced by the laminate as the orientation moves from ±30 to ±45°. Figure 25.8 indicated a significant contrast in strain to failure between these two orientations and insight into why this occurs is shown in Fig. 25.9. The ±30° laminate experiences tensile stress in the transverse direction for the axial compressive loading condition; the ultimate failure stress and strain correspond closely to the thresholds set by F_{2t} and ε_{2t}. A key observation is that the residual stress for this laminate is quite large (~26 MPa) compared to F_{2t} (76.4 MPa).

25.6 Angle-Ply Rate Dependence

The yield points were obtained from the angle-ply stress-strain curves (Fig. 25.8) and are provided in Table 25.4.

Due to premature failure, the yield points for the $[\pm15]_{14s}$ laminate were not obtained; in the remaining cases, the yield point stress was seen to increase with increasing strain rate. In relation to the unidirectional lamina, the angle-ply counterparts are shown to yield at a lower applied axial stress. This behavior is due to the biaxial state of stress as well as the 'locked-in' residual stresses from curing. The axial stress at yielding was transformed to the associated lamina stresses using laminate mechanics, and the residual stress was incorporated for $\Delta T = -150\ K$. At the extremes, the tested angle-ply laminates range from $[90]_{54}$ in compression, through $[\pm\theta]_{14s}$, to $[90]_{16}$ in tension. The Northwestern lamina Yield Criteria presented previously were applied and the yield data were plotted and compared to the classical failure theories. The static angle-ply yield envelope is depicted in Fig. 25.10, and all three envelopes are shown in Fig. 25.11.

As a result of the noted boundary layups, the strain rate dependence of the angle-ply laminates is the same as that of the lamina according to relation (25.4) with $m = 0.092$. The data was recast in terms of the static strain rate and plotted in Fig. 25.12 with the master NU Yield Criteria envelope. The master NU envelope fits the data quite well. Importantly, the

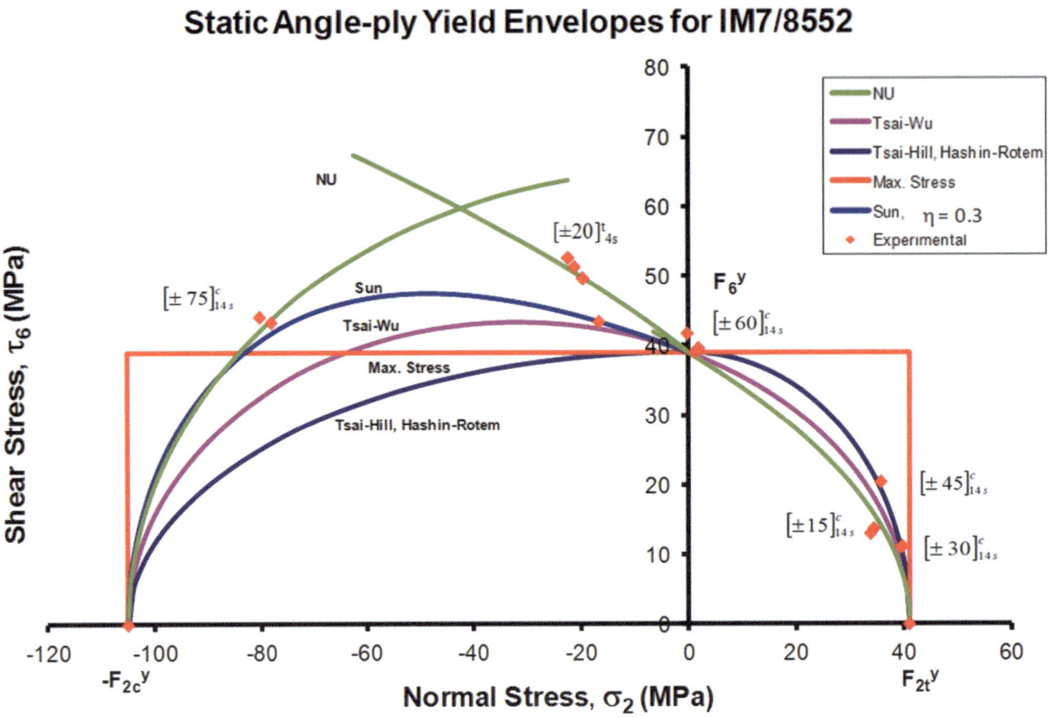

Fig. 25.10 Comparison of static angle-ply yield envelopes for IM7/8552

Fig. 25.11 Angle-ply yield envelopes at static, intermediate and dynamic strain rates for IM7/8552

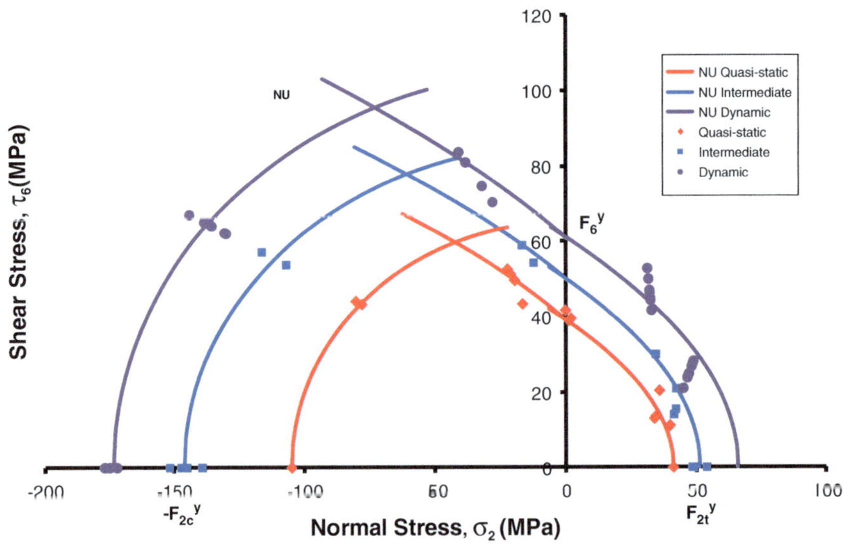

lamina and angle-ply data may be combined into a single yield envelope, as shown in Fig. 25.13. Thus, from a single set of lamina tests the strain-rate-dependent yield behavior of angle-ply laminates may be predicted with enhanced accuracy using a mitigated testing framework.

25.7 Conclusion

This investigation provides a direct comparison by which to evaluate the Northwestern Yield Criteria for lamina and the validity of its application to composite laminates. The results indicate that the failure-mode-based Northwestern Failure Theory may be recast as a set of yield criteria to effectively predict the yielding of both lamina and angle-ply laminates.

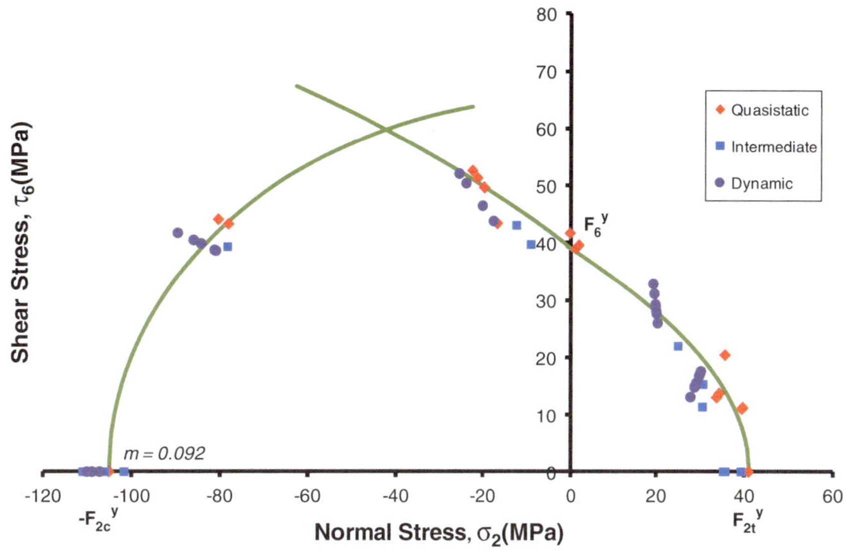

Fig. 25.12 Master angle-ply laminate yield envelope

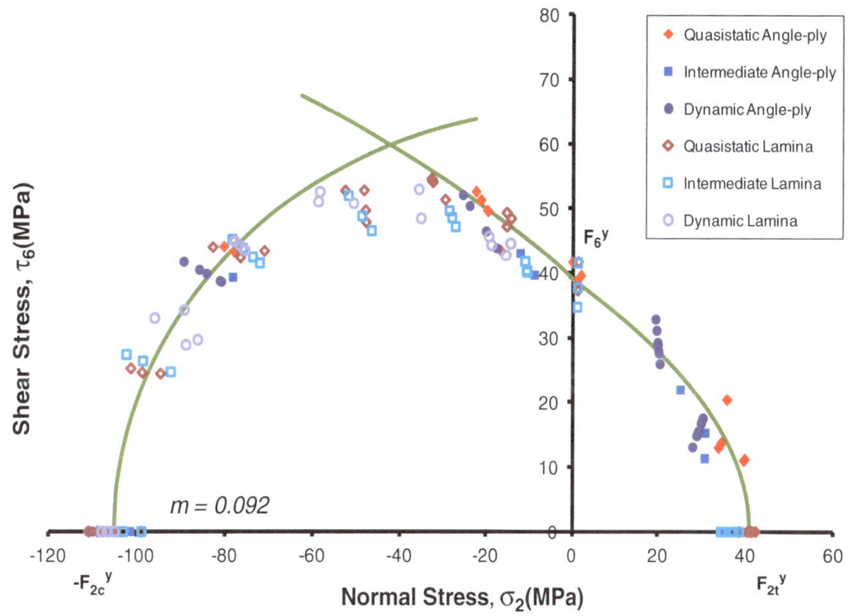

Fig. 25.13 Master NU lamina and angle-ply laminate yield envelope for IM7/8552

This reduced the number of required characterization tests by an order of magnitude; further, provides both the design engineer and structural analyst with key insight into composite laminate behavior. Investigating neat matrix strain-rate-dependence in light of the NU Yield Criteria is currently in development.

Acknowledgement This work was supported by the Office of Naval Research (ONR). We are grateful to Dr. Y. D. S. Rajapakse of ONR for his encouragement and cooperation.

References

1. ASME: Guide for Verification and Validation in Computational Solid Mechanics, pp. 1–36. ASME V&V 10, New York (2006)
2. Hart-Smith, L.J.: The role of biaxial stresses in discriminating between meaningful and illusory composite failure theories. Compos. Struct. **25**, 3–20 (1993)
3. Argüelles, A., Vina, J., Canteli, A., Lopez, A.: Influence of the matrix type on the mode I fracture of carbon-epoxy composites under dynamic delamination. Exp. Mech. **51**(3), 293–301 (2011)

4. Costa, M.L., Botelho, E.C., Paiva, J.M.F., Rezende, M.C.: Characterization of cure of carbon/epoxy prepreg used in aerospace field. Mater. Res. **8**(3), 317–322 (2005)
5. Eibl, S.: Observing inhomogeneity of plastic components in carbon fiber reinforced polymer materials by ATR-FTIR spectroscopy in the micrometer scale. J. Compos. Mater. **42**(12), 1231–1246 (2008)
6. Lee, J., Soutis, C.: A study on the compressive strength of thick carbon fibre–epoxy laminates. Compos. Sci. Technol. **67**(10), 2015–2026 (2007)
7. Marasco, A.I., Cartié, D.D.R., Partridge, I.K., Rezai, A.: Mechanical properties balance in novel Z-pinned sandwich panels: out-of-plane properties. Compos. A: Appl. Sci. Manuf. **37**(2), 295–302 (2006)
8. Murri, GB.: Evaluation of delamination onset and growth characterization methods under mode I fatigue loading. NASA/TM-2013-217966 (2013)
9. Tsotsis, T.K., Keller, S., Lee, K., Bardis, J., Bish, J.: Aging of polymeric composite specimens for 5000 hours at elevated pressure and temperature. Compos. Sci. Technol. **61**(1), 75–86 (2001)
10. Wolfrum, J., Eibl, S., Lietch, L.: Rapid evaluation of long-term thermal degradation of carbon fibre epoxy composites. Compos. Sci. Technol. **69**(3–4), 523–530 (2009)
11. NIAR, NCAMP.: Hexcel 8552 IM7 unidirectional prepreg 190 gsm & 35% RC qualification material property data report. CAM-RP-2009-015(Rev A), 22 April 2011
12. Daniel, I.M., Luo, J.-J., Schubel, P.M., Werner, B.T.: Interfiber/interlaminar failure of composites under multi-axial states of stress. Compos. Sci. Technol. **69**(6), 764–771 (2009)
13. Daniel, I.M., Cho, J.-M., Werner, B.T., Fenner, J.S.: Characterization and constitutive modeling of composite materials under static and dynamic loading. AIAA J. **49**(8), 1658–1664 (2011)
14. Daniel, I.M., Werner, B.T., Fenner, J.S.: Strain-rate-dependent failure criteria for composites. Compos. Sci. Technol. **71**(3), 357–364 (2011)
15. Werner, B.T., Daniel, I.M.: Characterization and modeling of polymeric matrix under static and dynamic loading. In: SEM XII International Congress & Exposition on Experimental & Applied Mechanics 2012, Costa Mesa (2012)
16. Schaefer, J.D., Werner, B.T., Daniel, I.M.: Strain-rate-dependent failure of a toughened matrix composite. Exp. Mech. **54**(6), 1111–1120 (2014)
17. Karim, M.R., Fatt, M.S.H.: Rate-dependent constitutive equations for carbon fiber-reinforced epoxy. Polym. Compos. **27**(5), 513–528 (2006)
18. Koerber, H., Camanho, P.P.: High strain rate characterisation of unidirectional carbon–epoxy IM7-8552 in longitudinal compression. Compos. A: Appl. Sci. Manuf. **42**(5), 462–470 (2011)
19. Thiruppukuzhi, S.V., Sun, C.T.: Models for the strain-rate-dependent behavior of polymer composites. Compos. Sci. Technol. **61**(1), 1–12 (2001)
20. Catalanotti, G., Camanho, P.P., Marques, A.T.: Three-dimensional failure criteria for fiber-reinforced laminates. Compos. Struct. **95**, 63–79 (2013)
21. Werner, B.T., Schaefer, J.D., Daniel, I.M.: Effect of ply dispersion on failure characteristics of multidirectional laminates. In: SEM 2013 Annual Conference & Exposition on Experimental and Applied Mechanics, Lombard, 3–5 June 2013
22. Cuntze, R.G.: The predictive capability of failure mode concept-based strength criteria for multi-directional laminates—part B. Compos. Sci. Technol. **64**(3–4), 487–516 (2004)
23. Jadhav, A., Woldesenbet, E., Pang, S.-S.: High strain rate properties of balanced angle-ply graphite/epoxy composites. Compos. Part B **34**(4), 339–346 (2003)
24. Schaefer, J.D.: Matrix dominated failure of fiber-reinforced composite laminates under static and dynamic loading. Dissertation, Northwestern University (2014)
25. Daniel, I.M., Ishai, O.: Engineering Mechanics of Composite Materials. Oxford University Press, Oxford (2006)

Chapter 26
Experimental Fatigue Specimen and Finite Element Analysis for Characterization of Dental Composites

Dhyaa Kafagy and Michael Keller

Abstract Dental composites are becoming more popular due to their semi white color and appearance. Mechanical damage such as cracks are causing the majority of short-term failures of dental composites. Using self-healing materials most of these failures can be prevented. Fatigue loads are a proper method to characterize the crack initiation and propagation. As healing makes uncertainty about the location of the crack tip, samples of tapered double cantilever beam (TDCB) are frequently used for their crack length independent in the measurements of healing efficiency and fracture toughness of self-healing. Due to the high cost of dental composite materials, tiny, inexpensive TDCB samples, about 30 % of the standard size, were developed and optimized with Rapid Prototyping (Objet 3D printer). FEA is also performed in order to visualize the stress field of the crack tip.

Keywords Dental materials • Composites • Microcapsules • Self-healing • Fatigue

26.1 Introduction

Resin-based dental composites are an increasingly popular restoration option when compared to conventional amalgams for dental restorations [1]. This increasing popularity is mainly driven by the improved esthetics of composites as the amalgam restorations are more noticeable. While composite restorations poses improved esthetics, they are susceptible to mechanical damage in the form of crack initiation and propagation.

Mechanical failures of composites increase the frequency dental procedures that must be performed to prevent further tooth decay. Repairs of resin composites typically require the removal and replacement of the original repair. One potential method to reduce these problems is to incorporate an ability for the resin to self-heal any mechanical damage that may be introduced during the lifespan of the restoration. One approach is to adopt the microcapsule-based system originally developed by White et al. [2], this approach has been show to be capable of autonomic retardation and arrest of propagating cracks. As part of the synthesis of all self-healing materials, an accurate testing method is critical. Since self-healing event of a crack with a statistically distributed healing component is by nature itself randomly distributed, determining the precise location of the healed crack-tip is often difficult. One approach is solve this problem is to fracture specimens which provide a crack-length-independent measure of fracture toughness such as provided by a tapered, double cantilever beam (TDCB) specimen.

The conventional size of TDCB specimens is unfeasible for use with restorative resin materials as the cost for a large testing sample is prohibitive. Therefore, a new compact TDCB or CTDCB was developed, using rapid prototyping, specifically for use in characterizing self-healing dental composites. Since the loading environment of composite restorations is cyclic fatigue, we investigate the fatigue performance of these sub size specimens.

D. Kafagy • M. Keller (✉)
Department of Mechanical Engineering, The University of Tulsa, 800 S. Tucker Dr., Tulsa, OK 74104,, USA
e-mail: dhk542@utulsa.edu

A.M. Beese et al. (eds.), *Fracture, Fatigue, Failure and Damage Evolution, Volume 8*, Conference Proceedings of the Society for Experimental Mechanics Series, DOI 10.1007/978-3-319-21611-9_26

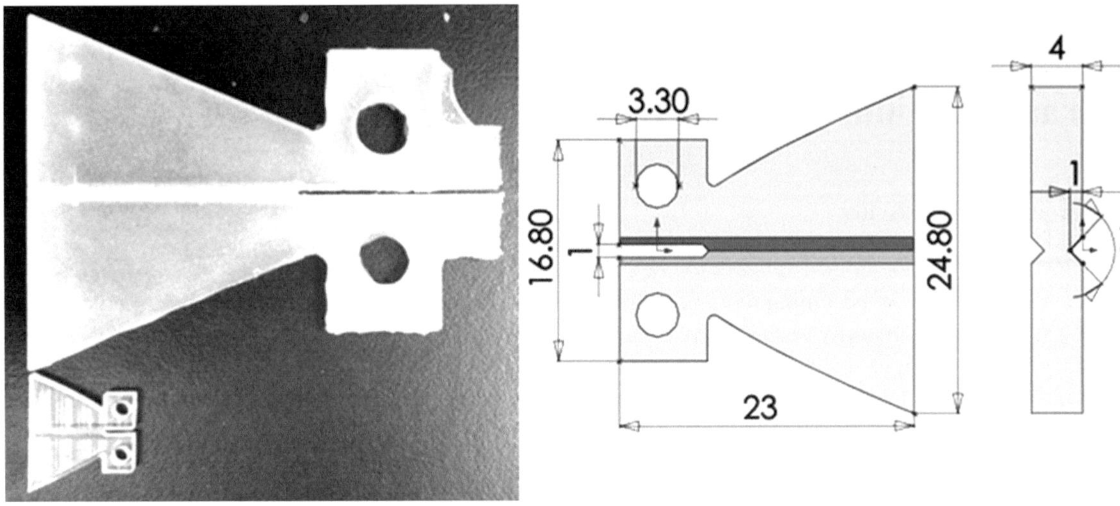

Fig. 26.1 Standard TDCB shape (*larger*) and new CTDCB dimensions (*compact*) [5]

26.2 Experimental Method

26.2.1 Specimen Design and Rapid Prototyping

TDCB specimens provide a crack-length-independent fracture specimen testing by tapering the height of the beam such that the change in compliance over change in crack length (dc/da) is constant. The beam shape can be defined using the following relationship [3],

$$3a^2/h^3 + 1/h \propto dC/da \tag{26.1}$$

whereas is the crack length and h is the width of the beam. The complex curve defined by the left side of (26.1), is often approximated by a linear taper in traditional TDCB designs. Figure 26.1 below shows a conventional TDCB as described in [4] along with the compact design used in this work.

For this study, the fatigue behavior of these miniature specimens was investigated using a prototype specimens fabricated using polyjet rapid prototyping (Objet Eden). This allowed for a rapid and accurate fabrication of the complicated curved geometry derived from (26.1). Samples were printed using high-resolution mode from a rigid polymer (RGD 720). Then were cleaned using high-pressure water and a dilute NaOH solution.

26.2.2 Fatigue Testing

The TDCB specimens were pin-loaded by a servo-hydraulic load frame (MTS) using a 100 N load cell (MTS). Cyclic loading was applied in load control, with a maximum load of 50 % of the quasi-static fracture load, which was previously determined. Before testing, each specimen was pre-cracked using razor sawing in order to ensure a sharp crack tip. The specimens were loaded using a sine wave at a frequency of 4 Hz, while collecting force, time, and displacement data. Load ratio for this testing was 0.1.

26.3 Finite Element Analysis

An initial study of the specimen was performed using Finite Element Analysis (FEA) (Abaqus) to investigate the displacement and stress fields. As an initial set of boundary conditions loading was applied using shear on the front edge of the loading “ears.” The specimen was meshed with six node brick elements and near the crack tip the mesh elements were refined. Asymmetry boundary condition was applied at the symmetric line between both sides of the specimen. Material properties were used from the manufacturers data sheet for the rapid prototype material.

26.4 Results and Discussion

26.4.1 *FEA Visualization of Crack Tip*

FEA visualization study of the crack tip was performed to see the stress and displacement fields about the crack tip. Figure 26.2 shows the stress field.

As we can see in Fig. 26.2, as the crack tip stress fields develop and grow as the crack length extends. For the simulation results shown in Fig. 26.2, the applied load was held constant for all crack lengths. We are currently extracting the predicted stress intensity factors to understand when end effects begin to dominate the stress field.

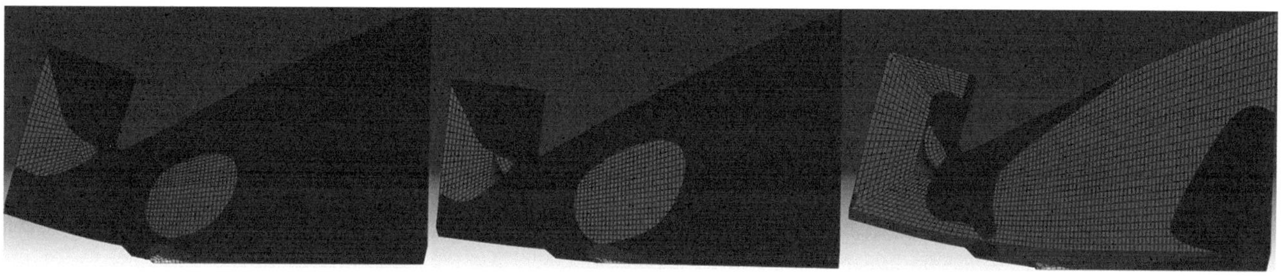

Fig. 26.2 FEA predicted stress fields in the tensile direction of the new TDCB compact model for three different crack lengths (1, 2, and 3 mm)

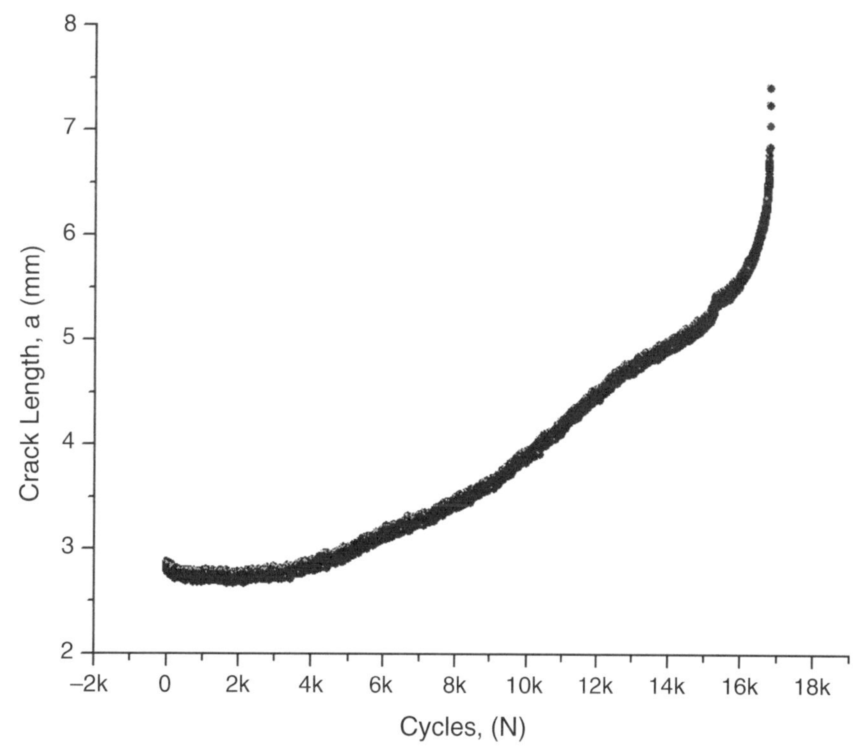

Fig. 26.3 Crack length vs. number of load cycles for a compact TDCB specimen

26.4.2 *Fatigue Testing*

A representative crack length versus cycle plot for the specimen is shown in Fig. 26.3. The crack-tip location in this graph was determined using a compliance calibration that was performed on s series of quasi-static tests [5]. Optical crack length measurements are in process to confirm that the predicted crack length from the compliance of the specimen is accurate. As we can see in Fig. 26.2, there is an incubation period at the beginning of the test where there is no apparent crack growth. This is likely due to a razor sawing process that did not initiate a sharp precrack into the starter-notch. We are currently investigating the pre-cracking procedure to ensure that a sharp crack is initiated.

After the initial fatigue plateau, a region of broadly linear crack-growth per cycle begins. This feature is expected for constant compliance specimens, as the stress intensity factor should be constant in this region. From the slope of this curve, the crack growth rate can be determined and was found to be approximately 215 μm/cycle. Catastrophic crack growth initiates after 2.5 mm of fatigue crack growth

Based on this initial fatigue test, the sub sized fatigue specimen appears to possess the required crack-length-independent crack-growth rates necessary for the characterization of self-healing composite materials. Based on the fatigue test shown above, the constant K region appears to extend for approximately 3 mm from the root of the starter notch for the geometry shown in Fig. 26.1b. Further testing and finite element simulations are ongoing to confirm this initial experimental finding.

26.5 Conclusions

A new, sub sized tapered double cantilevered beam specimens were design and fabricated. An initial fatigue test was performed and the sample geometry was shown to possess a linear crack growth rate region that corresponds to a region of constant stress intensity K. The linear crack-growth region indicates that these specimens will be appropriate for the evaluation of the self-healing of fatigue growth in restorative resins.

References

1. Watanabe, H., Khera, S.C., Vargas, M.A., Qian, F.: Fracture toughness comparison of six resin composites. Dent. Mater. **24**(3), 418–425 (2008)
2. White, S.R., Sottos, N.R., Geubelle, P.H., Moore, J.S., Kessler, M.R., Sriram, S.R., Brown, E.N., Viswanathan, S.: Autonomic healing of polymer composites. Nature **409**, 794–797 (2001)
3. Mostovoy, S., Crosley, P.B., Ripling, E.J.: Use of crack-line-loaded specimens for measuring plane-strain fracture toughness. J. Mater. **2**(3), 661–681 (1967)
4. Brown, E.N., Sottos, N.R., White, S.R.: Fracture testing of a self-healing polymer composite. Exp. Mech. **42**(4), 372–379 (2002)
5. Adams, K., Ivanoff, D., Khajotia, S., Keller, M.: Compact fracture specimen for characterization of dental composites. In: Tandon, G. (ed.) Composite, Hybrid, and Multifunctional Materials: Proceedings of the 2014 Annual Conference on Experimental and Applied Mechanics. Conference Proceedings of the Society for Experimental Mechanics Series, vol. 4. Springer, Heidelberg (2014)

Chapter 27
Fracture Toughness and Impact Damage Resistance of Nanoreinforced Carbon/Epoxy Composites

Joel S. Fenner and Isaac M. Daniel

Abstract In this study, the objective was to develop, manufacture, and test hybrid nano/microcomposites with a nanoparticle reinforced matrix and demonstrate improvements to damage tolerance via Mode-II fracture toughness and impact damage absorption. The material employed was a woven carbon fiber/epoxy composite, with multi-wall carbon nanotubes as a nano-scale reinforcement to the matrix. A direct-mixing process, aided by a block copolymer dispersant and sonication, was employed to produce the nanoparticle-filled epoxy matrix. Fracture toughness was tested by several different Mode-II and mixed Mode-I/Mode-II specimens to determine the toughness improvement. Testing and material difficulties were overcome by this approach, showing a Mode-II toughness improvement of approx. 35 % in the hybrid material. Impact tests were performed in a falling-weight drop tower at different energies to introduce interlaminar damage in samples of both materials. Impact damaged specimens were imaged by ultrasonic c-scans to assess the area of the damage zone at each ply interface. Post-mortem optical microscopy confirmed the interlaminar nature of the impact damage. These tests showed a consistently smaller absorbed energy and smaller total damage area for hybrid composite over reference material, translating to a nominally higher 'effective impact toughness' in the hybrid composite (approx 42 %) regardless of specific impact energy.

Keywords Nanocomposites • Hybrid nano/microcomposites • Test methods • Fracture toughness • Impact behavior

27.1 Introduction

The effect of nanoparticles on the mechanical properties of composite materials is an area of continuing interest in recent materials research. Such research is often motivated by the surprising enhancements that can be obtained from the introduction of relatively small quantities of nanoparticles [1–10]. The result of work in this area has considerable bearing on the possible applications of these materials and thus promotes their wider use.

Damage tolerance enhancements are to be expected from the addition of nanoparticles to a composite material owing to additional energy-absorbing mechanisms that arise. In the case of carbon nanotubes, mechanisms such as nanotube pullout from matrix, nanotube fracture, and nanotube stretching cause additional energy absorption at the nanoscale (Fig. 27.1) [2]. Mechanisms such as these contribute to bulk macroscopic energy absorption, which in turn improves overall fracture toughness and impact damage tolerance.

This objective of this study was to examine the effect of nanoparticles, namely carbon nanotubes (CNT), on the damage tolerance and energy-absorption properties of composites, especially impact damage and associated Mode-II fracture toughness, which often dominates interlaminar damage in impact [8–10].

27.2 Material Processing

The material employed in this investigation was a woven carbon fiber/epoxy composite with matrix reinforcement provided by short multi-wall carbon nanotubes. The major mechanical reinforcement was provided by a 5-harness satin weave carbon fabric dry preform (Hexcel, AGP370-5H, AS4 fibers, 6 k tows). The matrix was a typical Bisphenol-A epoxy resin (DGEBA, Huntsman GY 6010) cured with an anhydride hardener (Methyltetraphthalic anhydride, Huntsman HY 917)

J.S. Fenner • I.M. Daniel (✉)
Center for Intelligent Processing of Composites, Northwestern University, 2137 Tech Drive, Evanston, IL 60208, USA
e-mail: imdaniel@northwestern.edu

A.M. Beese et al. (eds.), *Fracture, Fatigue, Failure and Damage Evolution, Volume 8*, Conference Proceedings of the Society for Experimental Mechanics Series, DOI 10.1007/978-3-319-21611-9_27

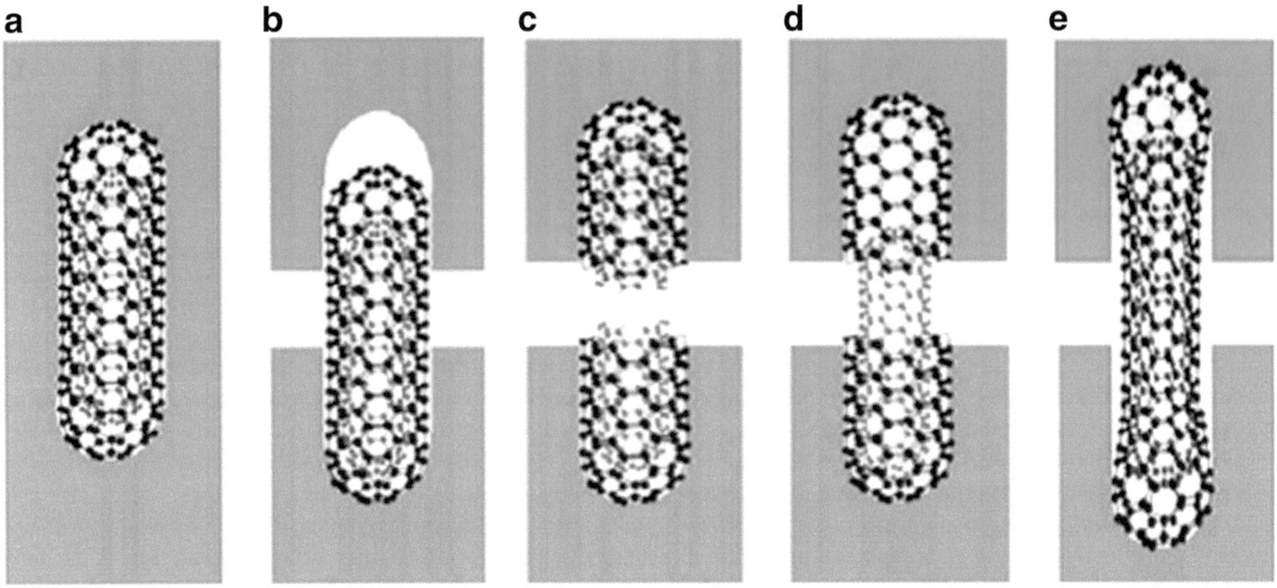

Fig. 27.1 Illustration of possible energy-absorbing mechanisms of CNTs embedded in matrix: (**a**) initial state, (**b**) pullout following CNT/matrix debonding, (**c**) fracture of CNT, (**d**) telescopic pullout-fracture of outer layer and pullout of inner layer, (**e**) partial debonding and stretching [2]

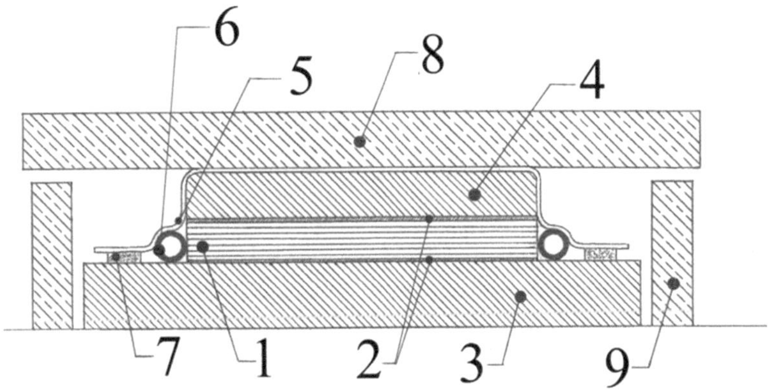

Fig. 27.2 Schematic diagram of composite fabrication mold

and an additional amine accelerator (1-Methylimidazole, Huntsman DY 070). The nominal mixture ratio was 100:90:1 (resin:hardener:accelerator) by weight. The resin was reinforced with short multi-wall carbon nanotubes (Helix Material Solutions) of 1–2 μm in length and 10–30 nm in outer diameter. A polyester block copolymer supplied in solution with a weakly volatile organic solvent (Disperbyk-2150, BYK Chemie) was used to facilitate dispersion of the CNTs [5].

Preparation of the nanoparticle enhanced matrix was carried out primarily by simple mixing and sonication. A weighed amount of DGEBA was mixed with 0.5 %wt of nanotubes and 0.5 %wt of copolymer solution. The materials were mixed together thoroughly and then vacuum degassed at an elevated temperature (80 °C) to remove the organic solvent. The anhydride hardener was then added and mixed thoroughly, followed by a repeat vacuum degassing. The resulting mixture was subjected to sonication (90 W at 20 kHz for 30 min.) to disperse the nanotubes. Just prior to composite infusion, the accelerator amine was then added, and the resin mixture infused into the carbon fiber preform, layer by layer, in a wet layup process. The wetted perform stack was subjected to final degassing and then brought to elevated temperature (150 °C) in a mold to achieve full cure. The construction of the mold (Fig. 27.2) allowed for careful control of finished laminate thickness and controlled removal of excess resin during curing. This process was developed and adapted to avoid previously encountered problems related to a marked increase in resin viscosity due to the presence of nanotubes and a filtration effect on nanotubes encountered in VARTM processing. The same procedure was used for the reference and nano-reinforced (hybrid) composites. After curing, composite plates were rough cut into specimens by means of diamond-abrasive cutting wheels, and wet-polished with SiC abrasive papers to final dimensions and smoothness.

27.3 Mode-II Fracture Testing

Initial Mode-II fracture toughness tests were performed by three-point End Notched Flexure (ENF), which was chosen for its overall simplicity in testing and analysis (Fig. 27.3) [11]. Specimens were prepared with nominal dimensions $L = 25$ mm, $a_0 = 7.6$ mm, $b = 12$ mm, $2h = 4$ mm (8 plies).

Tests were carried out on numerous specimens of each material type (reference & hybrid composite), giving curves of the form shown in Fig. 27.4. From these curves, calculation of the Mode-II strain energy release rate was made by various methods. For the ENF specimen, the sample compliance may be given as

$$C = \frac{\delta}{P} = \frac{1}{8bh}\left[\frac{2L^3 + 3a^3}{E_1 h^2} + \frac{12L}{5G_{13}}\right] \tag{27.1}$$

where a is the crack length
b is the specimen width
$2h$ is the specimen thickness
E_1 is the elastic modulus along the beam span
G_{13} is the shear modulus in the plane of bending

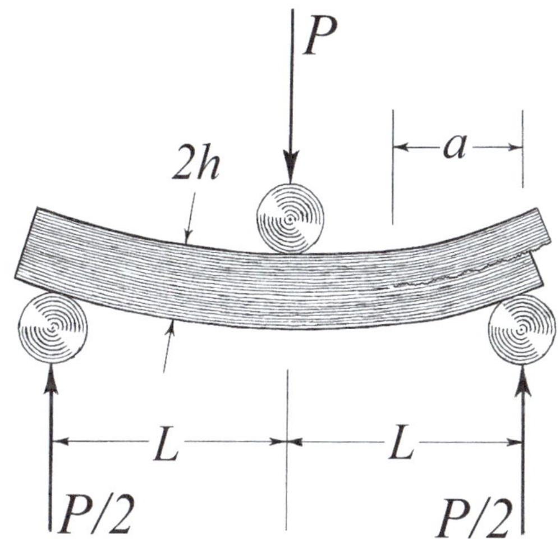

Fig. 27.3 Illustration of End Notched Flexure (ENF) test

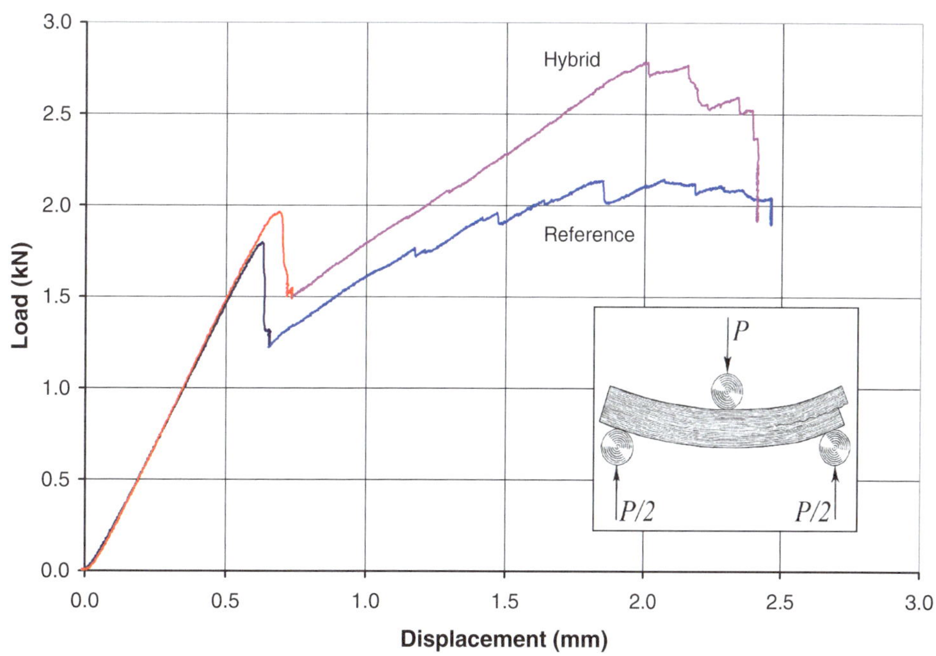

Fig. 27.4 Representative curves of load-displacement behavior for Mode-II ENF tests

Table 27.1 Mode-II fracture toughness from ENF tests, calculated by various methods

Method of calculation	Reference composite (J/m^2)	Hybrid composite (J/m^2)
Peak load	553 (±200)	729 (±239)
Peak load & individual specimen compliance calibration	590 (±178)	789 (±144)
Work of fracture at first release, calculated crack lengths	1610 (±531)	1850 (±144)
Work of fracture at first release, measured crack lengths	3870 (±1340)	3730 (±215)

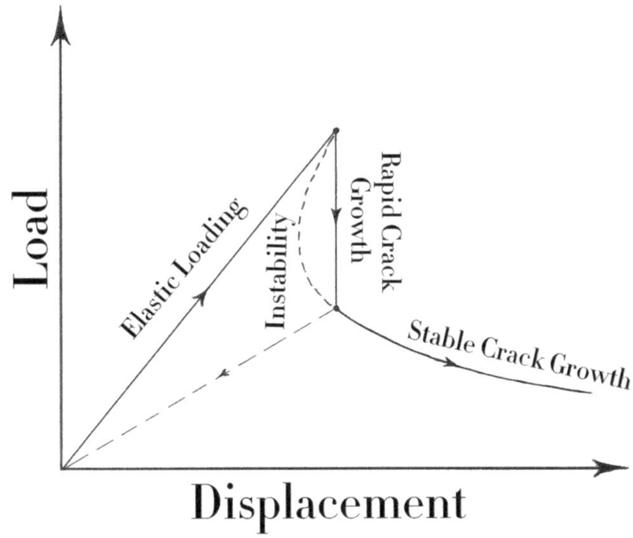

Fig. 27.5 Theoretical behavior of a three-point end notched flexure specimen with a short ($a < a_c$) starting crack

Hence, by the compliance method, the Mode-II fracture toughness at peak load at crack advance (G_{II}) is given by the relation

$$G_{II} = \frac{9a^2P^2}{16E_1b^2h^3} \tag{27.2}$$

where P is the applied load at the central roller

Alternatively, the fracture toughness was computed making use of the loading (until first peak) and unloading (immediately after first peak) compliances of the specimen to determine the crack length. Neglecting shear effects in (27.1), the effective crack length may be calculated as

$$a \cong \left[\frac{8\left(\delta/P\right)E_1bh^3 - 2L^3}{3} \right]^{1/3} \tag{27.4}$$

This permitted simple calculation of the crack length a pre- and post-peak from the measured specimen compliances. Fracture toughness was then calculated by the area method as

$$G_{II} = \frac{P_1\delta_2 - P_2\delta_1}{2b\left(a_2 - a_1\right)} \tag{27.5}$$

Calculation of the Mode-II fracture toughness purely by peak load or by peak load with initial compliance correction showed an increase of 30 % in G_{II} in hybrid composite over reference composite (Table 27.1). Depending on the method of calculation, however, some disparate "apparent" values of G_{II} were obtained from the data, such as those obtained using post-peak compliance or manually-measured crack lengths.

The discrepancy between various calculation methods served to indicate an obvious experimental inconsistency. In prior work on Mode-I fracture toughness with identical material [12], differing methods of calculation typically resulted in very good agreement. Furthermore, the load-displacement behavior (Fig. 27.4) of the samples did not exhibit the expected post-peak load decay that is to be expected in a typical ENF test (Fig. 27.5). These details suggested some mechanism impeding crack growth immediately after initial crack release.

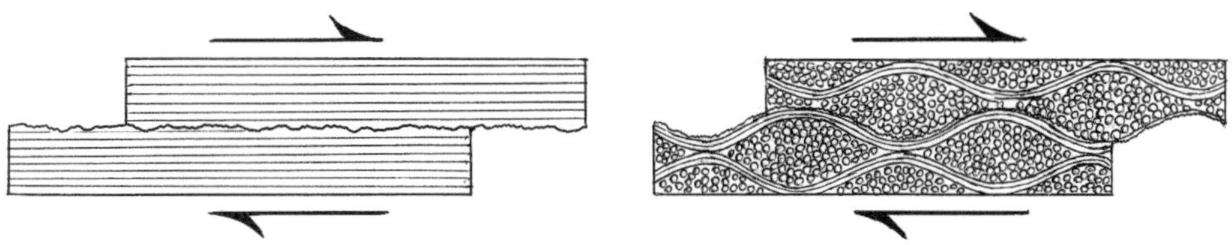

Fig. 27.6 Illustration of simple Mode-II sliding behavior (*left*) and Mode-II woven composite ply-on-ply interlock (*right*)

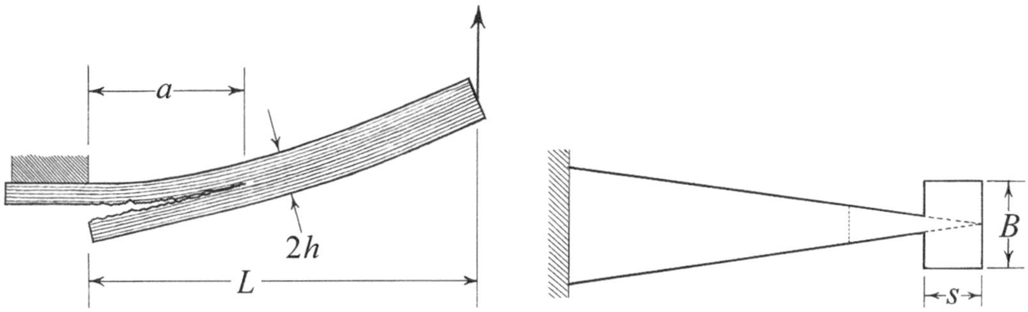

Fig. 27.7 Mixed-mode form of tapered cantilever beam specimen

The most likely cause is ply-on-ply friction or mechanical interference disturbs the outcome of this type of test, especially when employed on a woven composite material (as in Fig. 27.6) [13]. The extent of this effect is quite probably more evident in the unloading behavior rather than the loading behavior, as crack propagation involves more sliding motion than elastic loading, and as such, the values of G_{II} obtained by peak loads *only* should be the most accurate. Even if this rationale is correct, there is still significant experimental variation in numerical outcome of these tests, and hence the precise values of G_{IIc} for the two materials are difficult to ascertain by ENF test. The values with the highest confidence were

$G_{IIc} = 590\ (\pm 178)\ \mathrm{J/m^2}$ for the reference material

$G_{IIc} = 789\ (\pm 144)\ \mathrm{J/m^2}$ for the hybrid nano-reinforced material showing a nominal improvement of 34 % due to incorporation of carbon nanotubes.

While these tests show a marked qualitative difference between the reference and hybrid materials in Mode-II interlaminar fracture, the calculated values of Mode-II fracture toughness are unrealistically high. Most likely, this is due to a fundamental problem in testing woven composite material in pure Mode-II fracture, wherein adjacent fabric plies at an interlaminar crack plane can mechanically interfere with one another through 'interlocking' woven fiber crimps (Fig. 27.6). Because this interference impedes (or prevents) sliding of the crack faces, it causes such tests to exhibit artificially high forces at crack propagation, and hence gives artificially high values for calculated Mode-II fracture toughness. The effect also worsens with increasing crack length.

27.4 Mixed-Mode Testing

As a means of overcoming the difficulties encountered in pure Mode-II testing of woven composites, mixed-mode fracture tests were attempted following the general form of Fig. 27.7. In such tests, the presence of an opening-mode component alleviates interlaminar crimp-on-crimp mechanical interference by separating the crack faces.

Assuming a linear superposition of modes applies to such a specimen (Fig. 27.8) the total specimen strain energy release rate may be given as

$$G_T = \frac{21P^2k^2}{4E_1h^3} \tag{27.6}$$

where $k = a/b$ is the width taper ratio

For the complete specimen, the total fracture toughness is a merely sum of the strain energy release rates of the component modes, expressed as

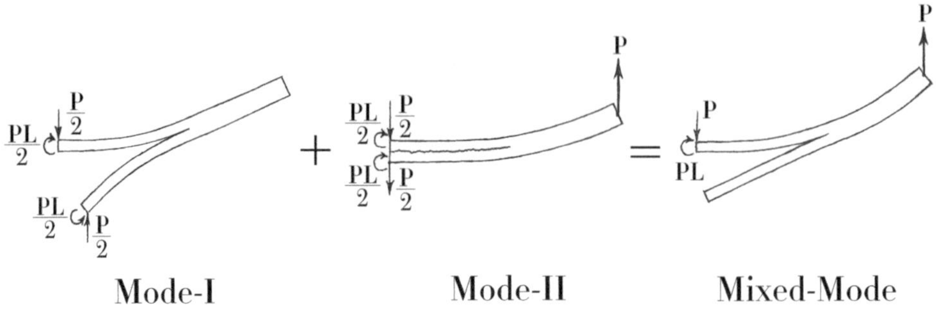

Fig. 27.8 Linear superposition components of mixed-mode cantilever beam specimen

$$G_T = G_I + G_{II} \tag{27.7}$$

Thus, the fractional contributions of the different modes (I & II) in this specimen are essentially

$$\frac{G_I}{G_T} = \frac{4}{7} \tag{27.8}$$

and

$$\frac{G_{II}}{G_T} = \frac{3}{7} \tag{27.9}$$

regardless of crack length. This makes the width-tapered specimen very convenient, as it possesses constant fractional contributions from the different modes throughout a test.

However, mixed-mode testing then requires the introduction of an interaction criterion to describe the contribution of the two fracture modes (I & II) to the overall specimen load-displacement behavior at crack advance. At the present, there is no fundamentally superior theory for addressing mixed-mode fracture. For the purposes of this study, three of the most significant criteria were selected for the purpose of extracting the Mode-II component from the mixed-mode specimen response: those of Wu & Reuter [14], Benzeggagh & Kenane [15], and Reeder [16]. These criteria may be expressed as

$$\left(\frac{G_I}{G_{Ic}}\right)^{\alpha} + \left(\frac{G_{II}}{G_{IIc}}\right)^{\beta} = 1 \quad \text{Wu \& Reuter} \tag{27.10}$$

$$G_{Tc} = G_{Ic} + (G_{IIc} - G_{Ic})\left(\frac{G_{II}}{G_T}\right)^{m} \quad \text{Benzeggah \& Kenane} \tag{27.11}$$

$$G_{Tc} = G_{Ic}\left[1 - \left(\frac{G_{II}}{G_T}\right)^{m}\right] + G_{IIc}\left(\frac{G_{II}}{G_T}\right) \quad \text{Reeder} \tag{27.12}$$

All mixed-mode interaction criteria also depend on semi-empirical parameters to define the extent of interaction of the different fracture modes. Ideally, these parameters are calibrated experimentally for a given material by performing pure Mode-I and Mode-II tests followed by mixed Mode-I/Mode-II tests where the individual contributions of the different modes are known. Since the motivation for resorting to a mixed-mode attack in this case is the inability to obtain reliable values for the pure Mode-II behavior, published parameters from other studies were necessarily a resort in being able to apply these criteria. For Wu & Reuter, values of $\alpha = 1$ and $\beta = 2$ were proposed as generally valid parameters [14] for orthotropic plates. These constants were also validated experimentally by O'Brien [17], showing their applicability to carbon-fiber composite materials. For the Benzeggagh & Kenane criterion, the authors [15] suggested that a value of $m = 2$ described most brittle materials fairly well, and that a value of $m = 1.557$ fit their experimental data for a carbon/epoxy composite especially well. For the Reeder criterion, a similar value of $m = 2$ was chosen.

From prior testing by Mode-I DCB specimens [12] on identical materials, the Mode-I interlaminar fracture toughness was found to be $G_{Ic} = 176\ (\pm 14)\ \text{J/m}^2$ for the reference material for the reference composite. This permitted calculation of the Mode-II fracture toughness by the mixed-mode criteria as (Table 27.2)

Table 27.2 Results of calculated Mode-II fracture toughness for reference composite from mixed-mode testing by different interaction theories

	Wu & Reuter (J/m^2)	Benzeggagh-Kenane (J/m^2)	Reeder (J/m^2)
Reference composite	475 (±255)	556 (±90)	312 (±56)

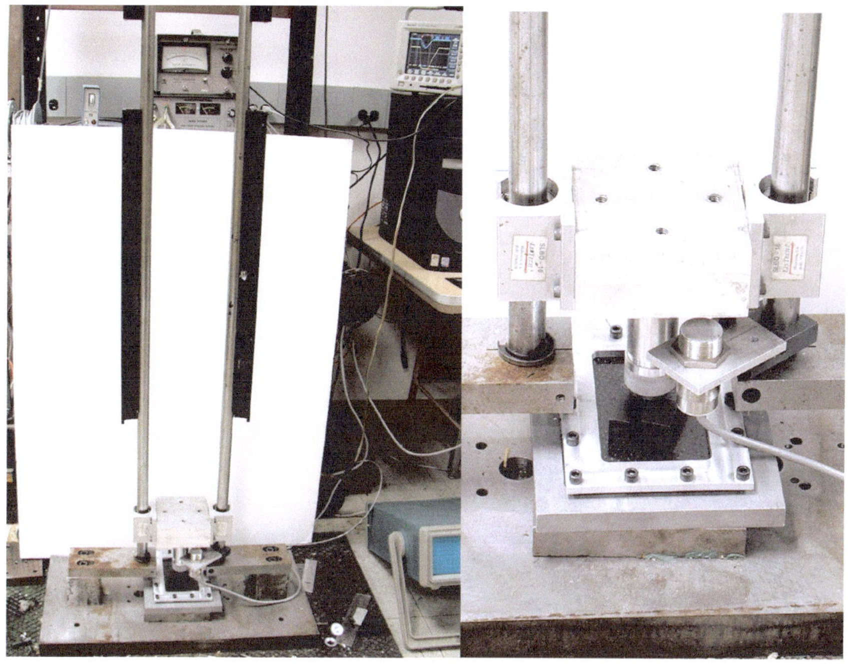

Fig. 27.9 Overview photographs of falling weight drop tower used in impact damage tests (*left*: photograph of tower and instrumentation, *right*: close view of impactor and inductive position sensor with sample in fixture prior to test)

While there was some disagreement between the result obtained based on the choice of interaction criterion, all these values were somewhat close to the value of $G_{IIc} = 590$ J/m^2 obtained from pure Mode-II testing. The value of 556 J/m^2 obtained from Benzeggagh-Kenane calculations was also closest, which is perhaps due to the interaction parameter $m = 1.557$ being suggested experimentally rather than just a *generally* valid numerical parameter as in the case of the other theories. This reinforces the validity of the Mode-II fracture toughness results, which were uncertain due to testing complications.

27.5 Impact Testing

Impact tests were carried out in a falling-weight drop tower to introduce interlaminar damage in composite plate samples [18]. Samples were of nominal dimensions l = 150 mm (6 in), w = 100 mm (4 in), h = 5 mm (0.200 in) (16 plies), prepared by abrasive machining from larger cast plates. The drop tower was instrumented with an inductive position sensor (Fig. 27.9) to allow accurate measurement of impactor velocity immediately before and after impact, thereby allowing calculation of the absorbed impact energy as

$$E_a = {}^1\!/_2 m \left(v_f^2 - v_i^2 \right) \tag{27.13}$$

where m is the impactor mass

v_i and v_f are the pre- and post-impact velocities, respectively

From tests on multiple samples of both materials (reference and hybrid nanocomposite) at varying initial impactor heights, sufficient data were generated to produce plots of relative energy absorption as a result of impact (Fig. 27.10). These tests showed a generally lower proportion of absorbed energy in the hybrid composite. The two materials only converged at higher impact energies due to a substantial amount of obvious fiber breakage at the sample face opposite the impact site, which shows those tests to be dominated by fiber failure rather than interlaminar damage.

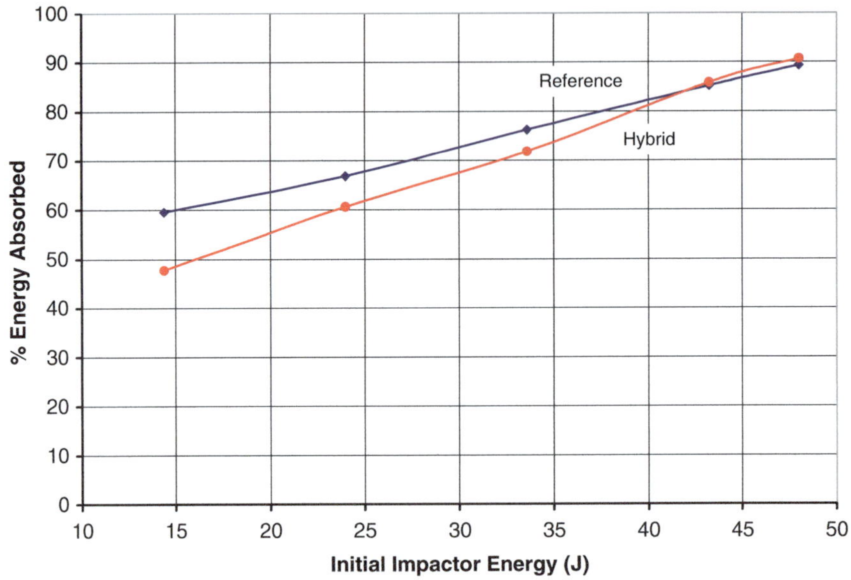

Fig. 27.10 Plot of relative energy absorption through impact as a function of initial impactor potential energy for the two materials

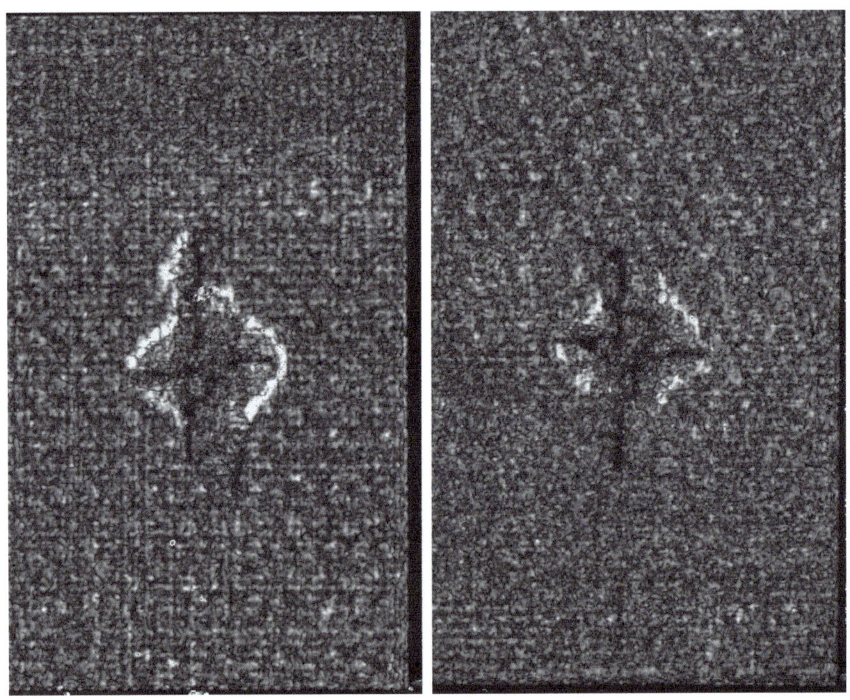

Fig. 27.11 Ultrasonic C-scan images of sample midplane (ply 8), 40 J impact, *left*: reference material, *right*: hybrid material

Following impact, each sample was imaged ultrasonically in a water-immersion scanning tank utilizing a 5 MHz focused probe. Ultrasonic C-scan timing gates were specified to collect reflected wave signals separately for each of the 16 plies in the composite, permitting through-thickness imaging of the damage zone. These various C-scan images (Figs. 27.11 and 27.12) were then examined to assess the size of the damage zone at each of the 16 plies.

By digital analysis of the C-scan images (16 for each specimen tested) it was then possible to compute an individual damage are for each ply of a tested sample (Fig. 27.13), defined by the contrast boundary in the image. This in turn gave a means of quantifying the 'total damage area' for each sample, representative of the total crack surface area formed due to delamination. Computation of this quantity for each sample yielded sufficient data to produce plots showing the variation of the total damage area with absorbed impact energy (Fig. 27.14). The tests then showed a consistently smaller total damage area in the hybrid composite, again suggesting some reinforcing effect from the presence of the nanotubes.

To confirm the ultrasonic assessment of the samples, additional work was performed by cutting post-mortem cross sections and examining them by optical microscope. These studies showed undamaged material to be effectively free of interlaminar cracking (Fig. 27.15), while samples subjected to impact damage showed extensive and random cracking within the fiber

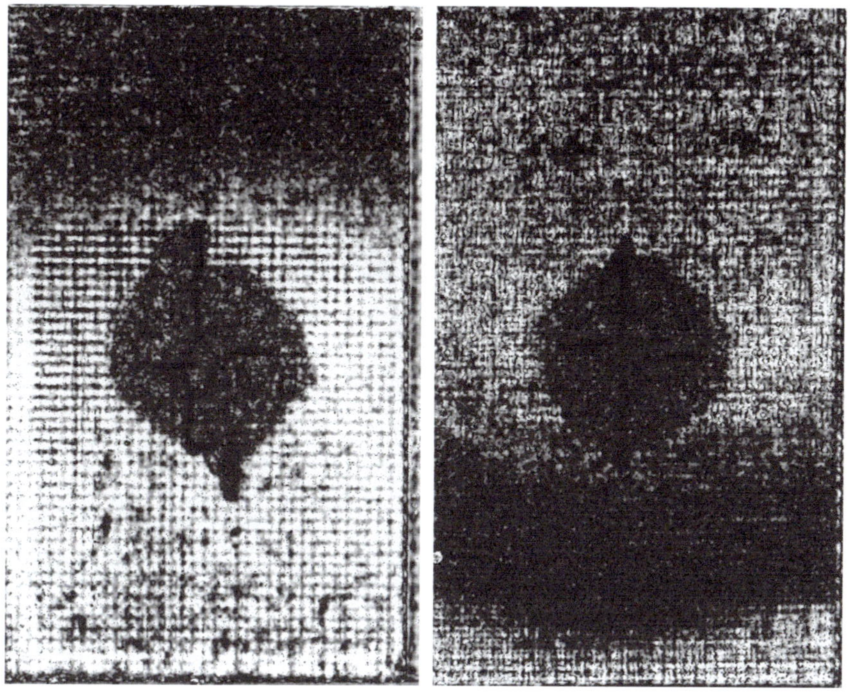

Fig. 27.12 Ultrasonic C-scan images of sample backwall (ply 16), 40 J impact, *left*: reference material, *right*: hybrid material

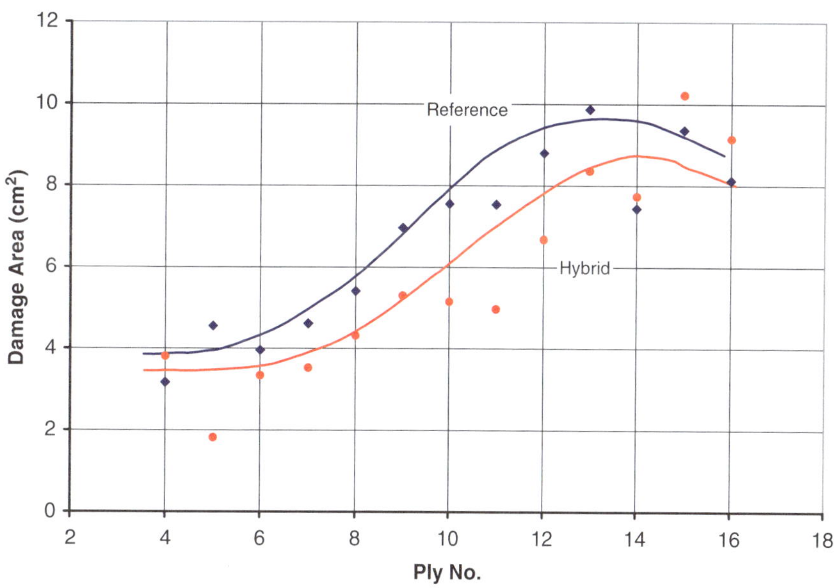

Fig. 27.13 Representative measured damage area from ultrasonic C-scan tomographs at individual ply locations

tows of the woven material. This both confirmed the findings of ultrasonic scanning, and showed that crack formation was not merely inter-ply but also had an intra-ply component as well that was not discernable by ultrasonic imaging. As such, it implies more than one mode of fracture is present under impact conditions (Fig. 27.16).

As a means of quantifying the overall damage tolerance behavior of these materials under impact, an *effective* fracture toughness was then defined as

$$G_T \cong \frac{E_a}{A_T} \tag{27.14}$$

where E_a is the absorbed impact energy for a given test

A_T is the total measured delamination area (across all plies) for a given impact specimen.

While this definition possess an inherent limitation in that it does not precisely quantify *all* of the internal fracture surface produced by impact, as only the 2-D projection of that damage onto the ultrasonic C-scans is quantified, it is fairly easy to

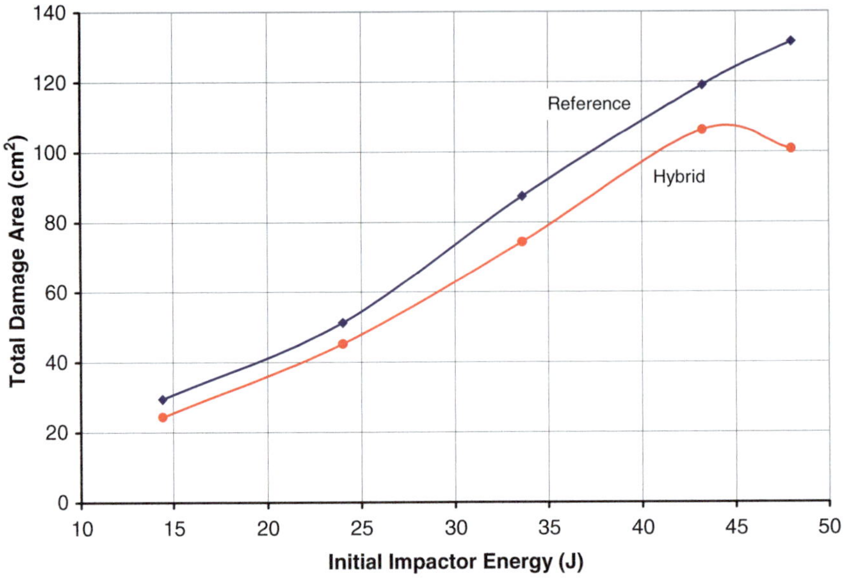

Fig. 27.14 Plot of measured total damage area as a function of impactor initial potential energy for the two composite materials

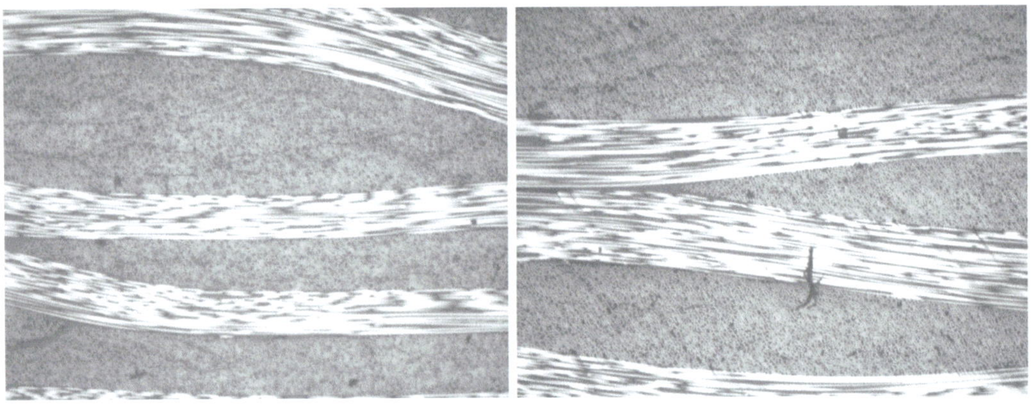

Fig. 27.15 Optical microscope images of cut surfaces from undamaged impact testing specimens, 100× magnification (*left*: reference material, *right*: hybrid composite)

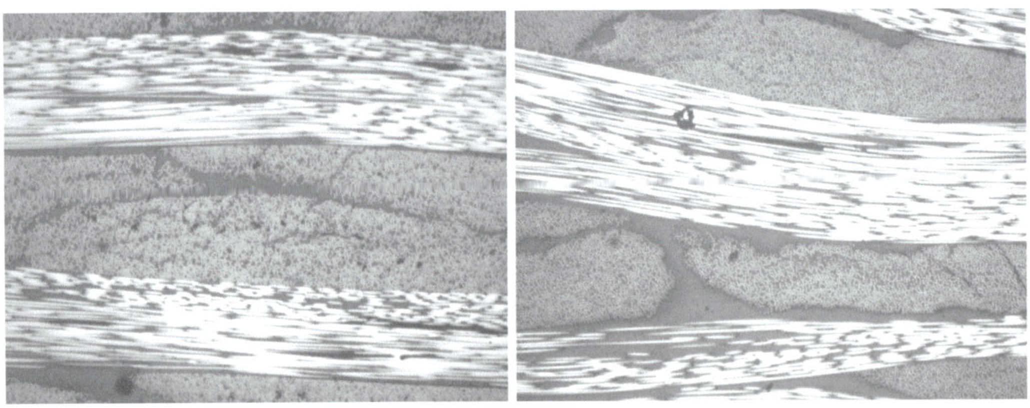

Fig. 27.16 Optical microscope images of cut surfaces from tested impact specimens, 19 J imparted energy, 100× magnification, (*left*: reference material, *right*: hybrid composite)

calculate from the available data. By plotting absorbed impact energy as a function of total measured damage area, the slope readily gives this effective fracture toughness (Fig. 27.17)

By this method, the *effective* total fracture toughnesses were determined to be

$G_T = 3030\ J/m^2$ for the reference material

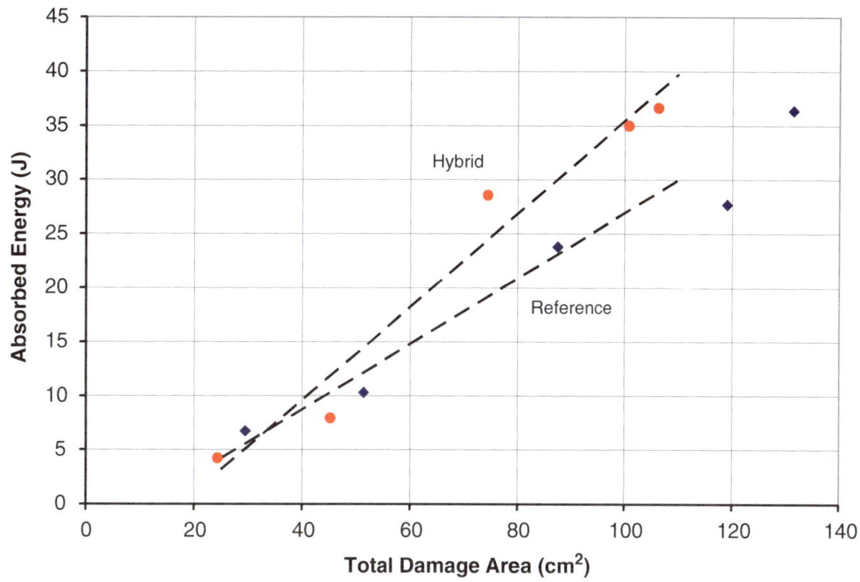

Fig. 27.17 Plot of absorbed impact energy as a function of total damage area showing approximate linear fits

$G_T = 4300$ J/m^2 for the hybrid nanoreinforced material showing an improvement of 42 % due to incorporation of carbon nanotubes.

27.6 Conclusions

This study demonstrated that damage tolerance of fiber-reinforced composite materials can be significantly improved by nanoparticle reinforcement of the matrix at low concentrations. The presence of nanoparticles in a hybrid nano/microcomposite can offer a significant increase in interlaminar Mode-II fracture toughness, improving resistance to internal sliding delamination. Furthermore, nanoparticles also can reduce the size of internal delamination developed from impact damage to composites, reducing the overall size of an internal delamination in both diameter and through-thickness volume. In addition, the increase in Mode-II fracture toughness correlates well on a relative basis with the overall reduction in impact damage observed through the addition of carbon nanotubes to a composite.

The result of this study is significant in that it shows how overall damage tolerance of a composite, and hence its resistance to impact damage as is seen in realistic service, may be enhanced by introduction of a fairly modest quantity of nanoparticles and simple additional material processing. This not only translates to the possibility of developing more durable or robust composite materials for a broader range of applications, but also for materials that will survive significantly longer in a given application.

Acknowledgment This work was supported by the Office of Naval Research (ONR). We are grateful to Dr. Y. D. S. Rajapakse of ONR for his encouragement and cooperation.

References

1. Iwahori, Y., Ishiwata, S., Sumizawa, T., Ishikawa, T.: Mechanical properties improvements in two-phase and three-phase composites using carbon nano-fiber dispersed resin. Compos. A: Appl. Sci. Manuf. **36**, 1430 (2005)
2. Gojny, F.H., Malte, M.H.G., Fiedler, B., Schulte, K.: Influence of different carbon nanotubes on the mechanical properties of epoxy matrix composites-a comparative study. Comp. Sci. Technol. **65**, 2300–2313 (2005)
3. Dean, D., Obore, A.M., Richmond, S., Nyairo, E.: Multiscale fiber-reinforced nanocomposites: synthesis, processing and properties. Compos. Sci. Technol. **66**, 2135 (2006)
4. Bekyarova, E., Thostenson, E.T., Yu, A., Kim, H., Gao, J., Tang, J., Hahn, H.T., Chou, T.W., Itkis, M.E., Haddon, R.C.: Multiscale carbon nanotube-carbon fiber reinforcement for advanced epoxy composites. Langmuir **23**, 3970 (2007)
5. Cho, J.-M., Daniel, I.M.: Reinforcement of carbon/epoxy composites with MWCNTs and dispersion enhancing block copolymer. Scr. Mater. **58**, 533–536 (2008)

6. Siddiqui, N.A., Woo, R.S.C., Kim, J.K., Leung, C.C.K., Munir, A.: Mode I interlaminar fracture behavior and mechanical properties of CFRPs with nanoclay-filled epoxy matrix. Compos. A: Appl. Sci. Manuf. **38**, 449 (2007)
7. Davis, D.C., Whelan, B.D.: An experimental study of interlaminar shear fracture toughness of a nanotube reinforced composite. Compos. Part B **42**, 105–116 (2011)
8. Lin, J.C., Chang, L.C., Nien, M.H., Ho, H.L.: Mechanical behavior of various nanoparticle filled composites at low-velocity impact. Compos. Struct. **74**, 30 (2006)
9. Iqbal, K., Khan, S.U., Munir, A., Kim, J.K.: Impact damage resistance of CFRP with nanoclay-filled epoxy matrix. Comp. Sci. Technol. **69**, 1949 (2009)
10. Hosur, M.V., Chowdhury, F.H., Jeelani, S.: Processing and low-velocity impact performance of nanophased woven carbon/epoxy composite laminates. Proceedings of the Twelfth US-Japan Conference on Composite Materials, 114 (2006)
11. Carlsson, L.A., Gillespie, J.W., Pipes, J.R., Pipes, R.B.: On the analysis and design of the end notched flexure (ENF) specimen for mode II testing. J. Compos. Mater. **20**, 594 (1986)
12. Fenner, J.S., Daniel, I.M.: Hybrid nanoreinforced carbon/epoxy composites for enhanced damage tolerance and fatigue life. Compos. Part A **65**, 47 (2014)
13. Argüelles, A., Viña, J., Cantelli, A.F., Bonhomme, J.: Influence of resin type on the delamination behavior of carbon fiber reinforced composites under mode-II loading. Int. J. Damage Mech. **20**(7), 963 (2011)
14. Wu, E.M., Reuter, R.C.J.: Crack extension in fiberglass reinforced plastics and a critical examination of the general fracture criterion. Theoretical and Applied Mechanics Report No. 275 (Bureau of Naval Weapons, Contract No. NaW-64-0178-s) Urbana, (1965)
15. Kenane, M., Benzeggagh, M.L.: Measurement of mixed-mode delamination fracture toughness of unidirectional glass/epoxy composites with mixed-mode bending apparatus. Compos. Sci. Technol. **56**(4), 439 (1996)
16. Reeder, J.R.: 3D mixed mode delamination fracture criteria: an experimentalist's perspective. Proceedings of American Society for Composites, 21st Annual Technical Conference, Dearborn, (2006)
17. O'Brien, T.K.: Interlaminar fracture toughness: the long and winding road to standardization. Compos. Part B. **29**(1), 57–62 (1998)
18. ASTM Standard D7136: Standard test method for measuring the damage resistance of a fiber-reinforced polymer matrix composite to a drop-weight impact event. ASTM International: West Conshohocken (2005)

Chapter 28
Compression Testing of Micro-Scale Unidirectional Polymer Matrix Composites

Torin Quick, Sirina Safriet, David Mollenhauer, Chad Ryther, and Robert Wheeler

Abstract This project builds on work done by Lu et al. An experimental study is carried out to characterize the failure behavior of a fiber reinforced polymer matrix composite at the micro-scale using the same test methodology. In order to address the issue of catastrophic failure observed in the previous effort, a physical stop for the indenter that limits maximum displacement to a predetermined value is integrated into the specimen design. Micron-sized specimens of IM7/BMI unidirectional composite with an integrated indenter displacement control were fabricated using Focused Ion Beam (FIB) milling. The specimens were compression tested using a custom built, SEM-based in-situ micro-testing device. During compression, SEM images are acquired continuously between displacement intervals so the deformation phenomena can be observed. Initial results showed that the integrated indenter displacement control prevents complete destruction of the specimen after the onset of failure. Damage observed includes interface failure, broken fibers, and general crushing. Parallel efforts on larger-scale compressive testing are conducted on millimeter-sized specimens using an in situ mechanical test frame located in an X-ray micro computed tomography (μCT) system. Failure response includes longitudinal splitting or brooming and kinking. A quantitative comparison of the compressive strength and modulus obtained from the two size scales specimen shows that there is no indication of a size effect. The experimental results will be used to validate the numerical models of micro-compression behavior.

Keywords Failure mechanism • Composite • In-situ compression testing • Kink bands • X-ray micro computed tomography

28.1 Introduction

In spite of significant improvements in composite resins and fibers during the last few decades, the compressive strength of polymer matrix composites still remains at 50–60 % of their tensile strength. This significantly reduces the advantage position of these materials in structures in which compressive strength is a primary design requirement. The mechanisms of composite failure in compression are not fully understood. Therefore, there is the need for a better understanding of the physics and mechanics of compressive failure, which can be achieved by micro-scale experimental studies.

The failure mechanisms in unidirectional composites under compression load have been examined extensively [1–5]. Four common failure mechanisms are micro-buckling, kinking, fiber failure and splitting [6]. Each of these failure modes may be observed in a single specimen, or a specific mode may dominate for the same composite material tested under different conditions.

Under the USAF Summer Faculty Fellowship Program, a methodology for performing in-situ compression test to characterize the failure behavior of a fiber reinforced composite at the micro-scale was pioneered by Lu et al. [7]. The material investigated was unidirectional carbon fiber reinforced polyimide matrix composite (IM7/BMI). The specimens were prepared by two steps in the micro fabrication process: ion sputtering (coarse cutting) and ion milling FIB (fine cutting). Speckle patterns were created on the specimen surface for further use in Digital Image Analysis (DIC). In-situ compression tests were conducted using a custom micro-mechanical testing device placed inside the SEM. The micron-size specimens

T. Quick • D. Mollenhauer • C. Ryther
U.S. Air Force Research Laboratory, WPAFB, Dayton, OH 45433, USA

S. Safriet (✉)
University of Dayton Research Institute, Dayton, OH 45469, USA
e-mail: Sirina.Safriet.ctr@us.af.mil

R. Wheeler
Micro Testing Solutions LLC, Hilliard, OH 43026, USA

A.M. Beese et al. (eds.), *Fracture, Fatigue, Failure and Damage Evolution, Volume 8*, Conference Proceedings of the Society for Experimental Mechanics Series, DOI 10.1007/978-3-319-21611-9_28

were compressed by a conical indenter with a flat end surface. High resolution SEM images and DIC were used to monitor specimen deformation, strains, and fracture during loading. While this methodology demonstrated the feasibility of the testing of micro-compression specimens, significant limitations were encountered. Excessive compliance of the micro-load frame in the SEM leads to difficulties in precisely controlling the applied loads and strains during and immediately following the onset of failure. The result is a nearly instantaneous destruction of the specimen.

This study aims to address this challenge by the integration of a physical stop for the indenter that limits maximum strain to a predetermined value. Micro-compression specimens with integrated indenter displacement control will be fabricated and compression tested. With the integrated indenter displacement control, it is expected that the onset and propagation of damage can be observed without the catastrophic failure of the specimen.

Parallel efforts on larger-scale compressive testing will be conducted on millimeter-sized specimens using a Deben mechanical test frame within the Xradia X-ray microscope, similar to work accomplished by Wang et al. [8]. In-situ micro tomography allows detailed observation of the local internal deformations at different strain levels. The X-ray CT data sets will be complimented with digital volume correlation (DVC) to obtain a full-strain data at the loading increments to hopefully provide insight into the sequence of events that lead to the failure of the specimen. Comparison of quantitative force/displacement data and failure processes observed in micro- and meso-scale specimens will help in the understanding of "size" effect and the role of imperfections/defects in the compressive failure of composite laminates. The experimental results will be used to validate numerical models for micro-compression behavior.

The goals are: (1) to understand the failure mechanism of composites at the micro-level by observing the sequence of events of deformation phenomena, and (2) to use U.S. Air Force Research Laboratory composite analysis codes to model micro-compression behavior.

28.2 Experimental

28.2.1 Materials and Sample Preparation

The material used in this study is a carbon fiber reinforced bismaleimide composite (IM7/BMI) from Cytec Engineering Materials. This unidirectional prepreg tape has a ply thickness of 125 μm and a fiber volume fraction of 60 %. Composite panels of 1 ply and 24 plies of the material were manufactured and autoclave cured following the company recommended cure cycle [9].

For fabrication of a micro-compression specimen, a small piece (1 × 1 cm) of material was cut from the single ply composite panel. The specimen surfaces were carefully polished using a series of silicon-carbide paper and diamond lapping pads to thin the bulk material down to reduce the FIB milling time. The specimen was mounted to an SEM stub and sputter coated with a carbon layer to prevent electrical charging. Focused ion beam milling was used to create the pillar and arresting shoulders that act as the physical stop for the indenter. This was conducted in a TESCAN LYRA 3 FEG-SEM × FIB. The apertures and beam currents were adjusted depending on the process steps. Material was removed on all four sides of the pillar to create an approximately 20 × 15 × 60 μm pillar. The top surface of the material was cut independent of the pillar to create the arresting shoulders for the 130 μm diameter indenter. The pillar was cut to its final height in order to set the vertical offset between the top of the pillar and shoulders. For this effort, the ideal difference in height between the two surfaces is the amount the material would strain before failure. This height can be adjusted depending on the desired outcome of the test. The specimen was prepared for DIC analysis by sputter coating a thin layer of Cr followed by Pt. The speckle pattern was created on the front surface by FIB milling away small areas of the Pt coating such that the underlying material was not affected.

The specimens used in the meso-scale testing were small cylindrical solid rod specimens (8 mm long × 1.4 mm diameter) and would be used for in-situ micro tomography. They were fabricated following these steps. A rectangular section (8.5 × 25 × 3.4 mm) of the 24 ply panel was cut and polished on the top and bottom surfaces. Care was taken to ensure that the top and bottom of the specimens were parallel and flat. A special aligning jig was fabricated to hold the specimen during machining using a diamond core-drill specially designed for preserving the core. Multiple cylindrical solid rod specimens of uniform cross section were obtained. Their ends were surface ground parallel to within 2 μm tolerance. The rod specimens were ultrasonically cleaned for 5 min to remove debris. They were air-dried and oven-dried at 38 °C overnight. The specimens were kept in a desiccator until ready to test.

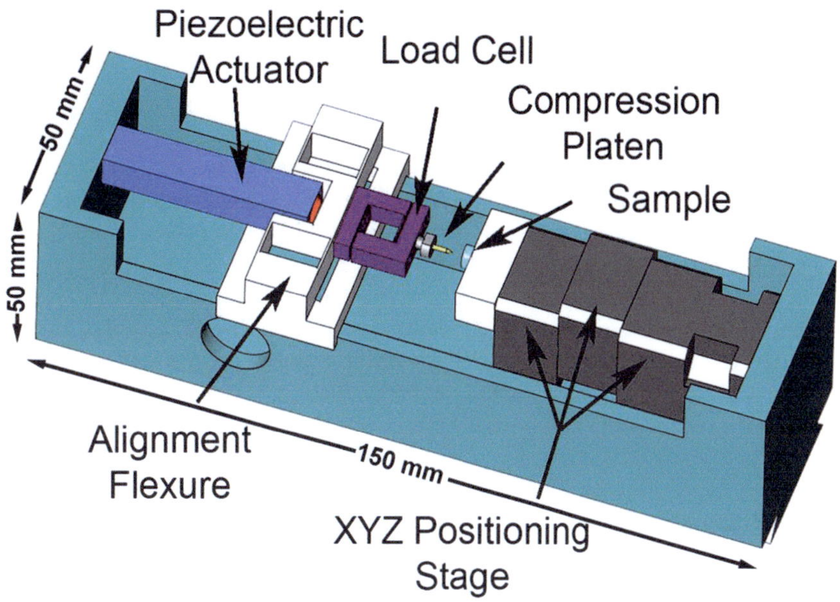

Fig. 28.1 Schematic illustration of the in-situ micro-load fixture used in the experiments

28.2.2 In-Situ Micro-Compression Set Up

A custom built, SEM-based loading fixture was used to conduct the compression tests on the micron-size specimens. The schematic illustration of the fixture is shown in Fig. 28.1. A piezoelectric actuator was used to impose sub-nanometer displacement over a total stroke of 40 μm. A strain-gage load cell, with a maximum load capacity of 100 g, was used to measure the forces applied to the sample. An alignment flexure was employed to ensure uniform axial motion of the load train. A polished 130 μm sapphire fiber was used as the compression platen. The contact surface was polished with FIB milling to ensure the surface was perpendicular to the loading direction. At the other end of the test device, a piezoelectric controlled slip-stick motion x-y-z positioning stage was used to precisely place the specimen underneath the compression platen. This stage based its movement on a piezoelectric inertial force mechanism that provided nanometer scale positioning resolution with zero backlash. The fixture is 50 mm wide × 50 mm tall × 150 mm long and thus fit well inside the SEM chamber without disturbing or inhibiting the SEM's function. Instrumentation control and data acquisition were achieved by using the Labview software from National Instrument (NI). The compression tests were conducted in a quasi-static manner, with a typical displacement rate of 100 nm/s. The test was separated by periods where the piezoelectric actuator was held at a constant voltage in order to optimize the collection of high-resolution SEM images.

28.2.3 In-Situ μCT Compression Set Up

Small cylindrical composite rods were used for in-situ compression testing inside the X-ray μCT. The Deben CT5000 testing stage was used in the Xradia X-ray microscope Versa XRM-520 for a real time observation of the damage evolution of the materials which were subjected to load. The specimens were positioned perpendicularly to the beam direction on the hardened steel disks (M50). The built-in controller software *Scout and Scan* was used for setting the imaging parameters. A scintillator CCD detector (obj 4×) was used. The specimen was first scanned without load with a voxel size of 1.5 μm. Then, the specimen was loaded in compression with a loading speed of 0.1 mm/min. The projection (2D X-ray) images were continuously acquired to observe damage of the specimen while under compressive loading. They were taken at 1 s intervals of exposure time. The load and displacement were displayed and automatically recorded via Microtest software. After the specimen failed, CT scan was performed to obtain volumetric images of damaged specimen. A source voltage of 60 kV was used at a power of 5 W. Thousand six hundred and one projections were taken over a rotation of 360°, and the exposure time for each projection was 45 s.

28.3 Results and Discussion

28.3.1 Micro-Scale Compression Testing

Two micron-size specimens (designated FP-1 and FP-2) were fabricated and tested. Figure 28.2 shows an SEM image of the FP-2 specimen with an integrated indenter displacement control. The final dimensions of the FP-2 specimen are 19 μm (width) × 63 μm (height) × 15 μm (thickness). The arresting shoulders are ~25 μm from the pillar, with a total width of 137 μm, offset ~3 μm below the pillar height.

The cross sectional area of the FP-2 specimen is shown in Fig. 28.3. Yellow circles were drawn to outline the fibers (diameter ~6 μm). The fibers are not equally distributed inside the polymer matrix: some are closed packed together while some are isolated with large resin rich regions in between them. Partial fibers are also seen near the specimen edges.

The specimen was compressed to ~0.51 N. The aim of this test was to observe the deformation process at the micro-level without the catastrophic failure of the specimen. Figure 28.4 shows a sequence of SEM images of the FP-2 specimen under compression from a single scan. The failure of the specimen was instantaneous and occurred midway through one of the SEM imaging scans (Fig. 28.4b) as observed by the sudden platen movement. The integrated indenter displacement control was effective in preventing catastrophic failure of the specimen as evident in Fig. 28.4c.

The stresses, strains and moduli were calculated from test data and images recorded. The total strain was computed from measuring three image-based distances; the starting distance from platen to arresting shoulder, total length of the pillar, and failure jump distance. From these measurements the displacement of the pillar before failure could be calculated and total strain could be obtained. Note that there is potential for an error in the strain measurement. It has shown that SEM scanning process introduces small shifts in imaging position, which results in an error in the strain measurement [10]. This error can be minimized through implementation of an image integration process that combines image data from multiple scans. The SEM images acquired are from single, non-integrated scans. Image integration will be used to eliminate step changes in SEM images for the future test.

The compressive strength and the modulus of the FP-2 specimen are 1.72 GPa and 81 GPa, respectively. An additional test was conducted on the FP-1 specimen (20 × 50 × 13 μm). The stress and modulus are 1.18 GPa and 41.6 GPa, respectively. Manufacturer's values for a generic IM7/epoxy composite are 1.69 and 150 GPa for compressive strength and modulus, respectively [11]. Both the stresses and modulus from our test are lower than the values previously reported from the

Fig. 28.2 SEM image of a micro-compression specimen with integrated indenter displacement control. FP-2 specimen dimensions: 19 μm (width) × 63 μm (height) × 15 μm (thickness) with ~3 μm displacement control

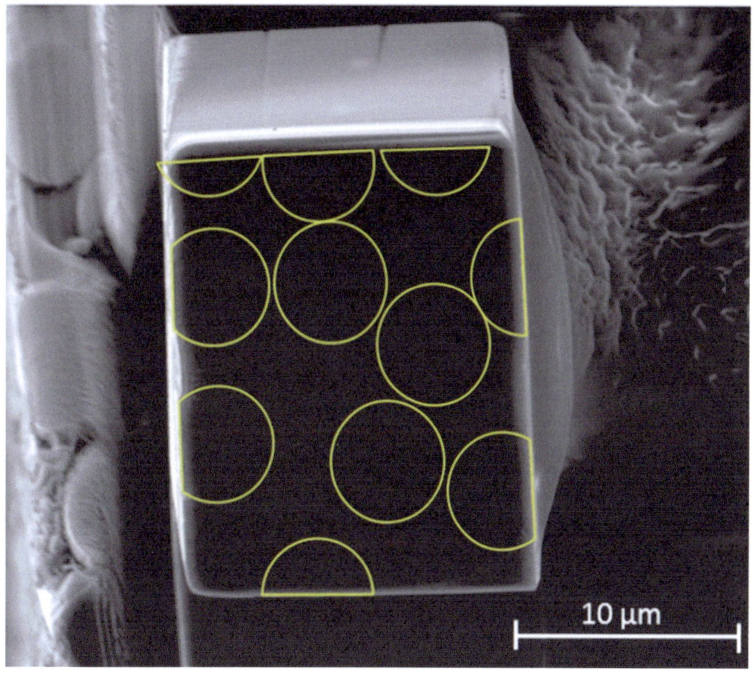

Fig. 28.3 SEM image of the cross sectional area of FP-2. Fiber distributions in matrix are shown

same test method by Lu [7]. This may be attributed to the difference in the processing technique of the same material. The composite material tested in this study was autoclave-cured, while the material investigated from Lu et al. was compression molded. The failed FP-2 specimen was unloaded and the morphology of the fractured surface was examined to identify failure mechanism. Damage observed are fiber-matrix debonding, broken fibers, and general crushing as shown in Fig. 28.5.

The SEM images of the composite specimen under compression of the last five load steps before the onset of failure were used for DIC analysis. It is noted that a small amount of misalignment is always present in the fibers of a composite specimen. The failure was sudden. There is not much change in the displacement and strain field until the last load step before the onset of failure, as evident in the last two load steps of the FP-1specimen (Fig. 28.6). The image size was 1510×4095 pixels and DIC was performed using a subset size of 35 pixel and step 5. Displacement along the length of the specimen ($\times$ direction) is consistent with the compressive loading of the specimen. High shear deformation occurs in matrix between fibers. The strain field is consistent with a highly inhomogeneous material response, which localized strain that can lead to fiber-matrix debonding or matrix failure, as observed in this specimen.

28.3.2 Meso-Scale Compression Testing

Compression tests on the meso-scale specimens were conducted using a Deben mechanical test frame located in an Xradia X-ray microscope. Hardened steel disks (M50) were used as compression platens. The M50 plates were adhered to the Deben compression plates with epoxy. The tests were performed on cylindrical specimens of diameter $\sim$1.4 mm and a gage length of 8.2 mm. Three specimens were end–loaded to failure. The first specimen used a thin Nylon support at the bottom for stabilization. The specimen failed at the bottom area inside the nylon. To avoid any effects of the nylon on the test, the remaining two specimens were free-standing during loading. For all specimens, the failure was sudden and occurred at the bottom end; due to stress concentration. The specimens exhibited broom-like fracture with extensive longitudinal splitting as shown in Fig. 28.7. End loading unidirectional composite specimens has been found to lead to failure by splitting and brooming of fibers at loading interface [1].

This test method introduces the stress concentrations at the loading interface, which causes the failure to occur at the specimen ends rather than the middle of the specimen. To minimize end effects, aluminum 6061 disks with a precise hole for the specimen were made. Epoxy was used to adhere the ends of the specimen to the aluminum disks. An additional test was performed on the confined specimen, CD-4. Confinement of the test specimen is a standard practice in the testing of

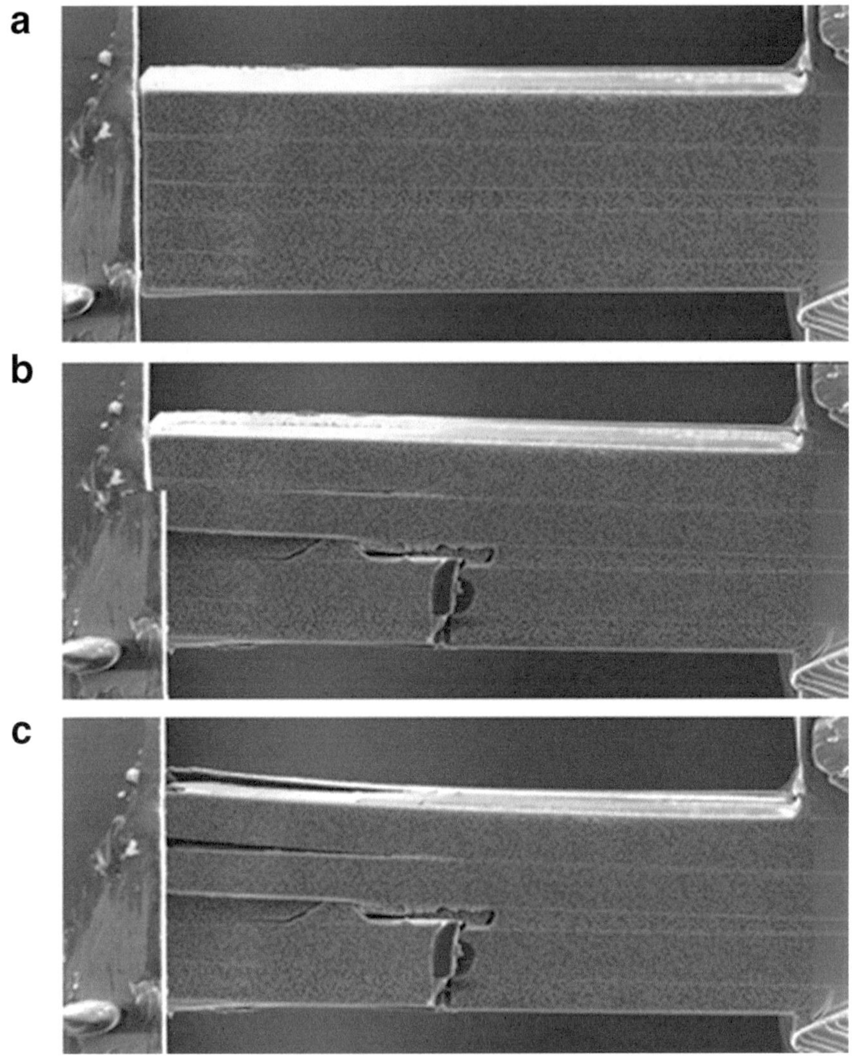

Fig. 28.4 SEM images showing the failure process of the IM7/BMI composite under compression; (**a**) before failure, (**b**) during failure, and (**c**) after failure

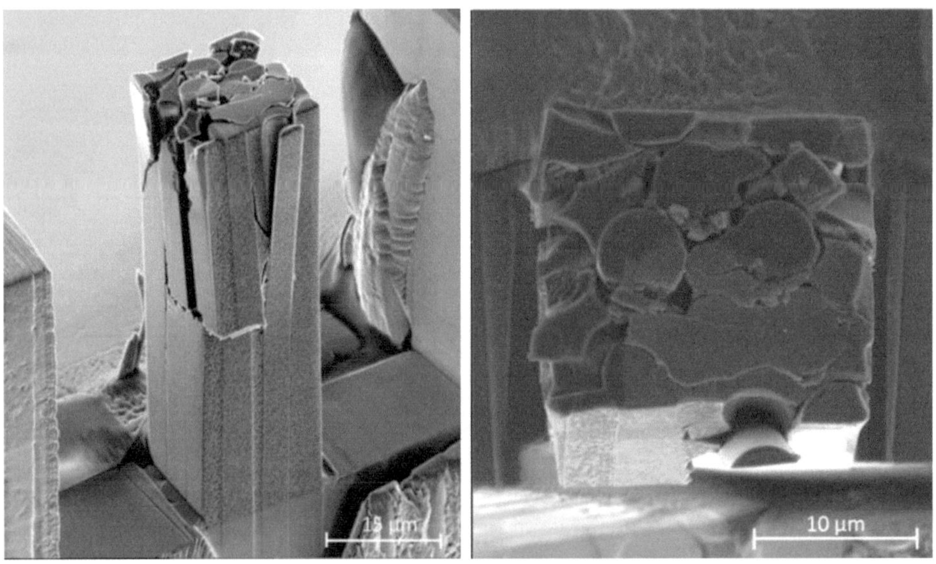

Fig. 28.5 SEM image of post-failure FP-2 specimen

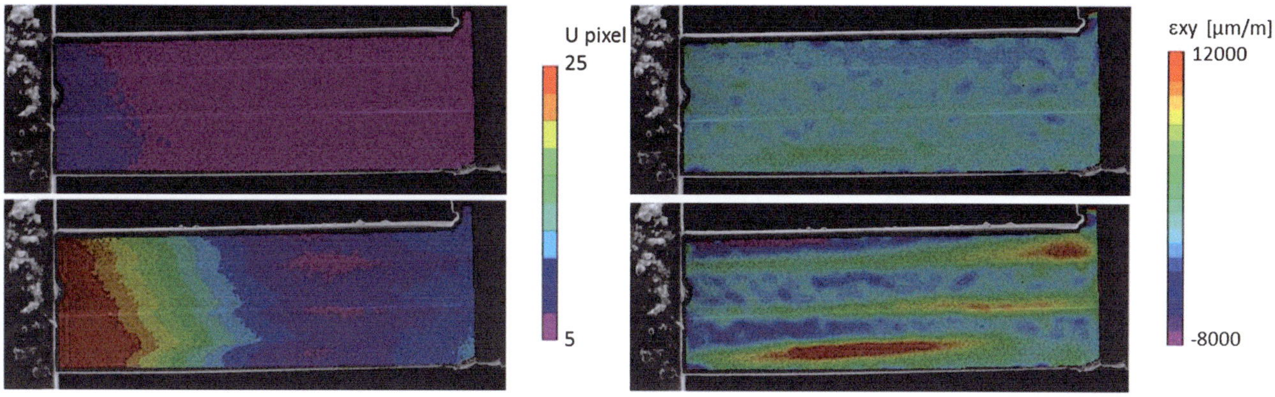

Fig. 28.6 Displacement (**a**) (× direction) and strain fields (**b**) (ε_{xy}) in the FP-1 composite specimen under compression from the last two steps before the onset of failure

Fig. 28.7 Virtual tomographic slice showing brooming of failed specimen, CD-3

materials with a high degree of anisotropy. The specimen failed at the top of the lower disk, but at a higher load (2100 N) than the previous tests. Several layers of kink bands were observed with splitting occurring near the outer radius of the specimen, as seen in Fig. 28.8.

The compressive failure was instantaneous. It is desired in this work to capture the initiation and propagation of damage so that failure mechanism can be further studied. Work is in progress to fabricate and test more micron-size specimens with the goal of limiting indenter travel at precisely the moment that initiation of damage occurs. For meso-scale specimens, the loading rate must be reduced to limit failure after initiation. The current ability to arrest the compressive force is limited by the rate of data acquisition and the rate of loading. The lowest loading rate of the current loading fixture used is 0.1 mm/min.

The compressive strength, strain and modulus from the meso-scale specimens are listed in Table 28.1. Note that the sample displacements, which are used to compute the strain, were directly obtained from test data recorded via Microtest software.

The compressive strength obtained from the confined specimen is higher than the free standing specimens. Also included in Table 28.1 for comparison are the stress, strain and modulus from the tests performed on micron-size specimens.

Two test methods were used for measuring the compressive stress-strain response. The first method involved testing rectangular shaped-specimens of micron size scale using a custom built, SEM-based load fixture. The second method involved testing cylindrical specimens of diameter ~1.4 mm and a gage length of 8.2 mm (meso-scale) using the Deben testing stage inside an X-ray microscope. The stress and strain obtained from both test methods were in reasonably good agreement, with the exception of FP-2 and CD-4 having higher stresses. Specimen configurations, size, and test method do not seem to have any major effect on the measured response.

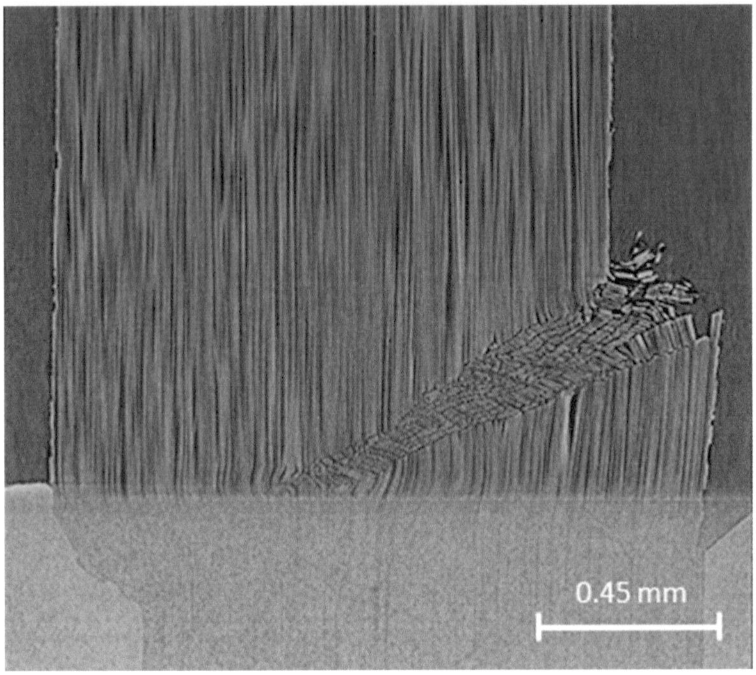

Fig. 28.8 Virtual slice in the xy plane of the confined specimen, CD-4

Table 28.1 Maximum stress and modulus of unidirectional IM7/BMI composite from different size scales

	Dimension	Stress (GPa)	Strain (%)	Modulus (GPa)
Micro-scale				
FP-1	20 × 13 × 50 μm	1.18	2.8	41.6
FP-2	19 × 13 × 65 μm	1.72	2.1	81.0
Average	–	1.45	2.5	61.3
Meso-scale				
CD-1	1.37 mm Ø, 8.19 mm long	1.10	2.9	54.6
CD-2	1.37 mm Ø, 8.19 mm long	1.01	2.2	68.2
CD-3	1.40 mm Ø, 8.19 mm long	1.17	2.6	74.3
CD-4	1.38 mm Ø, 8.20 mm long	1.41	2.6	71.9
Average	–	1.17	2.6	67.3
Results from Lu et al.				
Micro	18 × 15 × 53 μm	1.98	1.04	190
Macro	6.25 × 3.17 × 108 mm	0.95	1.33	71

The modulus obtained from the micron-size specimens is much lower than the one reported from Lu's work. This may be attributed to the difference in processing technique between the two materials or due to differences in obtaining specimen displacement data. Overall, the modulus obtained from micro, meso and macro scales are on the same order of magnitude. Based on our results shown to date, there is no clear indication of a size effect.

28.4 Summary

The failure mechanism of a fiber reinforced polymer matrix composite at the micro and meso-scales were investigated through an experimental study. Micron-sized specimens of IM7/BMI unidirectional composite with an integrated indenter displacement control were fabricated and compression-tested using a custom built, SEM-based in-situ micro-loading fixture. The integrated indenter displacement control is effective in preventing the complete destruction of the specimen after the onset of failure. Interface failure, broken fibers, and general crushing, were observed. The failure was sudden so it was not possible to capture the sequence of events that lead to failure under these test conditions. In future work, image integration will be used to eliminate step changes in SEM images during tests to improve the accuracy of DIC analysis.

Meso-size specimens were fabricated and in-situ compression tested inside an X-ray μCT. Splitting, brooming and kinking were observed for end loading specimens. A confined specimen shows kink bands with splitting occurring near the outer radius of the specimen. Loading rate limitations for this test method prevent the needed precision to arrest the failure immediately after its onset. We plan to continue testing using the confined specimen method with incremental loadings to obtain volumetric images.

Direct comparison of the compressive strength and modulus obtained from the two size scales specimen shows that there is no indication of a size effect.

Acknowledgements The authors gratefully acknowledge Dr. Michael Uchic of the AFRL for his informative technical discussions and suggestion. The authors thank Mr. Arthur Safriet from the University of Dayton Research Institute (UDRI) for his help in designing and fabricating a special aligning jig and supporting fixtures for specimen preparation and testing. Also thanks to Mr. Ron Trejo, Mr. Mike Nickel, Ms. Marlene Houtz from UDRI for their help with autoclave processing and X-ray CT scan.

References

1. Kyriakides, S., Arseculeratne, R., Perry, E.J., ad Liechti, K.M.: On the compressive failure of fiber reinforced composites. Int. J. Solids Struct. **32**(6–7), 689–738 (1995)
2. Pimenta, S., Gutkin, R., Pinho, S.T., Robinson, P.: A micromechanical model for kink band formation: Part I—experimental study and numerical modeling. Compos. Sci. Technol. **69**, 948–955 (2009)
3. Pimenta, S., Gutkin, R., Pinho, S.T., Robinson, P.: A micromechanical model for kink band formation: Part II—analytical modeling. Compos. Sci. Technol. **69**, 956–964 (2009)
4. Vogler, T.J., Kyriakides, S.: On the initiation and growth of kink bands in fiber composites: Part I. experiments. Int. J. Solids Struct. **38**, 2639–2651 (2001)
5. Waas, A.M., Schultheisz, C.R.: Compressive failure of composites, Part I: experimental studies. Prog. Aerosp. Sci. **32**(1), 43–78 (1996)
6. Schultheisz, C.R., Waas, A.M.: Compressive failure of composites, Part I: testing and micromechanical theories. Prog. Aerosp. Sci. **32**(1), 1–42 (1996)
7. Lu, Y.C., Wheeler, R., Tandon, G.P., Schoeppner, G.A.: In-situ micro-compression testing for characterizing failure of unidirectional fiber composites. Proceedings of the American Society for Composites, 28th Technical Conference, State College, 9–11 September 2013
8. Wang, Y., Soutis, C., Withers, P.: X-ray microtomographic imaging of kink bands in carbon fiber-epoxy composites. Proceedings of the 16th European Conference on Composite Materials, Seville, 22–26 June 2014
9. CYCOM® 5250-4 prepreg system technical data sheet, Argosy International, AECM_00008, 20 March 2012
10. Sutton, M.A., Li, N., Joy, D.C., Reynolds, A.P., Li, X.: Scanning electron microscopy for quantitative small and large deformation measurements Part I: SEM imaging at magnifications from 200 to 10,000. Exp. Mech. **47**, 775–787 (2007). doi:10.1007/s11340-007-942-z
11. HexTow Carbon Fiber brochure, Hexel Corporation (2013)

Chapter 29
Crack Analysis of Wood Under Climate Variations

Nicolas Angellier, Rostand Moutou Pitti, and Frédéric Dubois

Abstract The knowledge of crack driving forces such as energy release rate and stress intensity factors is very important in the assessment of the reliability of timber structures. This work deals with static and creep fracture tests in opening mode crack growth. They are performed in climatic chamber, which allows reproducing the real environment effects on the cracking of Double Cantilever Beam specimens machined in Douglas and White Fir species. The evolutions of the crack length are posted versus time. The aim of this work is to compare the obtained experimental results with numerical tools given by a finite element model. The numerical model is based on a new analytical formulation of the A-integral, generalized to viscoelastic orthotropic material, which allows taking into account the effect of thermal and hydric loads in the cracking process.

Keywords Wood • Fracture • Creep test • Energy release rate • Variable environment

29.1 Introduction

It is known that the mechanical behaviour of wood is highly dependent on environmental conditions and rheological effects. It is governed by mechano-sorptive effects, which result from a complex interaction between mechanical loading and moisture content variations [1]. In fact, due to the advantages provided by its mechanical behavior particularly under extreme loading conditions such as fire and seismic events, in addition to its aesthetic and environmental effects, timber is commonly employed in building and civil engineering structures [2]. As well known, timber elements exhibit micro-cracks, which can propagate due to fatigue, overload or creep loading and cause failure of the structure. In addition, timber is a hygroscopic material whose mechanical behavior is very sensitive to climatic changes such as temperature and moisture variations. For example, drying process accelerate the crack growth, while wetting process induce the delay of the crack propagation. This fact is accelerated with the time-dependent behavior of wood during long-term loading and creep tests.

In order to bring some responses to these notorious problems, the ANR JCJC2013 CLIMBOIS project is dealing with« effects of climatic and mechanical variations on the durability of timber structures » [3]. The first experimental results belong to the first task of the project on « fracture and viscoelastic behavior », and concerns experimental validation of analytical formalisms [4] of fracture mechanics and numerical modeling of fracture coupled with viscoelasticity [5]. In order to understand the effect of thermal process on crack growth process, numerical results by finite element calculations have already been obtained in the framework of the same project [6]. In the past, the literature background shows that viscoelastic tests have been performed in timber structure so as to monitor the strain under moisture content variations [7]. But, this work is devoted to fracture tests performed in climatic chamber in GEMH laboratory in France.

The first part of the paper presents the experimental set-ups with samples preparation. The wood sample geometry is a Double Cantilever Beam (DCB) in order to provide the stability of energy release rate during the crack growth process in opening mode. Hence, mechanical loadings for characterization static test in constant humidity environment and for creep

N. Angellier (✉) • F. Dubois
Heterogeneous Material Research Group, Civil Engineering Center, Université de Limoges, Egletons 19300, France
e-mail: nicolas.angellier@unilim.fr

R. Moutou Pitti
Université Clermont Auvergne, Université Blaise Pascal, Institut Pascal, BP 20206, Clermont-Ferrand 63000, France

CNRS, UMR 6602, Institut Pascal, Aubiere 63171, France

A.M. Beese et al. (eds.), *Fracture, Fatigue, Failure and Damage Evolution, Volume 8*, Conference Proceedings of the Society for Experimental Mechanics Series, DOI 10.1007/978-3-319-21611-9_29

test in variable humidity environment are presented. The last section shows results for two species used in Europe (Douglas and White Fir). Then, these results are discussed: in terms of force-displacement curve for static tests and of crack advance for creep tests.

29.2 Experimental Setup

In this section, we briefly describe wood samples and the experimental devices used to characterize the failure process in Douglas and White Fir under several mechanical and hydric loadings.

29.2.1 *Sample Preparation*

The experimental protocols are based on using a Double Cantilever Beam (DCB) with variable inertia beam specimen. The chosen species are Douglas and White Fir. Several samples of each species have been machined in a Radial-Longitudinal (RL) configuration. Samples are conditioned in climatic chambers in which temperature and relative humidity are regulated at 25 °C and 40 % RH or 90 % RH, respectively, corresponding to average moisture content levels around 9 and 17 %. After conditioning, a pre-crack with an initial length equal to 50 mm, is performed along the grain direction with a band saw (3 mm thick). The geometry and loading symmetry allow assuming an open mode configuration according to a particular choice of grain alignment with the crack, Fig. 29.1.

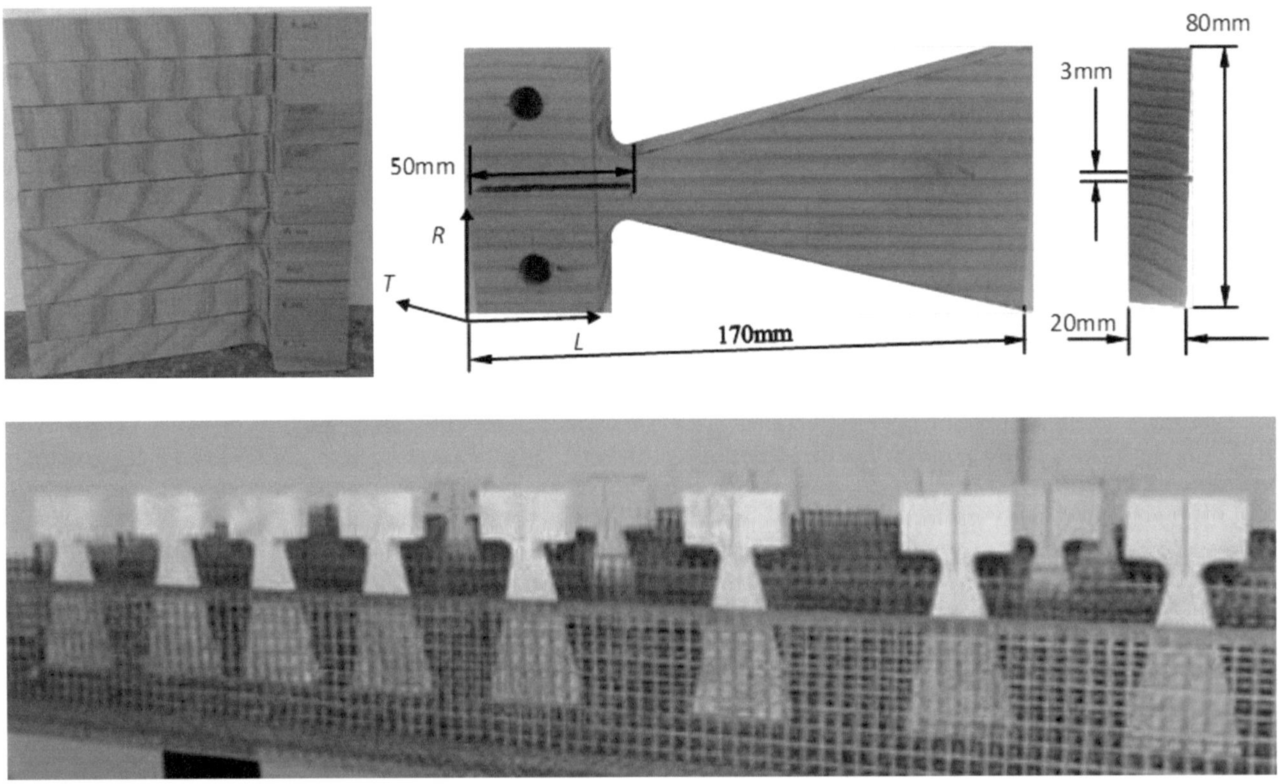

Fig. 29.1 Double cantilever beam with variable inertia specimen

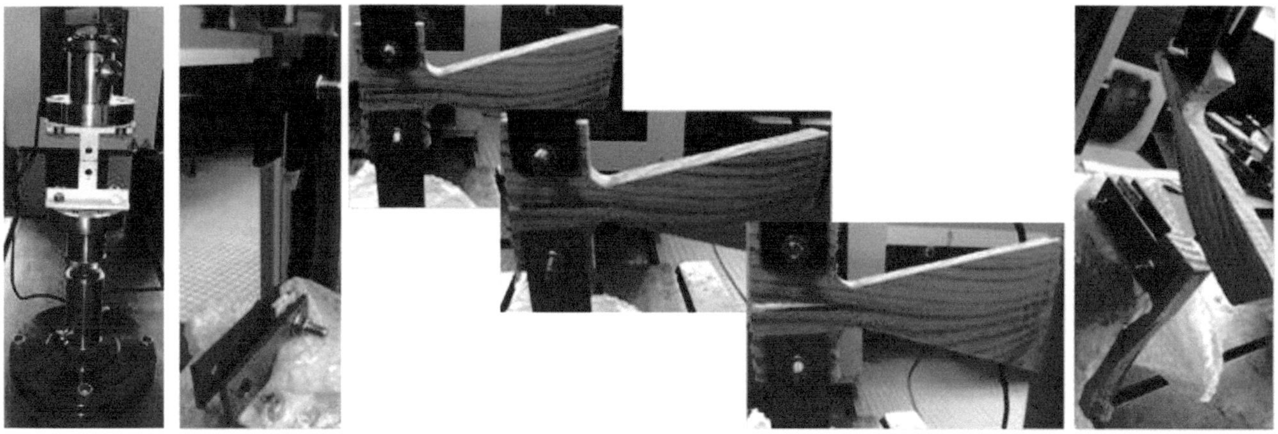

Fig. 29.2 Experimental testing machine

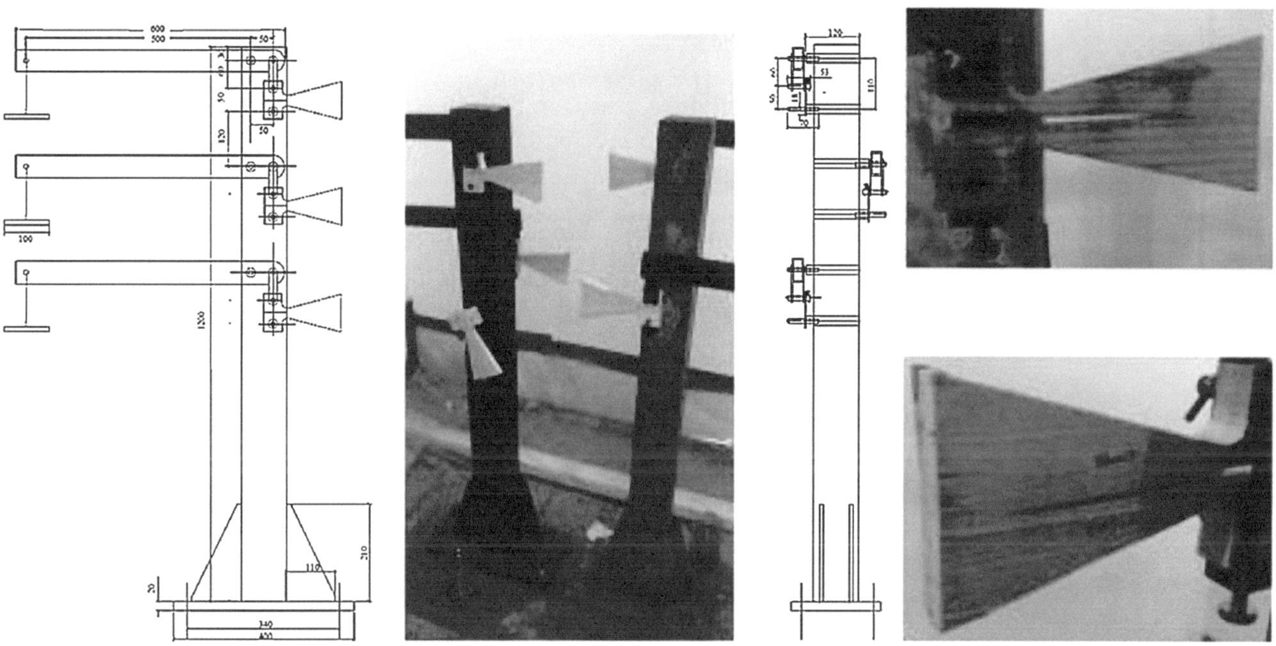

Fig. 29.3 Creep bench

29.2.2 Mechanical Loading for Characterization Static Test in Constant Humidity Environment

The Zwick electromechanical testing machine, with a 50 kN load capacity, is controlled in displacement, which allows forcing stable crack growth during the experimental test. This load is applied to the specimen by use of shafts pushed into holes drilled through the top and bottom cantilevers, Fig. 29.2. Specimens are tested at a constant displacement rate of 0.5 mm/min. During the tests on 2 wet and 2 dry specimens of each species, both force and displacement are measured and recorded, Figs. 29.7 and 29.8.

29.2.3 Mechanical Loading for Creep Test in Variable Humidity Evironment

The experimental device is a creep bench compound of four poles [5] that allow testing four DCB specimens for each specie together with an initial load (determined by the previous static test) applied through lever arm enabling to initiate cracking without exceeding the stability zone, Fig. 29.3.

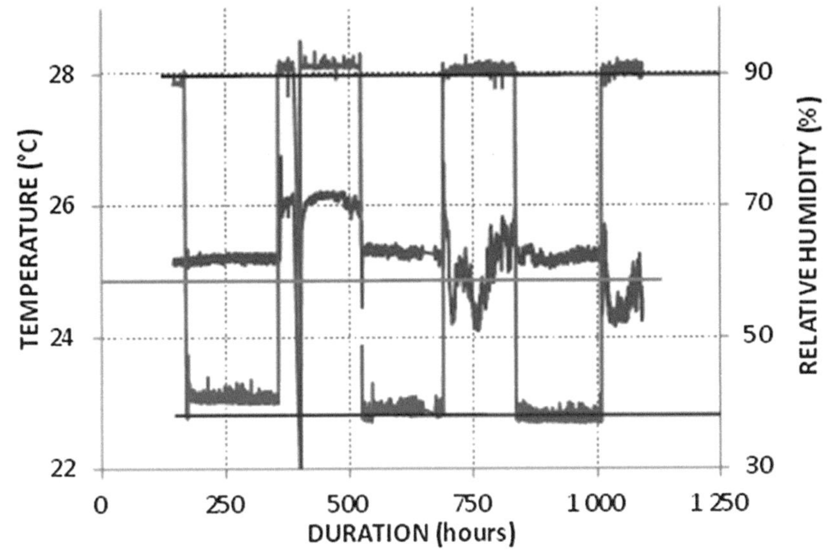

Fig. 29.4 Relative humidity and temperature monitoring

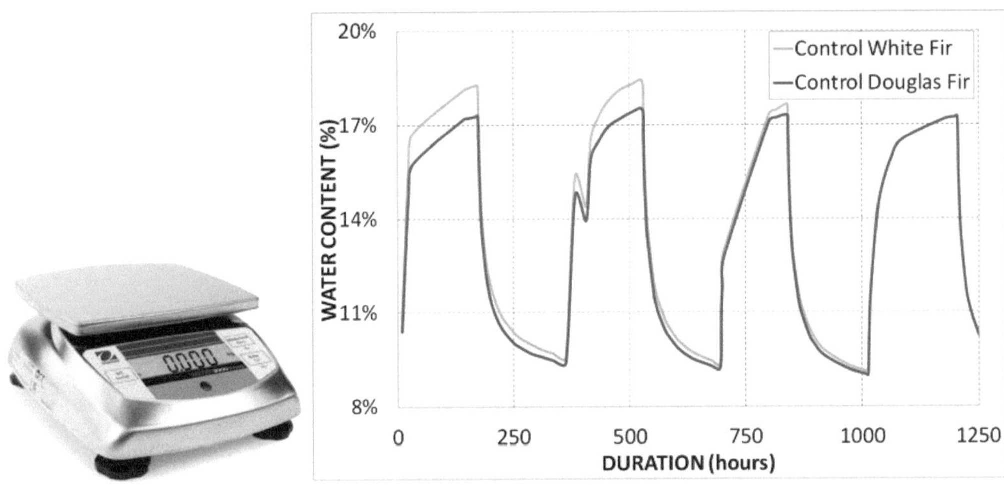

Fig. 29.5 Water content monitoring

The creep benches are into a 23 m^3 climatic chamber that provides hydric cycles between 40 and 90 % RH at a constant temperature of 25 °C, Fig. 29.4. Two supplementary control specimens are weighed regularly for monitoring the water content, Fig. 29.5, and the crack tip position is regularly measured on both sides of the tested specimens, Fig. 29.6.

29.3 Results and Discussions

In this section we discuss the results obtained for characterization static tests and creep tests for a better understanding of the failure process in Douglas and White Fir under variable humidity environment.

29.3.1 Characterization Static Test in Constant Humidity Environment

The opening mode crack tests are performed to assess the critical fracture force and the energy release rate of both Douglas and White Fir species. The comparison of the force-displacement curves, Fig. 29.7, emphasizes differences between Douglas and White Fir specimens' behavior: if initial rigidity (around 150 N/mm) and critical force (around 200 N) remain unchanged,

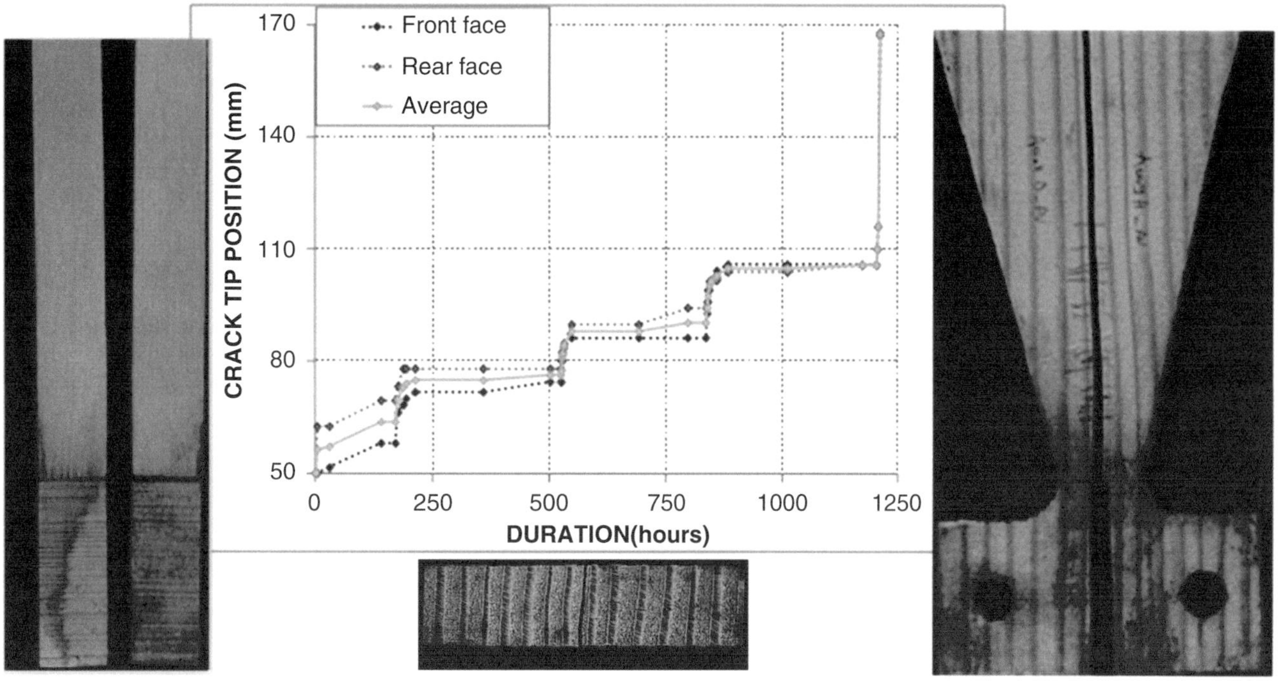

Fig. 29.6 Example of result: crack growth monitoring

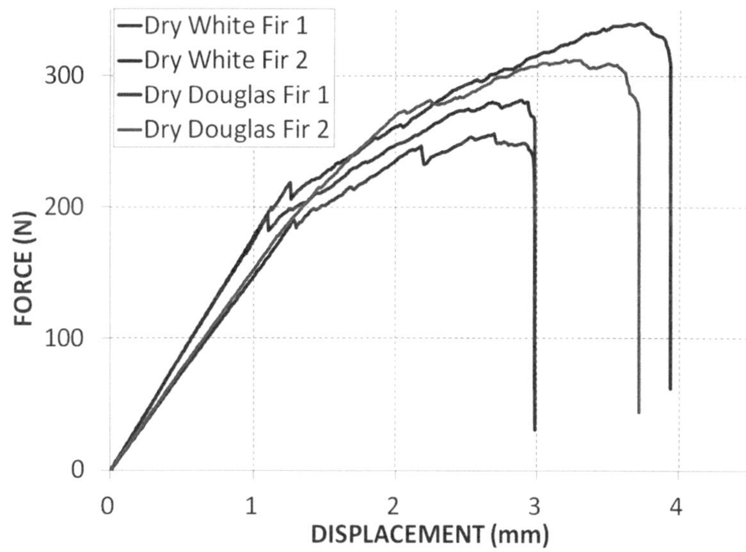

Fig. 29.7 Force-displacement curves for dry specimens (douglas and white fir)

the post fracture phase exhibits different displacements, maximal forces or rigidities as shown in Table 29.1. Simultaneously, the comparison of the force-displacement curves, Fig. 29.8, between wet and dry specimens behavior leads to the same observations. The average critical force value is used as the external force for forthcoming creep tests.

A thermodynamic approach allows illustrating the separation energies induced during the crack growth process. The first thermodynamic principle is based on the separation of external work in terms of a released energy (elastic energy) and a global dissipated fracture energy; this last energy at the same time, corresponds to the dissipation induced by the crack tip advance on new surface creations, but also the process zone development in a damaged and non linear zone and crack bridging. For all unloading phases, the crack is assumed to be closed once again without any particular interference regardless of the crack length. Figure 29.9 shows a load-displacement curve and corresponding calculated energies evolutions: external work, released energy and dissipated fracture energy.

According to an energy release rate concept, the energy release rate expresses the energy loss induced by the crack tip advance (on a created crack surface). The energy released is correlated with the dissipated energy, crack length and sample thickness, Table 29.2. For dry specimens values are around 310 J/m^2 for Douglas Fir and around 200 J/m^2 for White Fir. For wet specimens, values are around 400 J/m^2 for Douglas Fir and around 330 J/m^2 for White Fir.

Table 29.1 Samples force-displacement curves parameters: initial rigidity, critical force, maximal force, rupture displacement

Sample	Initial rigidity (N/mm)	Critical force (N)	Maximal force (N)	Rupture displacement (mm)
Dry douglas fir 1	153	178	312	3.73
Dry douglas fir 2	178	198	340	3.94
Dry white fir 1	148	188	256	2.98
Dry white fir 2	176	182	281	2.99
Wet douglas fir 1	151	230	301	4.75
Wet douglas fir 2	151	185	235	7.07
Wet white fir 1	149	228	313	3.96
Wet white fir 2	163	191	300	4.11

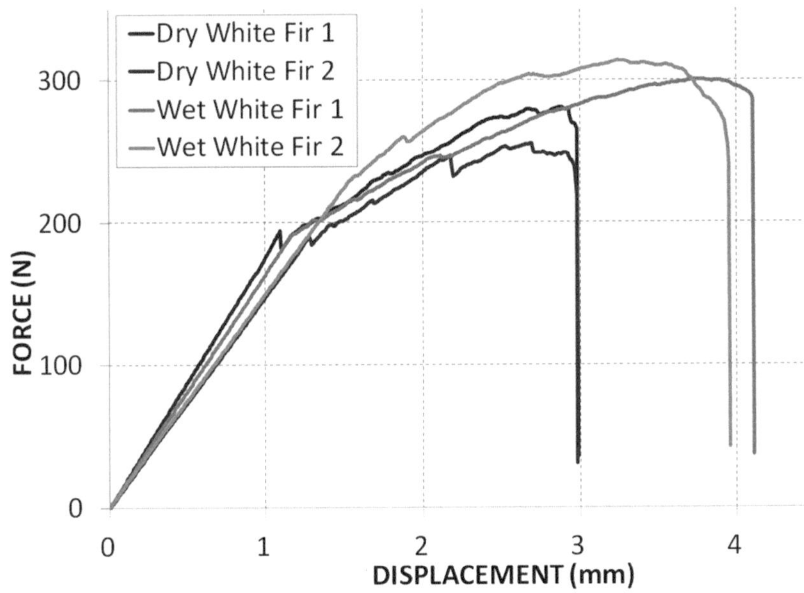

Fig. 29.8 Force-displacement curves for wet and dry specimens (white fir)

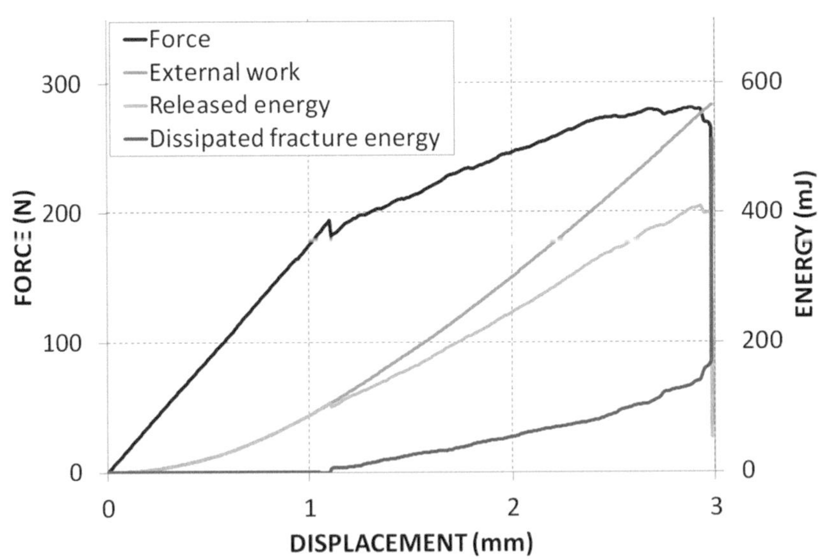

Fig. 29.9 Example of result: force and energies vs. displacement

Table 29.2 Samples energy release rate's determination

Sample	Thickness (mm)	Crack length (mm)	Dissipated fracture energy (mJ)	Energy release rate (J/m^2)
Dry douglas fir 1	20	120	719	300
Dry douglas fir 2			791	329
Dry white fir 1			468	195
Dry white fir 2			512	213
Wet douglas fir 1			954	398
Wet douglas fir 2			1170	487
Wet white fir 1			790	329
Wet white fir 2			805	335

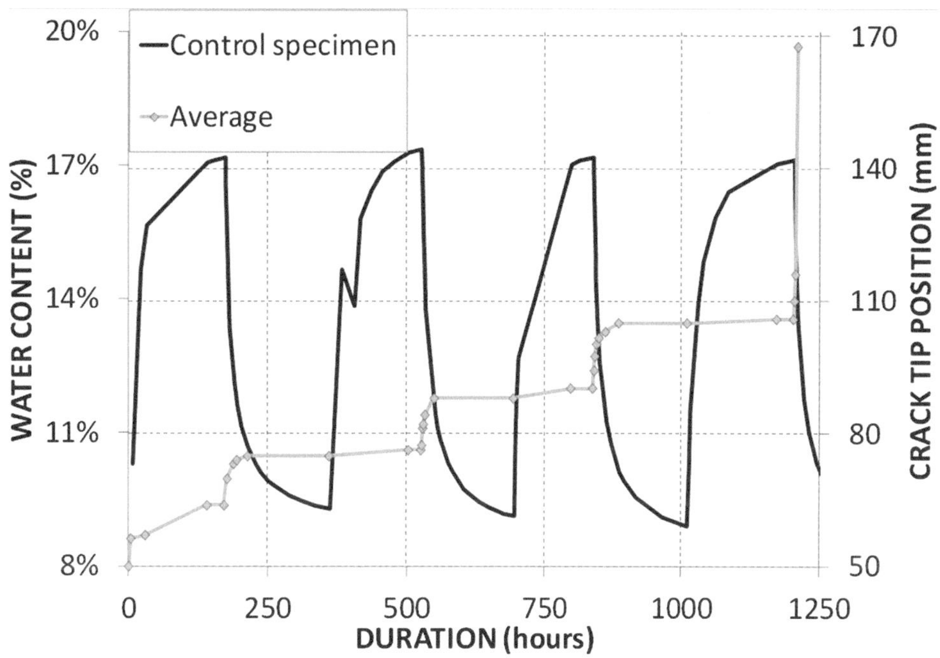

Fig. 29.10 Example of result: superposition of water content and crack growth evolutions

Table 29.3 Samples crack advance's parameters

Sample	Load applied (N)	Average crack advance step (mm)	Rupture crack tip position (mm)	Rupture time (h)	Number of hydric cycles to rupture
Douglas fir A	215.8	9	>88	–	>4
Douglas fir B	214.5	5	>75	–	>4
Douglas fir C	215.1	25	117	843.9	3
Douglas fir D	215.4	13	106	1210.9	4
White fir A	189.2	18	96	846	3
White fir B	191.8	17	87	857.6	3
White fir C	189.5	6	–	–	?

29.3.2 Creep Test in Variable Humidity Environment

By superposing water content evolution and crack tip position vs. time, we can observe that crack propagation principally occurs at the beginning of each drying phase, Fig. 29.10. In average, the crack advance step is between 5 and 25 mm at each phase, Table 29.3.

The complete rupture of the test specimens occurs after 3 or more cycles as shown in Fig. 29.11, when the crack tip position is between 90 and 120 mm (corresponding to a crack length of 40 mm and 70 mm respectively), Table 29.3: the stability zone has been exceeded.

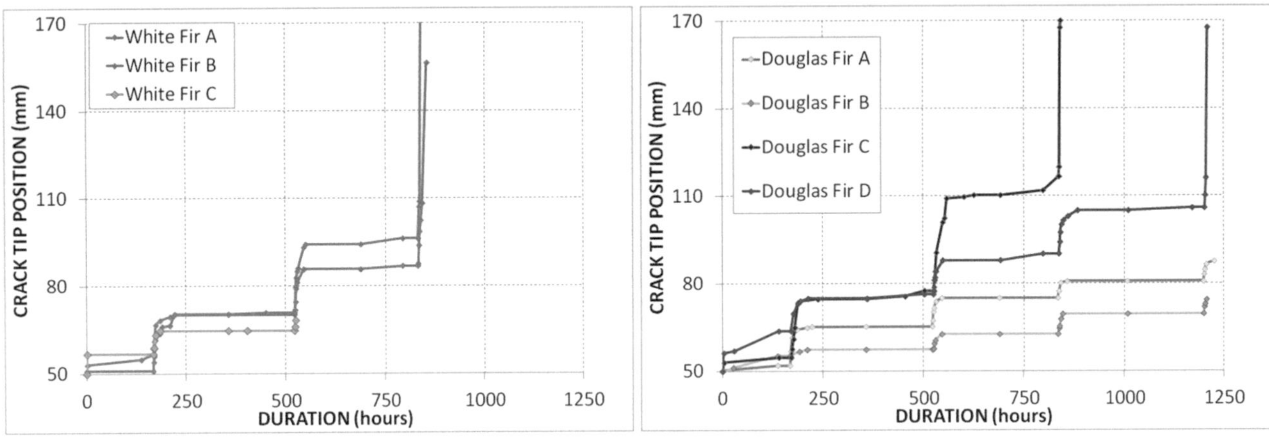

Fig. 29.11 Comparison of crack growth evolutions (white and douglas fir)

Conclusion and Outlook

These experimental results constitute a first database in order to validate numerical tools given by a finite element model based on a new analytical formulation of the A-integral. This independent integral allows taking into account the effect of thermal and hydric loads in the cracking process. Coming crack growth tests will be conducted with the miniaturized MMGC specimen with non destructive measurements (optical methods, acoustic emission). In fact these results will allow a better understanding of in-service behavior of timber structures subjected to mechanical loads and climatic variations. The final aim is to expand the use of wood material (characterized by low environmental impact), and in particular the studied Massif Central's (France) species in civil engineering structures.

Acknowledgment The authors wish to strongly acknowledge the National Agency of Research (ANR) for its financial support of this work through the project CLIMBOIS N° ANR-13-JS09-0003-01 labeled by ViaMeca.

References

1. Nguyen, S.L., Saifouni, O., Destrebecq, J.F., Pitti, R.M.: An incremental model for wood behaviour including hydro-lock effect. In: Final Cost Action FP0904 Conference, Recent Advances in the Field of TH and THM Wood Treatment, Skellefteå, Sweden, 19–21 May 2014
2. Riahi, H., Pitti, R.M., Chateauneuf, A., Dubois, F.: Stochastic analysis of mixed mode fracture in timber material using polynomial chaos expansion. In: Final Cost Action FP0904 Conference, Recent Advances in the Field of TH and THM Wood Treatment, Skellefteå, 19–21 May 2014
3. Pitti, R.M., Diakhaté, M., Arteaga, E.B., Aoues, Y., Angellier, N., Riahi, H., Chateauneuf, A., Dubois, F.: ANR JCJC-2013 Project CLIMBOIS: effects of climatic and mechanical variations on the durability of timber structures: use of acoustic emission tool to evaluate wood mechanical behaviour. In: COST Action FP1302 Wood Music, Opening Conference. Paris, 26–28 Feb 2014
4. Riahi, H., Pitti, R.M., Dubois, F.: On analytical formulation integrating mixed mode fracture and climate change in wood. In: COST Action FP0904 THMBP Workshop, Damage and Fracture Coupling with Thermo-Hydro-Mechanical effects, Bordeaux, 13–14 Apr 2014
5. Dubois, F., Chazal, C., Petit, C.: Viscoelastic crack growth process in wood timbers: an approach by finite element method for mode I fracture. Int. J. Fract. **113**, 367–388 (2002)
6. Riahi, H., Pitti R.M., Dubois, F., Fournely, E., Chateauneuf, A.: Numerical fracture analysis coupling thermo-hygro mechanical and viscoelastic behaviour. In: 9th International Conference on the Mechanics of Time-dependent Materials, Montreal, 27–30 May 2014
7. Dubois, F., Pitti, R.M., Pop, O., Fournely, E.: Numerical model about fracture analysis and mechano-sorptive behavior in wood. In: 9th International Conference on the Mechanics of Time-dependent Materials (MTDM), Montreal, 27–30 May 2014

Chapter 30
Numerical Fracture Analysis Under Temperature Variation by Energetic Method

Rostand Moutou Pitti, Seif Eddine Hamdi, Frédéric Dubois, Hassen Riahi, and Nicolas Angelier

Abstract It is known that temperature change can induce sudden crack propagation especially when the material is composed of fibers. In this fact, the crack growth process under mixed-mode coupling mechanical and thermal loads in orthotropic materials like wood is investigated in this work. The analytical formulation of *A* integral's combines the real and virtual mechanical and thermal stress/strain fields under transient diet in 2D. The Mixed Mode Crack Growth specimen providing the decrease of energy release rate during crack propagation is considered in order to compute the various mixed mode ratios. By using three specific routines, the analytical formulation is implemented in finite element software Cast3m. The efficiency of the proposed model is justified by showing the evolution of energy release rate and the stress intensity factors versus crack length and versus temperature variation in time dependent material.

Keywords Mixed mode crack growth • Thermal fields • Path independent integral • Finite element method • Wood material • Time-dependent material

30.1 Introduction

The micro-cracks commonly occur in many mechanical and civil engineering structures submitted to different loadings. But the main important fact for structural integrity remains the conditions of propagation of these small crack during the lifetime. Combining with mechanical solicitations as fatigue, overload or creep loading, the environmental actions like hydric or temperature play an important role in the propagation of these micro-cracks in the material. In the case of wood, due to its natural origin, its orthotropic and heterogeneous character with different defects such as knots, study these different approaches appear to be essential.

To predict the crack growth process many numerical methods were developed to characterize the mechanical fields around the crack tip. Among them, the background of energy methods come from invariant integrals which enables to evaluate the crack driving forces such as the crack growth rate and the stress intensity factors. The most popular is the J-integral proposed by Rice [1] based on the assessment of the strain energy density and Noether's theorem [2]. This method is inefficient when dealing with mixed mode crack growth problems because it is necessary to separate the displacement field into a symmetric and antisymmetric parts. To circumvent this difficulty, Chen and Shield [3] have developed the M-integral in order to separate fracture modes based on a bilinear form of the strain energy density with virtual mechanical fields.

In recent work, a new analytical formulation of A-integral developed by Moutou Pitti et al. [4] and implemented in finite element software by Riahi et al [5] is proposed. This formulation takes into account the effects of thermal load, induced by

R. Moutou Pitti (✉)
Université Clermont Auvergne, Université Blaise Pascal, Institut Pascal, BP 20206, Clermont-Ferrand F-63000, France

CNRS, UMR 6602, Institut Pascal, Aubiere F-63171, France

CENAREST, IRT, Libreville 14070, Gabon
e-mail: rostand.moutou_pitti@univ-bpclermont.fr

S.E. Hamdi
Université Clermont Auvergne, Université Blaise Pascal, Institut Pascal, BP 20206, Clermont-Ferrand F-63000, France

CNRS, UMR 6602, Institut Pascal, Aubiere F-63171, France

F. Dubois • N. Angelier
Heterogeneous Material Research Group, Civil Engineering Center, Université de Limoges, Egletons 19300, France

H. Riahi
LARIS, Université d' Angers, Angers F-49000, France

A.M. Beese et al. (eds.), *Fracture, Fatigue, Failure and Damage Evolution, Volume 8*, Conference Proceedings of the Society for Experimental Mechanics Series, DOI 10.1007/978-3-319-21611-9_30

temperature variation, and complexes boundaries conditions, such as contact between crack lips during crack growth process. However, this approach is limited to stationary crack coupled with a pressure on crack lips and is note suitable to viscoelastic materials.

The first part of this paper presents the mathematical formulation of the invariant integrals T and A with the part of crack growth process. Simultaneously, the energy release rate in mixed mode is proposed according to the real and virtual stress intensity factors. In the second section, the background of Mixed Mode Crack Growth (MMGC) specimen is proposed [6]. This specimen is combination between Double Cantilever Beam (DCB) and Compact tension Shear (CTS) specimens. The third section is dedicated to study the efficiency of the proposed approach in the case of orthotropic material, where crack growth analysis on wood MMGC specimen is performed.

30.2 Analytical Formulation of the Invariant Integrals *T* and *A*

In this section we recall the mathematical expression of T and A integrals parameters. According to the conservative and the Noether theorem defined with Arbitrary, Eulerian and Lagrangian conditions, if we consider a cracked body Ω and Γ a path which surrounds the crack tip oriented by the normal $\overrightarrow{n}$ of component n_j, as shown in Fig. 30.1a, the T-integral is given by:

$$\begin{aligned} T = & \int_\Gamma \frac{1}{2}\left[\sigma^v_{ij,k}\, u_i - \sigma^u_{ij}\, v_{i,k}\right] n_j\, d\Gamma + \int_\Gamma \frac{1}{2}\left[\gamma \Delta T_{,j}\,(v_k - \psi_k) - \gamma \Delta T\left(v_{k,j} - \psi_{k,j}\right)\right] n_j\, d\Gamma \\ & - \int_\Gamma \left[\sigma^v_{ij,k}\, u_{i,j} - \sigma^u_{ij,k}\, v_{i,j} + \beta\, \delta_{ij}\, u_{i,jk}\, \Delta T\right] n_j\, d\Gamma + \int_L \frac{1}{2}\left[p\, v_{1,1} - q\, v_{2,1}\right] dx_1 \end{aligned} \tag{30.1}$$

In the precedent expression, the first integral is the classical term of the T-integral which represent the effect of mechanical loads applied far from the crack, where σ^u_{ij} and σ^v_{ij} are stress tensor components deduced from the real displacement field u and the virtual displacement field v, respectively. The second integral represents the effect of thermal load induced by temperature variation ΔT, with ψ is a virtual displacement field as defined by Bui et al [7] and γ is a real coefficient function of material parameters. The third integral represents the pressure crack tip extension during the crack growth process. The last integral represents the effect of pressures p and q applied perpendicularly to the crack lips, where $L = OA_1 + OB_1$ is the integration path.

Although, experience have shown that integration through curvilinear path induce some inaccuracy on numerical results. To overcome this problem, the curvilinear path is transformed into surface domain by introducing a vector field $\overrightarrow{\theta}$ [8] as shown in Fig. 30.1b. This mapping function is continuously differentiable and takes these values: $\overrightarrow{\theta} = (1, 0)$ inside the ring V, and: $\overrightarrow{\theta} = (0, 0)$ outside it. Hence, the use of the Gauss–Ostrogradsky's theorem enables us to obtain the A-integral given by:

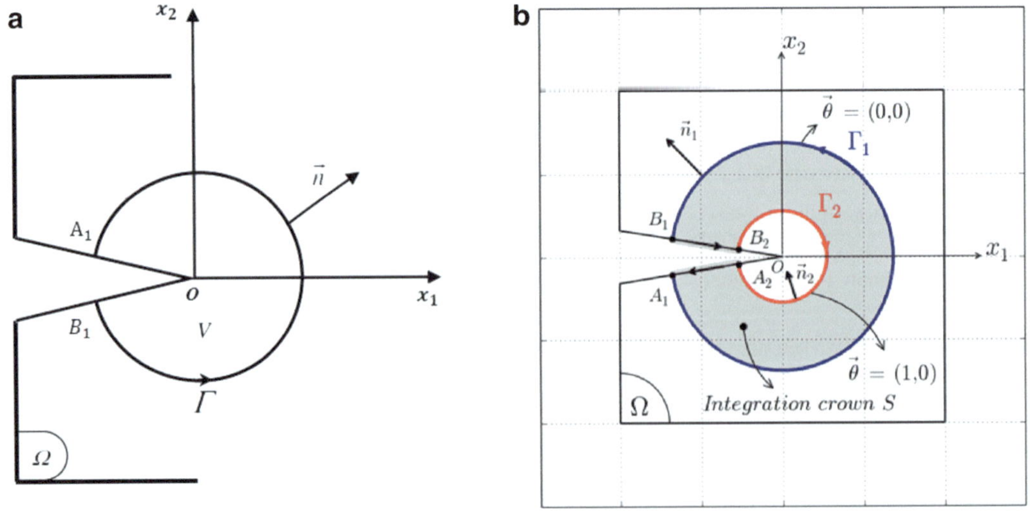

Fig. 30.1 (**a**) Curvilinear domain integral, (**b**) surface domain integral

$$A = \int_{\Omega} \frac{1}{2}\left[\sigma^{v}_{ij,k}\, u_i - \sigma^{u}_{ij}\, v_{i,k}\right]\theta_{k,j}\, dS + \int_{\Omega} \frac{1}{2}\left[\gamma \Delta T_{,j}\left(v_k - \psi_k\right) - \gamma \Delta T\left(v_{k,j} - \psi_{k,j}\right)\right]\theta_{k,j}\, dS$$
$$- \int_{\Omega}\left[\sigma^{v}_{ij,k}\, u_{i,j} - \sigma^{u}_{ij,k}\, v_{i,j} + \beta\, \delta_{ij}\, u_{i,jk}\Delta T\right]\theta_k\, dS + \int_{L} \frac{1}{2} F_i\, v_{i,j}\, \theta_j\, dx_1 \tag{30.2}$$

where $F_1 = p$ and $F_2 = q$ on the upper lip and $F_1 = -p$ and $F_2 = -q$ on the lower lip. Let us note that the virtual stress tensor components σ^{v}_{ij} are proportional to virtual thermal stress intensity factors ${}^{A}K^{v}_{I}$ and ${}^{A}K^{v}_{I}$ characterising virtual open and shear modes, respectively. Moreover, the A-integral, like M-integral, can be physically interpreted as a particular definition of real stress intensity factors ${}^{A}K^{u}_{I}$ and ${}^{A}K^{u}_{II}$. The mixed mode separation can be obtained by performing two distinct computations of ${}^{A}K^{u}_{I}$ and ${}^{A}K^{u}_{II}$ for particular values of ${}^{A}K^{v}_{I}$ and ${}^{A}K^{v}_{I}$, such as [9]:

$${}^{A}K^{u}_{I} = 8\frac{A\left({}^{A}K^{v}_{I}=1, {}^{A}K^{v}_{II}=0\right)}{C_1};\ {}^{A}K^{u}_{II} = 8\frac{A\left({}^{A}K^{v}_{I}=0, {}^{A}K^{v}_{II}=1\right)}{C_2} \tag{30.3}$$

In (30.3), C_1 and C_2 denote the elastic compliances in opening and shear modes, respectively. The thermal energy release rates in each specific fracture mode ${}^{A}G_I$ and ${}^{A}G_{II}$ are finally given by the following expression

$${}^{A}G_I = C_1\frac{\left({}^{A}K^{u}_{I}\right)^2}{8} \text{ and } {}^{A}G_{II} = C_2\frac{\left({}^{A}K^{u}_{II}\right)^2}{8} \tag{30.4}$$

The analytical formulation of the A-integral, presented above, is achieved assuming an elastic behavior. Unfortunately, timber has a viscoelastic behavior. In this case, the mechanical fields are time dependent and their computation is not a trivial task. Based on Boltzmann's superposition principle and considering a non-aging linear viscoelastic material, the strain and stress tensors, $\boldsymbol{\varepsilon}$ and $\boldsymbol{\sigma}$ respectively. The generalization to viscoelastic material is introduced by using a generalized Kelvin Voigt model presented in Fig. 30.2.

In this case, the characterization of the creep law only requires some relationship between the elastic moduli k^{p}_{ijkl} associated to the spring's and the dash-pot viscosities η^{p}_{ijkl}. In this context, each component J_{ijkl} of the creep tensor can be written as Dirichlet expansion and finally, the equation (30.2) is rewritten for each Kelvin Voigt cells as follow [10]

$$A_v = \sum_{m=0}^{N} A^{m}_{v} \text{ and } {}^{A}G^{v}_{\alpha} = \sum_{m=0}^{N} {}^{A}G^{m}_{\alpha} \text{ for } \alpha = 1,\ 2 \tag{30.5}$$

The quantities A^{m}_{v} are computed according to (30.2), where the real mechanical fields are obtained from the response of the mth Kelvin cell to the applied mechanical and thermal loading. The computation of viscoelastic energy release rate ${}^{A}G^{m}_{\alpha}$ in each fracture mode is obtained by (30.4) generalized to time dependent material. In the numerical process proposed in following section, the last term of (30.2) is not taking into account.

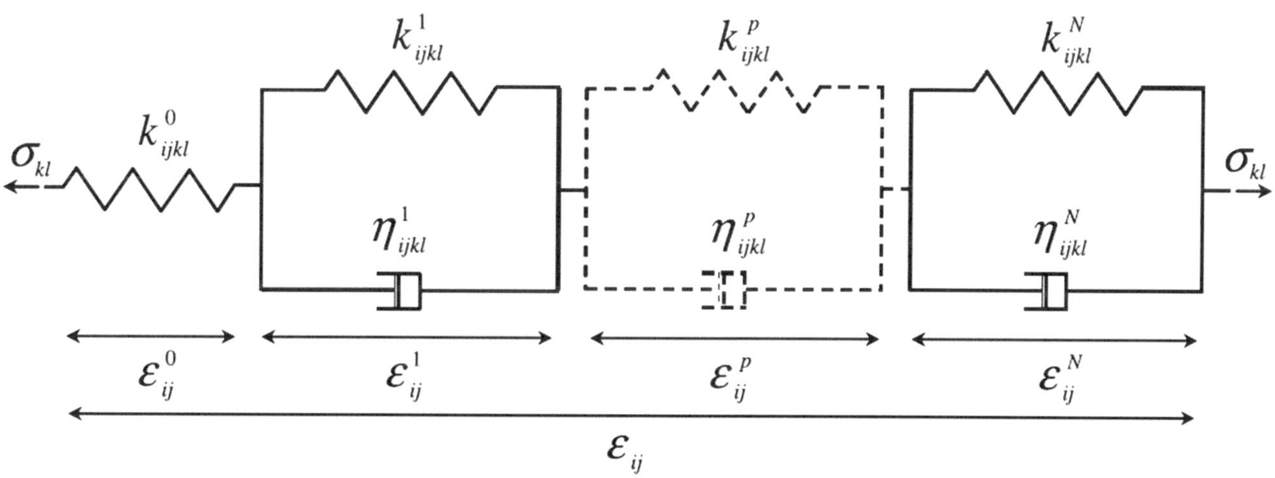

Fig. 30.2 Generalized Kelvin Voigt rheological model

30.3 Background of Mixed Mode Crack Growth specimen

30.3.1 MMCG Wood Specimen

Figure 30.3a shows the dimensions of wood specimen, with an overall size $210 \times 140 \times 25$ mm^3. The side of the specimen has inclinations with angles of 5° and 10° from the edge and the inferior hill, respectively, as shown in Fig. 30.1b. On both hills, four holes were perforated in order to fix the steel Arcans. The distance between the holes centers is 110 mm. The steel Arcans, were performed to allow for mixed-mode configurations. The type of wood used in this test is Douglas fir (*Pseudotsuga menziesii*) and the width of the annual ring varies between 3 and 5 mm and the crack is oriented following the (R/L) direction. An initial crack of 20 mm is machined in the wood specimen in the direction of the fibers.

The MMCG specimen, shown in Fig. 30.3b, is defined by the combination of a modified DCB specimen, and CTS specimen, on the one hand, and the observation of stability range of energy release rate versus crack length computed numerically with Mθ [6], on the other hand. The main objective of this specimen is to allow for the decrease in the energy release rate during the tests. The wood specimen presented in Fig. 30.3a, is fixed in a framed structure with Arc-formed handles containing symmetrical holes for load application according to various crack mode ratios β. The application of symmetric loads F_I with angle $\beta = 0°$ is equivalent to the opening mode, and the application of F_{II}, with angle $\beta = 90°$ corresponds to the shear mode as shown in Fig. 30.3b. The force F is separated into two forces F_x and F_y corresponding to the axis orientations (x, y) and the mixed-mode ratio in plane configuration. In this work, the mixed-mode fracture tests are obtained by applying the loads F_I and F_{II} with angles $\beta = 45°$.

30.3.2 Viscoelastic Crack Growth Routine with Thermal Fields

The analytical formulation of A-integral without taking into account the effect of pressure on the crack lips, was implemented in the finite elements software Cast3m. The MMCG specimen geometry variation versus time (crack evolution) and crack length is considered in the crack growth model. For symmetric geometries and loadings, this crack growth is modelled by moving boundary elements in a semi-mesh. In a mixed mode configuration, this symmetric loading is lost. Then, it is necessary to operate a re-mesh driven by the crack tip advance. In our cases, a fixed crack orientation is supposed. To overcome this difficulty, the hereditary mechanical fields have been projected in the transformed mesh. For a selected temperature, numerical modelling must separate the time and geometry variations. In a first step, all mechanical fields and the crack length a are supposed known at time t_n. Then, stress and external loading, named $\sigma_i(t_n)$ and $F_i(t_n)$, respectively, are defined in the initial mesh noted W_i characterized by a crack length a.

Firstly, the stress perturbation induced by a non-time dependent crack growth is evaluated. The first step consists of considering an instantaneous crack tip advance Δa. New mesh noted W_{i+1} is defined by re-meshing. With the same external loading, an elastic calculation provides the new stress state $\sigma_{i+1}(t_n)$. We note $\sigma(p)_{i+1}(t_n)$ the geometric projection of $\sigma_i(t_n)$ on mesh W_{i+1}. The second step involves calculating, with a differentiation, the stress perturbation $\Delta\sigma_{i+1}$ between the two

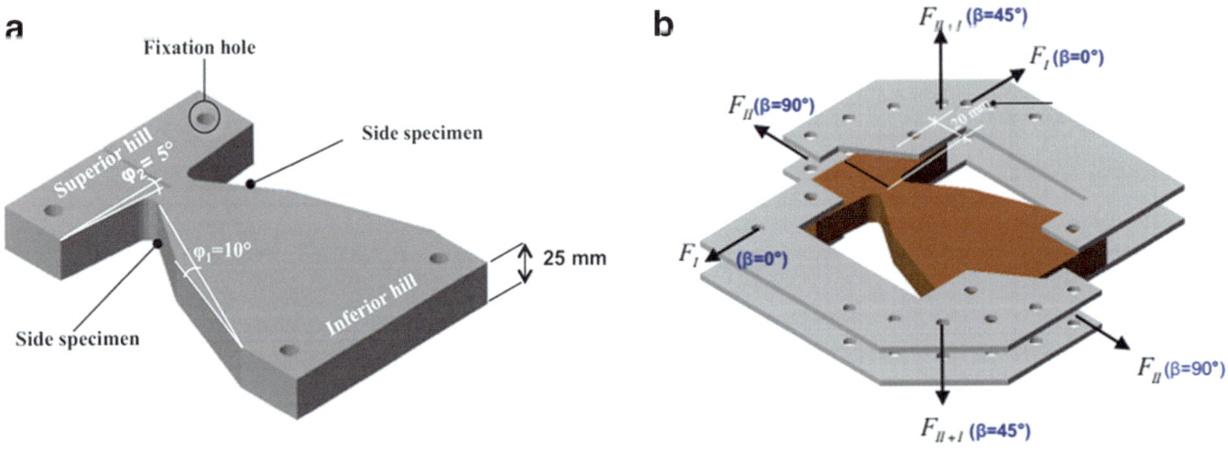

Fig. 30.3 (**a**) Dimension wood specimen. (**b**) MMGC specimen

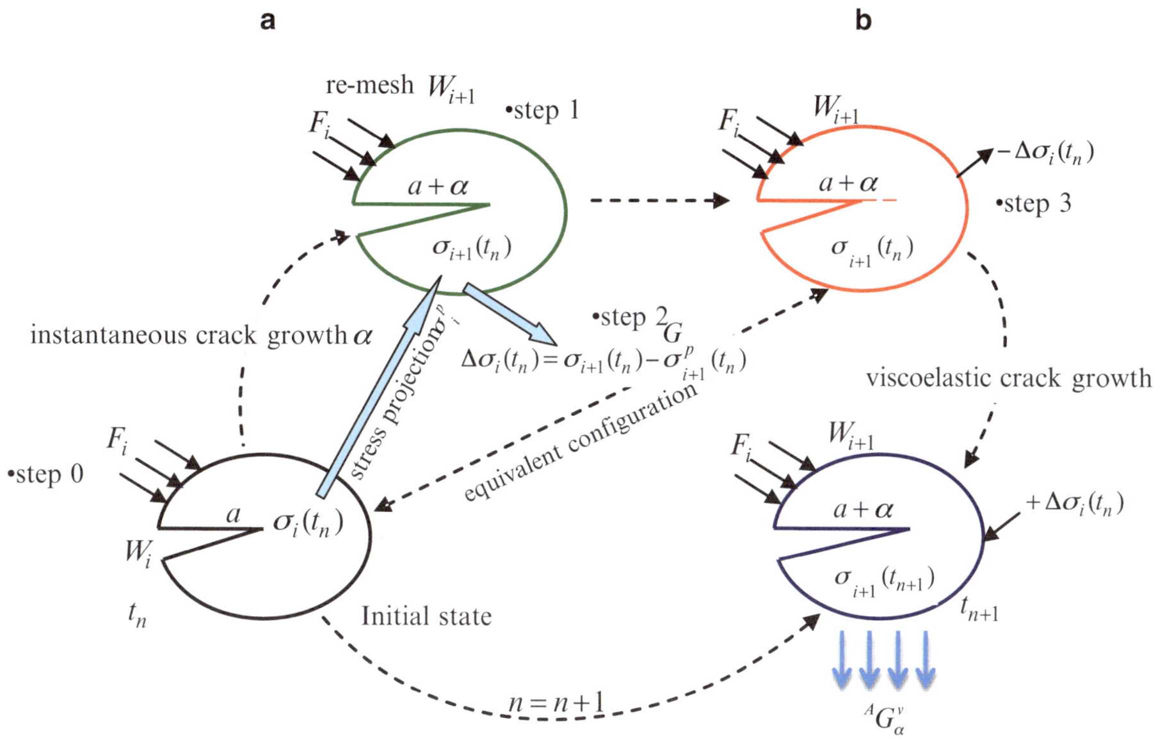

Fig. 30.4 Crack growth routine with thermal fields

mesh configurations. The final step involves the application of $-\Delta\sigma_{i+1}$ as a cohesion stress (equivalent external loading) by using the superposition principle. This supplementary loading allows to close the new crack on Δa. In this case, we obtain an equivalent configuration between the first and last steps (same mechanical state and same geometry), but with two different meshes. $\Delta\sigma_{i+1}$ can be interpreted as the stress cohesion in the process zone.

Secondly, the crack length advance is fixed (in our work we fixed Δa = 8 mm), and time is continuous. Employing the incremental viscoelastic procedure, the stress cohesion $\Delta\sigma_{i+1}$ is employed as an external load vector during the time increment Δt_n. In the time domain, the progressive un-cohesion of crack lips un-cohesion Δa is obtained. The true mechanical state can then be released. Finally, fracture parameters can be computed and the finite elements algorithm is incremented (Fig. 30.4).

30.4 Numerical Results

30.4.1 Finite Element Meshes

In this last section, we present numerical simulations, which prove the non-path dependence of the A-integral on hereditary behavior and discuss the crack growth process over time. In recent works, Moutou Pitti et al. [9], have shown that this approach is valid for a stationary crack. Now, we propose to generalize the validation by imposing a time dependent crack tip advance according to the numerical procedure presented in Sect. 30. 3. In order to validate A-integral expressions (30.5), it has been decided to simulate an orthotropic material. The geometry design is operated by the finite element method. The finite element mesh is shown in Fig. 30.5a in which a circular discretization around the crack tip allows us to define easily the temperature field and the integration crown using parametric elements (θ vector). The MMCG specimen is designed in order to obtain different mixed mode ratios and crack growth stability. The inferior hill is added in order to admit different mixed-mode loadings adding four holes allowing Arcan fixtures as shown in Fig. 30.5b. The A-integral is performed using virtual finite element displacement fields as shown in Fig. 30.5c for opening mode, as an example. The simulations integrate

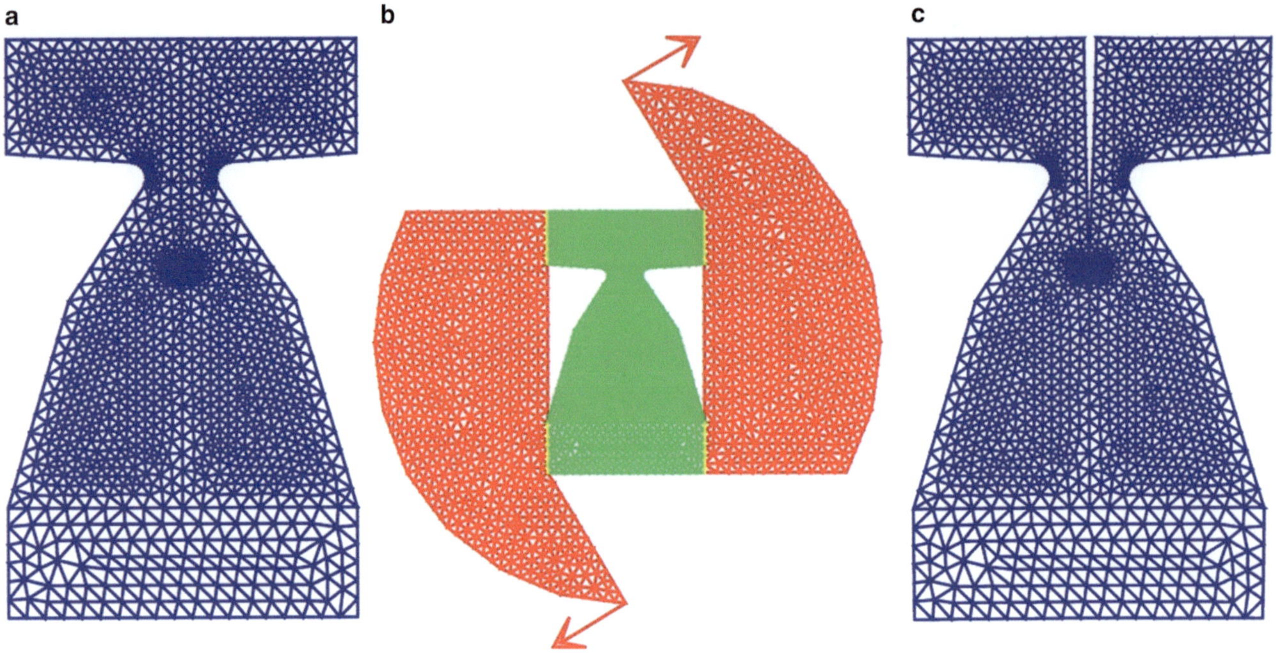

Fig. 30.5 (**a**) Numerical mesh of MMCG specimen with circular meshes around the crack tip. (**b**) Numerical mesh of MMCG specimen under loading and its fixation. (**c**) Virtual displacements for opening mode

orthotropic viscoelastic behavior for long-term loadings, with a longitudinal modulus $E_l = 1500$ MPa, transverse modulus $E_2 = 600$ Pa, normal modulus $E_3 = 600$ MPa, shear modulus $G_{12} = 700$ Pa, and Poisson's coefficient $\nu_{12} = 0.4$, $\nu_{13} = 0.4$, $\nu_{23} = 0.4$.

We note that, the numerical analysis is performed under plan stress conditions and based on the finite element mesh depicted in Fig. 30.5a. Moreover, in order to have good accuracy on the computation of the mechanical fields, a refined mesh is adopted in the neighborhood of the crack. In this work, we propose to deal with MMCG specimen with initial crack length ($a = 80\ mm$) subjected to tow temperature fields, $T_1 = 20\ °C$ and $T_2 = 30\ °C$, respectively. The numerical meshes of the MMCG specimen under the applied temperature fields is shown in Fig. 30.6.

30.4.2 *Viscoelastic Energy Release Rate*

For MMCG specimen, Figs. 30.7 and 30.8 show us the viscoelastic energy release rate evolution versus time and crack growth process for mixed mode using the A-integral concept. The crack growth process can be interpreted as brittle or ductile. If for all cases, the crack growth or damage evolutions in the process zone induce a rigidity decrease, we can note an energy release rate increase during the crack growth advance. With these considerations, for both mode 1 and mode 2 configurations, we can observe a progressive growing of the process zone (phase of energy release rate increasing) and, in the other hand, a stationary phase with a stabilization of its evolution. More precisely, we can observe, for the first mode G_1, a greater energy release rate then the second mode G_2, which correspond on brittle crack tip advance with an approximately peaks values around $8.5 \cdot 10^{-6}$ J/m^{-2} for G_1 and $7.8 \cdot 10^{-5}$ J/m^{-2} for G_2. Energy steady state evolution after these peaks illustrate crack growth stabilities for mixed mode calculated with A-integral without taking into account the effect of thermal load induced by a temperature field variation.

Figures 30.7 and 30.8 also shows the effect of temperature variation on the evolution of the viscoelastic energy release rate. We note that for both first mode G_1 and the second mode G_2, a gap appears between viscoelastic energy release values for $T_1 = 20$ °C and $T_2 = 30$ °C. In fact, for G_1 and G_2 modes, a greater energy release rate evolution can be observed, under the effect of temperature field $T_2 = 30$ °C then the temperature field of $T_1 = 20°$. Summarizing, we assimilate these evolutions as the critical values for G_1 and G_2, we conclude that the shear mode resistance is largest that the open mode resistance. According to the numerical results applied on the MMCG specimen, it is shown the capacity of A-integral to evaluate efficiently, viscoelastic characteristics and fracture parameters during crack growth and thermal load induced by a temperature field variation.

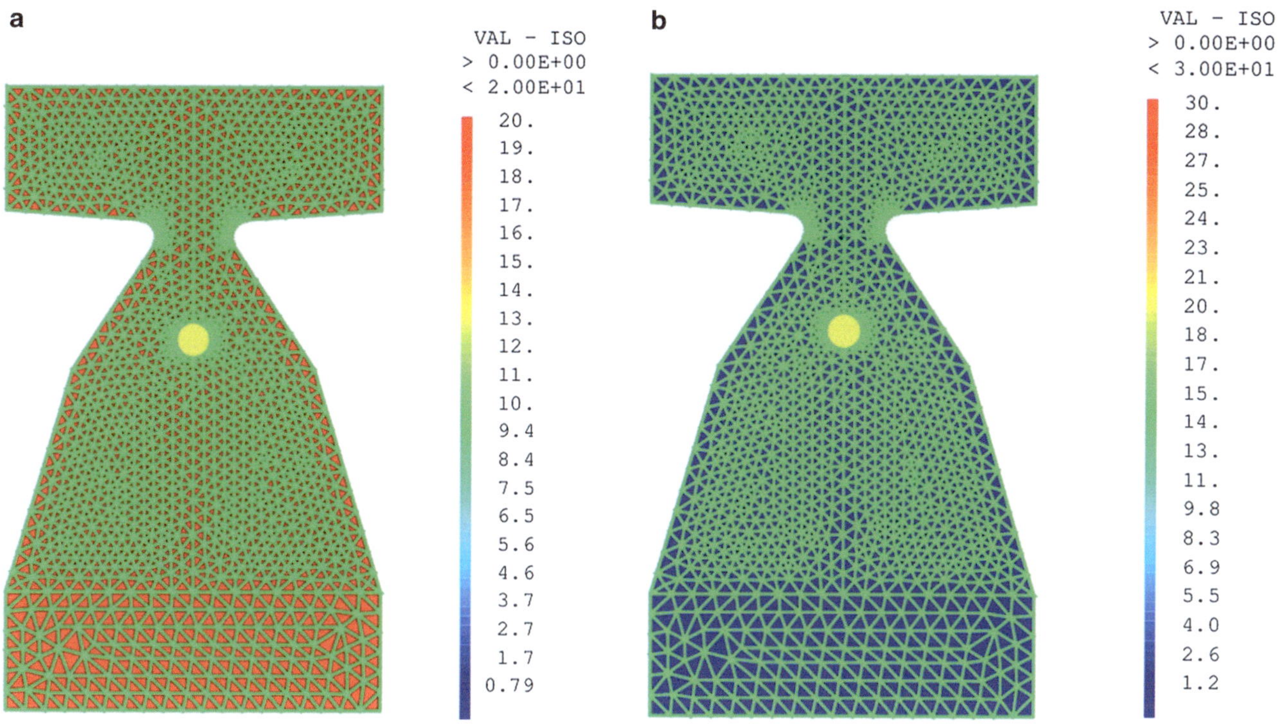

Fig. 30.6 Numerical mesh of the MMCG specimen under temperature field: (**a**) $T_1 = 20\ °C$ (**b**) $T_2 = 30\ °C$

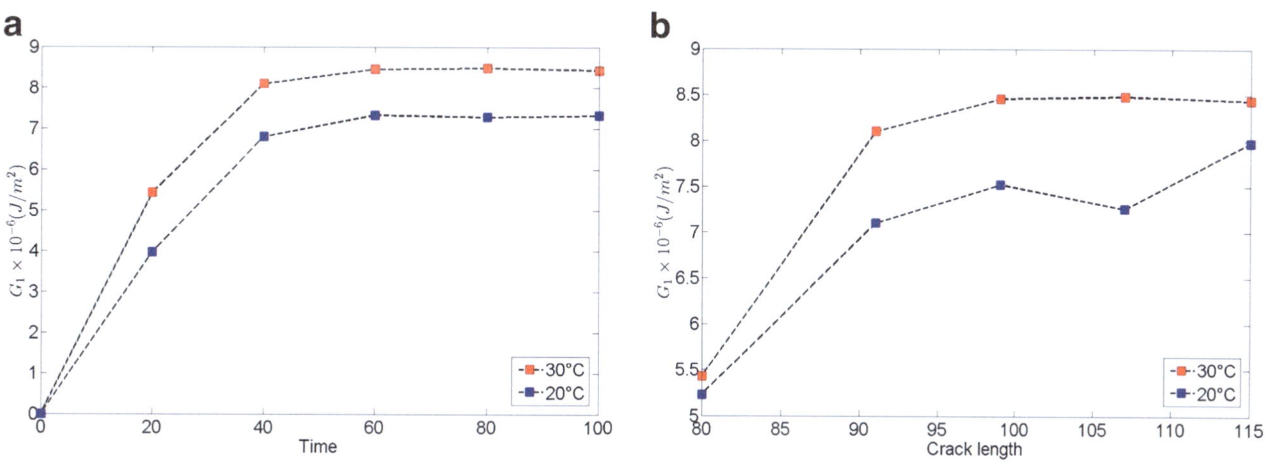

Fig. 30.7 (**a**) Viscoelastic energy release rate G_1 in mixed mode ($\beta = 45°$) versus time under temperature variation ($T_1 = 20\ °C$ and $T_2 = 30\ °C$). (**b**) Viscoelastic energy release rate G_1 in mixed mode ($\beta = 45°$) during crack growth process ($T_1 = 20\ °C$ and $T_2 = 30\ °C$)

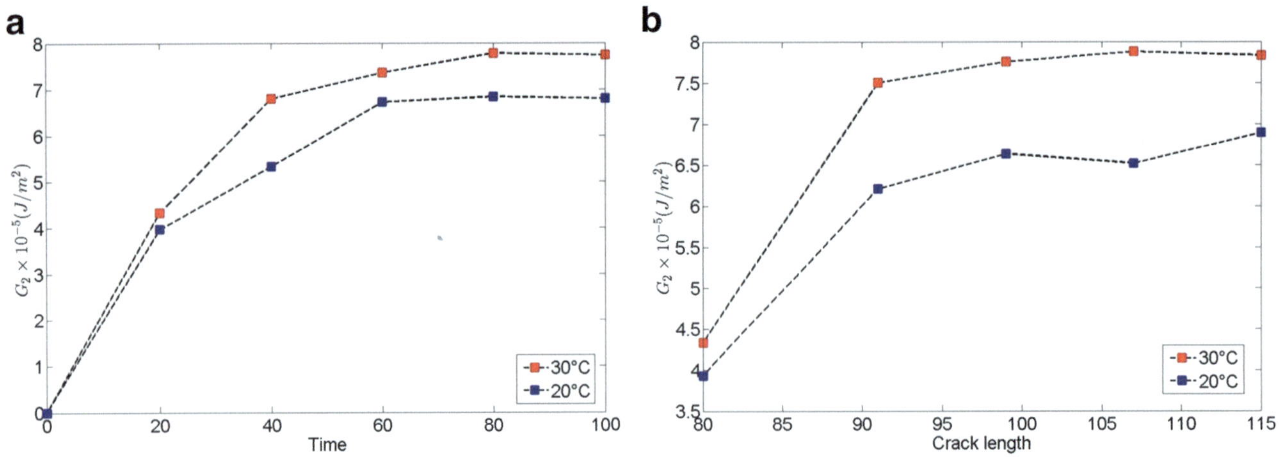

Fig. 30.8 (**a**) Viscoelastic energy release rate G_2 in mixed mode ($\beta = 45°$) versus time under temperature variation ($T_1 = 20$ °C and $T_2 = 30$ °C). (**b**) Viscoelastic energy release rate G_2 in mixed mode ($\beta = 45°$) during crack growth process under temperature variation ($T_1 = 20$ °C and $T_2 = 30$ °C)

Conclusion

The knowledge of crack driving forces such as energy release rate and stress intensity factors is very important in the assessment of the reliability of flawed structures. A new formulation of the *A*-integral is proposed, which allow to take into account the effect of thermal load induced by temperature variation, without taking into account the mechanical load applied on crack lips. The efficiency of the numerical implementation of crack propagation, in finite element software, is assessed by dealing with orthotropic materials. A good accuracy is observed on the estimates of the viscoelastic energy release rate in mixed mode for both time and crack length evolution. Moreover, the convergence is well achieved since the values of the viscoelastic energy release rate seem to be stabilized throughout the crack path.

In this paper, the analysis is achieved under the assumption of viscoelastic material under static crack propagation and temperature variation. In addition, the viscoelastic behavior of timber material was implemented in finite element software using spectral decomposition method of the creep tensor. However, automatic crack propagation needs to be achieved accordingly to a critic viscoelastic energy release rate values, which depends on the viscoelastic mechanical properties of timber material. This will be the subject of undergoing researches, in addition to the development of the *A*-integral in order to take into account the effect of moisture variation. Also, we hope to investigate the effect of uncertainties in the material and load parameters on the reliability of wood structure subjected to crack growth, by using stochastic computation methods.

Acknowledgment The authors wish to strongly acknowledge the National Agency of Research (ANR) for its financial support of this work through the project CLIMBOIS N° ANR-13-JS09-0003-01 labeled by ViaMeca.

References

1. Rice, J.R.: A path independent integral and the approximate analysis of strain concentrations by notches and cracks. J. Appl. Mech. **35**, 379–386 (1968)
2. Noether, E.: Invariant variations problem. Transp. Theory Stat. Phys. **1**, 183–207 (1918)
3. Chen, F.M.K., Shield, R.T.: Conservation laws in elasticity of the J-integral. J. Appl. Math. Phys. **28**, 1–22 (1977)
4. Moutou Pitti, R., Dubois, F., Petit, C.: Generalization of T and A integrals to time-dependent materials: analytical formulations. Int. J. Fract. **161**, 187–198 (2010)
5. Riahi, H., Pitti, R.M., Dubois, F., Fournely, E.: On numerical evaluation of mixed mode crack propagation coupling mechanical and thermal loads in wood material. In: Carroll, J., Daly, S. (eds.) Fracture, Fatigue, Failure, and Damage Evolution, vol. 5, pp. 21–26. Springer, Cham (2015). doi:10.1007/978-3-319-06977-7_3
6. Moutou, P.R., Dubois, F., Octavian, P.: A proposed mixed-mode fracture for wood under creep loadings. Int. J. Fract. **167**(2), 195–205 (2011). doi:10.1007/s10704-010-9544-z
7. Bui H.D., Proix, J.M.: Découplage des modes mixtes de rupture en thermoélasticité linéaire par des intégrales indépendantes du contour. In: Acte du Troisième Colloque Tendances Actuelles en Calcul de Structure, Bastia, pp. 631–643, 6–8 Nov 1985

8. Destuynder, P.H., Djaoua, M., Lescure, S.: Some remarks on elastic fracture mechanics. J de Mécanique Théorique Appliquée **2**, 113–135 (1983)
9. Moutou Pitti, R., Dubois, F., Petit, C., Sauvat, N.: Mixed mode fracture separation in viscoelastic orthotropic media: numerical and analytical approach by the Mθv-integral. Int. J. Fract. **145**, 181–193 (2007)
10. Riahi, H., Pitti, R.M., Dubois, F., Fournely, E., Chateauneuf, A.: Numerical fracture analysis coupling thermo-hygro mechanical and viscoelastic behaviour. In: 9th International Conference on the Mechanics of Time Dependent (MTDM), Montreal, 27–30 May 2014

Chapter 31
Use of a Multiplexed Photonic Doppler Velocimetry (MPDV) System to Study Plastic Deformation of Metallic Steel Plates in High Velocity Impact

Shawoon K. Roy, Michael Peña, Robert S. Hixson, Mohamed Trabia, Brendan O'Toole, Steven Becker, Edward Daykin, Richard Jennings, Melissa Matthes, and Michael Walling

Abstract High-velocity impact experiments with a gas gun pose unique challenges, in terms of experimental setup and computational simulation, since the projectile-target interaction creates extreme pressure and temperature within few micro seconds. The objective of this study is to accurately measure the plastic deformation of plates under projectile impact that does not lead to full penetration. In this work, free surface velocities at the back side of target plates are measured using the newly developed Multiplexed Photonic Doppler Velocimetry (MPDV) system, which is an interferometric fiber-optic technique which can measure velocity from the Doppler shift of the light reflected from the moving back surface of the target. The MPDV system allows measurements of velocity from different locations on the target plate using multiple optical fiber probes oriented in specific directions and patterns. Data are reduced using a Fast Fourier transformation (FFT) technique to obtain the free surface velocity profiles. These velocity profiles can present insights into the dynamic behavior of impacted materials under shock loading conditions.

Keywords MPDV • Multiplexing • High-velocity impact • Velocimetry • Data reduction

31.1 Introduction

Early experimental work in impact dynamics and shock physics included relatively simple diagnostic. Significant effort has taken place in the past few decades to develop new and accurate diagnostic systems to acquire data from this type of experiment. High-velocity impact experiments in modern days depend upon velocimetry data to represent the dynamic behavior of materials. A few of these diagnostic tools are already available to use in such dynamic impact experiments, for example, the velocity interferometer system for any reflector (VISAR) [1] and the photonic Doppler velocimeter (PDV) system [2]. PDV is an interferometric fiber optic technique to create Doppler induced beats where beat frequencies are related to surface displacement along a probe's beam axis. Ever since the design of modern PDV architecture, PDV systems have gained popularity for their relatively low system cost, ease of implementation, and robustness in both high-velocity and low-velocity dynamic impact experiments [3]. Many high-velocity impact studies have listed PDV as their primary diagnostic tool [4, 5]. Most of them focused on using single-probe PDV fiber to collect velocimetry data from a single point. Recent developments to PDV systems include frequency- and time-domain multiplexing with PDV (i.e. the MPDV system), which can record velocimetry data from several points onto a single digitizer channel. The concept of frequency upshift PDV or, heterodyne PDV (i.e. frequency-domain multiplexing) was first introduced by Daykin and Perez [6]. As described by Daykin [7], frequency-domain multiplexing used four channels into a single optical fiber, and then multiplexed this previous configuration twice in the time domain to get 8-velocity histories (i.e. velocimetry) in a single digitizer.

Multiplexed PDV (MPDV) is one of the most powerful tools invented in recent years and hence, researchers are currently focused on exploring the applications and abilities of such MPDV systems, including accuracy, and extraction analysis

S.K. Roy • M. Trabia (✉) • B. O'Toole
University of Nevada, Las Vegas, Las Vegas, NV, USA
e-mail: Mohamed.Trabia@unlv.edu

M. Peña • R.S. Hixson • S. Becker • E. Daykin • M. Walling
National Security Technologies, LLC, Las Vegas, USA

R. Jennings • M. Matthes
University of Nevada, Las Vegas, Las Vegas, USA

A.M. Beese et al. (eds.), *Fracture, Fatigue, Failure and Damage Evolution, Volume 8*, Conference Proceedings of the Society for Experimental Mechanics Series, DOI 10.1007/978-3-319-21611-9_31

of MPDV data [8–11]. The main objective of this article is to discuss the capability of MPDV systems in simple plate perforation experiment with high-velocity projectile impact i.e. a fundamental shock physics problem. The experiment was designed with different thickness steel plates to prevent complete penetration of the high-velocity projectile and to measure the resulting plastic deformation of the target plate. The use of an MPDV system in this kind of dynamic impact experiments provide a significant insight in the dynamic behavior of the target plate as a result of the penetration process.

31.2 Experimental Setup

31.2.1 Two-Stage Light Gas Gun

All high-velocity impact experiments were conducted with a two-stage light gas gun (Fig. 31.1a). A brief working principle of this gas gun is summarized here. The propellant gas in the powder breech drove a cylindrical piston into the pump tube. The moving piston then pressurized the light gas (hydrogen, helium, or nitrogen) in pump tube, which eventually burst a petal valve between the pump tube and the launch tube and accelerated the projectile down the launch tube and the drift tube. The projectile then impacted the target, which was bolted to a mounting frame inside the target chamber (Fig. 31.1b). A time interval system measured the projectile velocity by measuring the time between triggering two lasers at a specific location in the drift tube.

31.2.2 Materials

All gas gun experiments performed in this work used cylindrical Lexan projectiles with 5.58 mm diameter and 8.61 mm length. Three different types of steel plates were used as target materials: ASTM A36, HY100 and 304L. All target plates had a dimension of 152.4 × 152.4 × 12.7 mm. The thickness of target plates was chosen such that full penetration does not occur

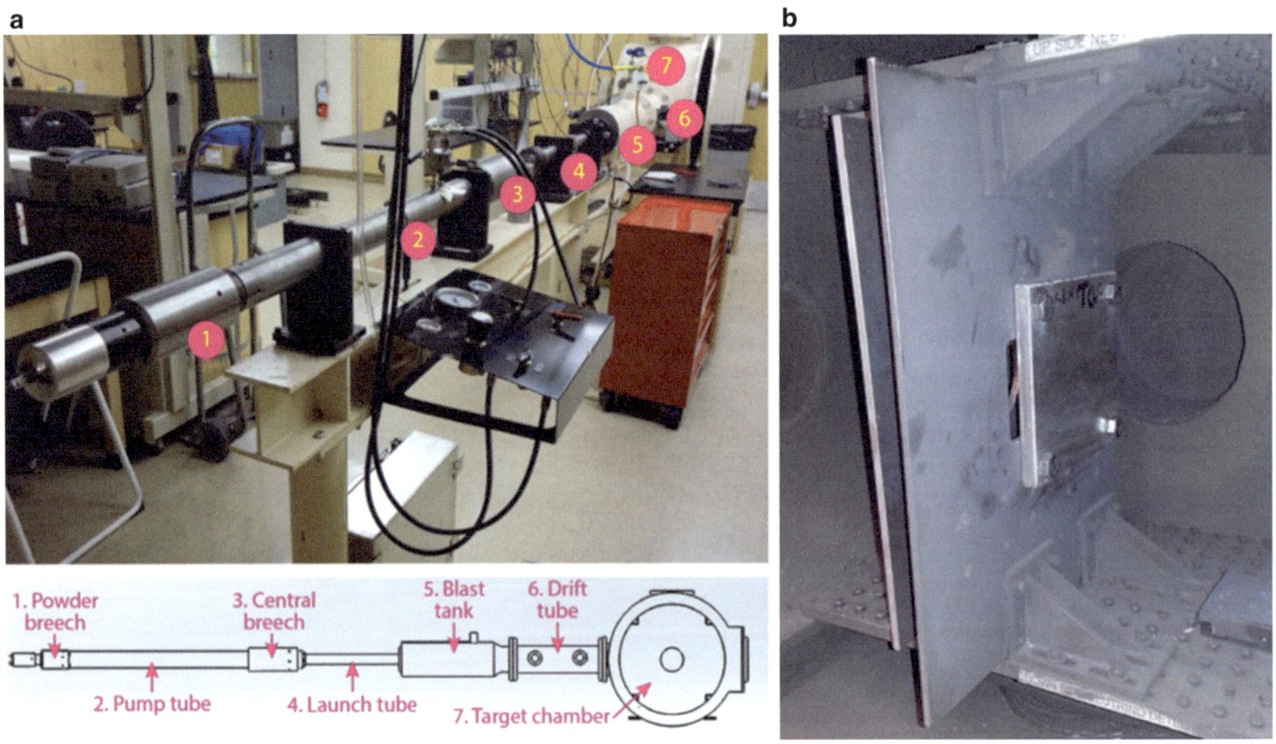

Fig. 31.1 (**a**) Schematic of two stage light gas gun, (**b**) bolted target plate

during the experiment but for which a significant bulge happens. This allowed the MPDV system to collect velocimetry data in a regime where significant plastic deformation was occurring. Depending upon the type of gas used and the fill pressure, the projectile impact velocity varied from about 4.5–6.8 km/s.

31.2.3 *Projectile Velocity Determination*

In all gas gun experiments, the projectile velocity was monitored by using a laser intervalometer system. The unit had two laser sources separated by a fixed distance. Each laser beam was passed through one port to a receiving station which had a narrow band pass filter to ensure that 32 photodiodes in the arrays were free from any external light. The flight of the projectile interrupted both laser beams within a time interval which was recorded by the digitizer.

31.2.4 *MPDV System*

As mentioned earlier, this work was focused on exploring the capability of MPDV system in dynamic impact experiments. The modern architecture of MPDV system was already explained E. Daykin [6] and hence, shown in Fig. 31.2 here. The architecture shown in Fig. 31.2 consists of eight lasers with two different wavelengths (defined by corresponding ITU bands). First, all four measurement lasers were multiplexed onto a single mode optical fiber, and then their signals were separated to their respective optical probes using a circulator and demultiplexer. Then measurement signals were multiplexed again (time-domain multiplexing) and sent to attenuators and combined with their respective local oscillators [3].

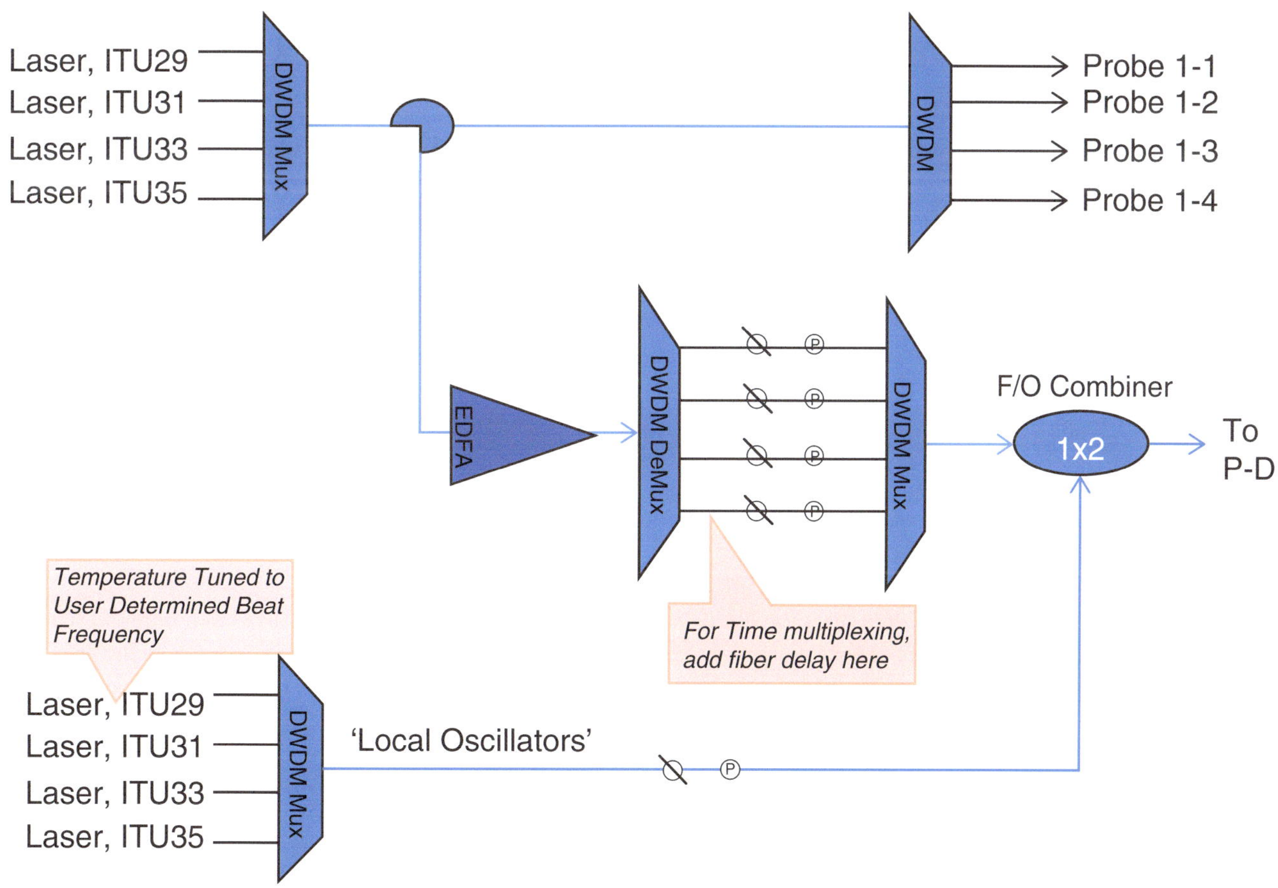

Fig. 31.2 The architecture of conceptual MPDV with 4× frequency multiplexing as described by E. Daykin [6]

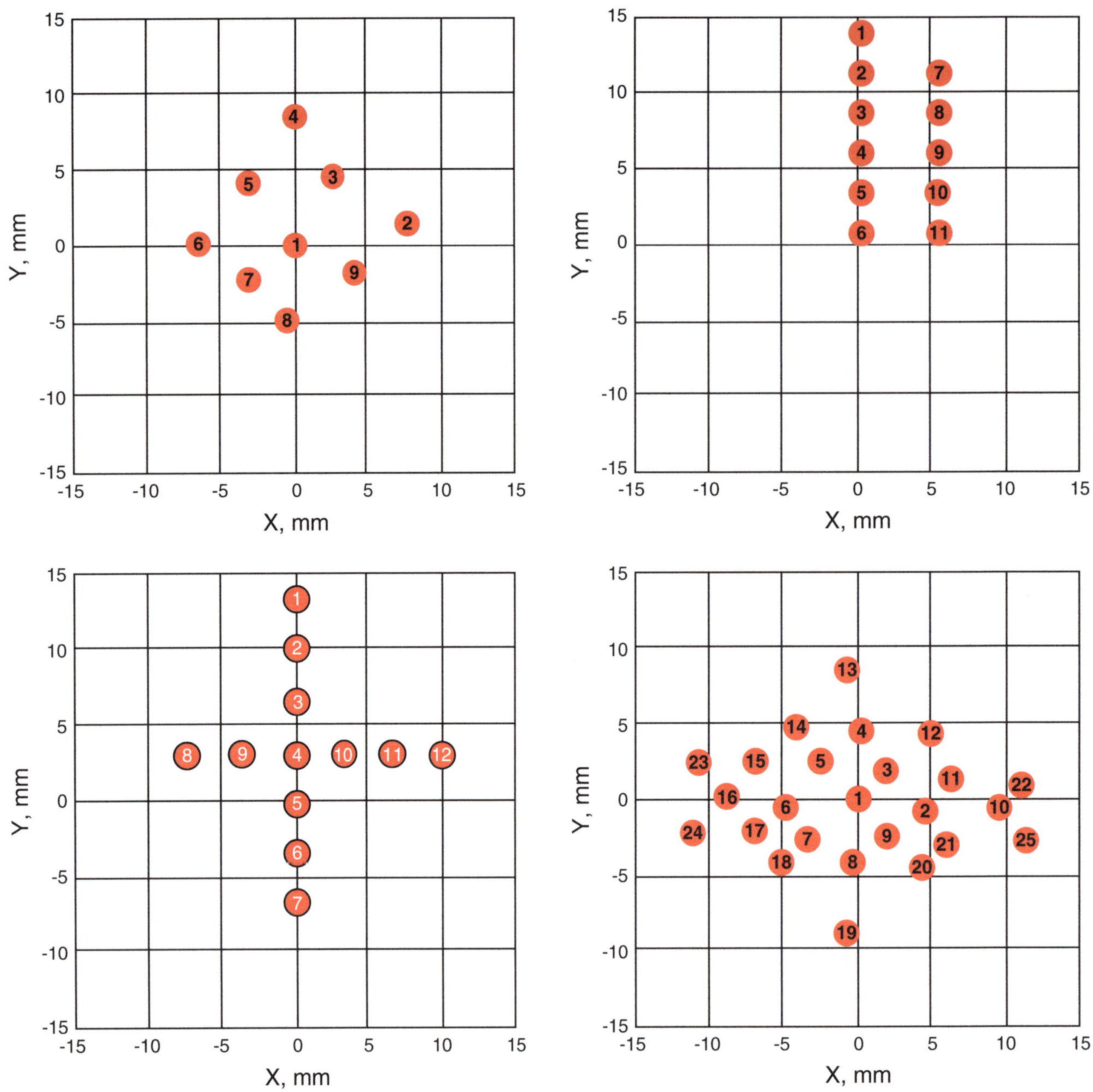

Fig. 31.3 Schematic of MPDV probe array arrangement: (**a**) 9-probe (**b**) 11-probe (**c**) 12-probe (**d**) 25-probe

All gas gun experiments performed to study the plastic deformation of steel plates in this work used four different types of MPDV probe array systems: 9-probe, 11-probe, 12-probe, 25-probe (Fig. 31.3). MPDV probes were focused on the back side of the target plate to capture velocimetry data at specific points around the expected center of impact. Probes were spread 10–15 mm approximately in both horizontal and vertical directions on the target plates. The MPDV system was triggered by the second laser in the intervalometer laser system assembly. A delay time was set in the MPDV system based on the projectile velocity to collect data from the back surface of the plate at the correct time. A schematic of the data acquisition by MPDV system is shown in Fig. 31.4.

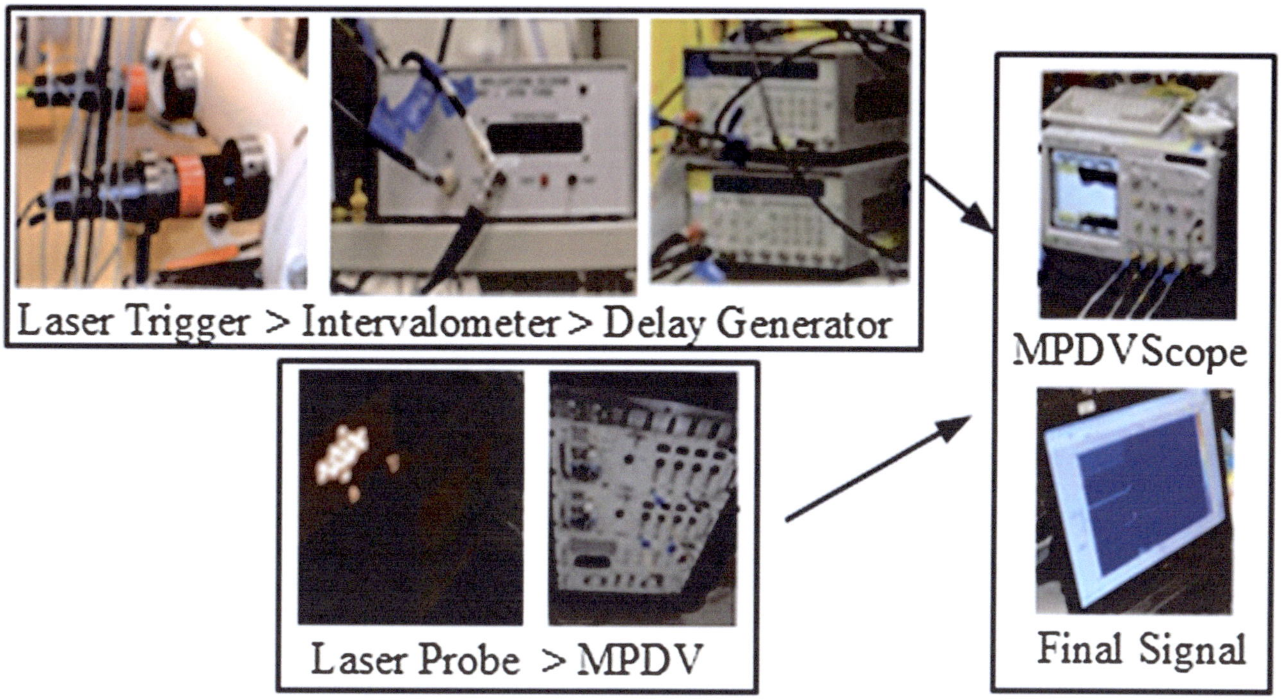

Fig. 31.4 Schematic of MPDV data acquisition system

31.2.5 Data Reduction Technique

As part of growing interest in MPDV systems, researchers have started to develop new tools to analyze the large amount of data collected from MPDV systems. A few of them have already been described [6, 12]. In the present work, the MPDV system collected data at a sampling rate of 10 GSa/s. All the raw data collected by the MPDV system were analyzed by software tools developed by National Security Technologies (NSTec), which uses a sliding Fast Fourier transformation (FFT) to get the free surface velocity profiles.

31.3 Results

31.3.1 Physical Observation of the Target Plates

In all gas gun experiments, impact of high-velocity Lexan projectiles created a crater on the front side and a bulge on the back side of target plates (Fig. 31.5). Lexan projectiles disintegrated due to the enormous pressure and heat generated upon impact. After each experiment, crater and bulge details for each target plate were measured. The distance between the flat rear surface of the plate and peak point of the bulge was taken to be the height of the bulge. An average value for multiple measurements of crater diameter, depth of penetration and bulge were taken as the final measurement. All the crater details and bulge dimensions are listed in Table 31.1. Sectioned plates showed spall due to release wave interactions. The results show that the size of spall cracks is proportional to impact velocity. Details of spall cracks from some of these experiments are also listed in Table 31.1.

31.3.2 Free Surface Velocity

Typically, all MPDV experiments captured free surface velocities from target plate for 30–40 μs whereas, the first 5 μs contains the most important features related to the dynamic properties of the materials used. Free surface velocity profiles

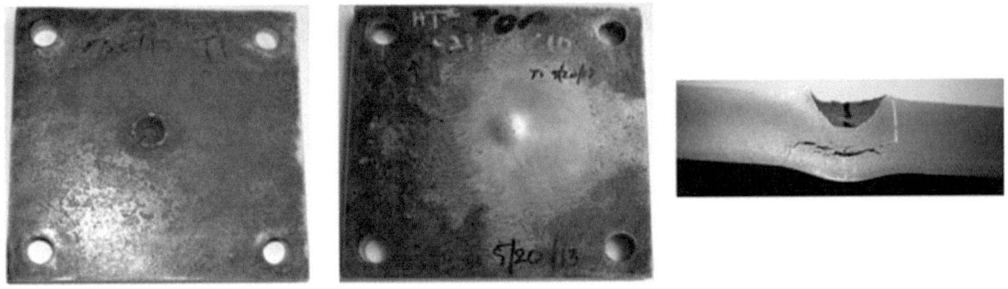

Fig. 31.5 Typical target plate after impact (Test ID 1000-024)

Table 31.1 Physical measurement of target plates after impact

Test ID	MPDV system details	Target Plates	Impact velocity, km/s	Crater diameter, mm	Penetration depth, mm	Bulge, mm	Spall crack details	
							Diameter, mm	Width, mm
1000-024	9 Probe	A36	5.708	17.2 ± 0.3	7.7 ± 0.3	3.1 ± 0.3	21.4 ± 0.2	1.9 ± 0.1
1000-025			4.763	15.4 ± 0.3	6.5 ± 0.3	1.4 ± 0.1	14.5 ± 0.2	0.2 ± 0.1
1000-088	11 Probe	304L	6.583	17.3 ± 0.2	6.4 ± 0.3	2.4 ± 0.1	n/a	n/a
1000-089		HY100	6.698	16.5 ± 0.3	6.4 ± 0.2	3.0 ± 0.3	n/a	n/a
1000-090	12 Probe	HY100	6.743	15.5 ± 0.1	4.5 ± 0.3	3.3 ± 0.1	n/a	n/a
1000-091		304L	6.758	15.5 ± 0.2	3.9 ± 0.1	3.2 ± 0.2	n/a	n/a
1000-026	25 Probe	A36	4.823	15.1 ± 0.2	6.5 ± 0.5	1.5 ± 0.1	n/a	n/a
1000-027			5.088	16.9 ± 0.8	7.0 ± 0.4	2.3 ± 0.2	n/a	n/a
1000-028			5.157	15.9 ± 0.4	6.5 ± 0.5	1.7 ± 0.2	18.5 ± 0.1	0.7 ± 0.1

at times up to 15–20 μs from typical 9-probe, 11-probe and 25-probe MPDV experiments are presented in Fig. 31.6. It should be noted that due to the similar type of velocity profiles for 304L and HY100 steels in 11-probe and 12-probe MPDV experiments, only 11-probe velocity profiles are presented in Fig. 31.6. In general, all A36 and 304L steel showed a two-wave structure velocity profile in compression: an elastic wave followed by a relatively sharp plastic wave. Because of known material properties differences between steel and stainless steel, subtle differences do exist in the measured wave profiles. The velocity peak for the plastic wave was proportional to the impact velocity. Velocity profiles also showed a second velocity peak and a spall signature in the form of a 'pullback velocity' signal within the first 5 μs. Supposedly, all low carbon alloy steels should show another wave rise after the plastic wave due to the $\alpha \leftrightarrow \varepsilon$ phase transition [13] similar to what was observed in shocked iron [14, 15]. However, in the case of HY100 steel, the signature of $\alpha \leftrightarrow \varepsilon$ phase transition is not prominent in our gas gun experiments at the rear surface. We suspect this could be due to the complex, asymmetric nature of the stress wave at that location in the target plate, and the relatively low stress amplitude of the wave at the rear surface of the target plate. Fundamental shock compression experiments that do observe the phase transition wave have all been done with a condition of uniaxial strain, which is not the case for the stress wave we observe. In all MPDV experiments, the impact center is within ±3.0 mm from the nearest PDV probe. Therefore, probes located closest to the impact center showed the earliest arrival of the free surface velocity signal and showed the highest peak velocities in general; but there were certain exceptions. While the exact reason of these anomalies is not yet understood completely, possible reasons may include the complex asymmetric nature of the stress wave.

31.4 Conclusion

Gas gun experiments were performed to measure the plastic deformation of steel plates during high-velocity impact and an MPDV system was used to measure free surface velocity during these experiments on the back of the target plate. Plastic deformation as represented by bulge of the target plates and crater damage were proportional to impact velocity. Velocimetry data captured by MPDV systems provide information on the comparative dynamic behavior of target plates made of different materials. Although, significant efforts are still needed to analyze these MPDV data accurately, implementation of MPDV system in high-velocity impact experiments certainly provides a new horizon for the researchers to explore. Work is ongoing to interpret the results of these experiments.

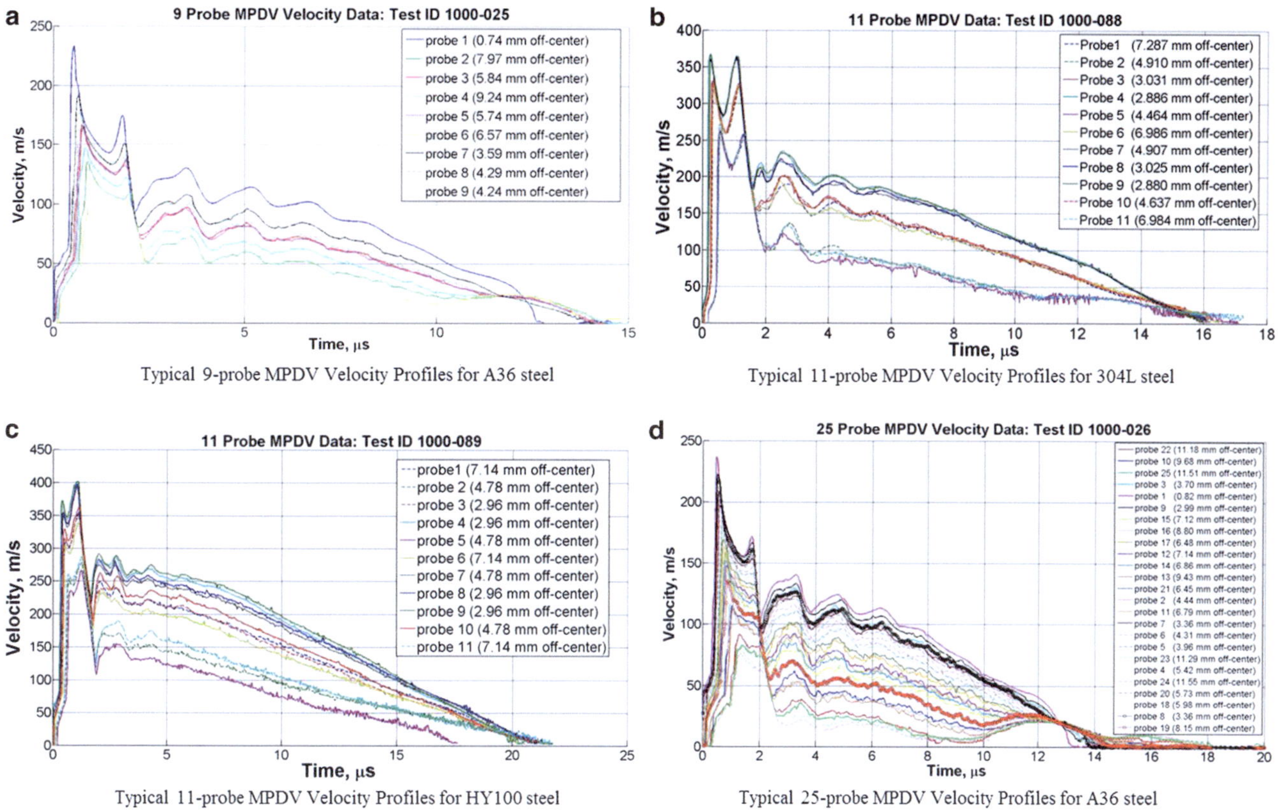

Fig. 31.6 Velocity profiles collected from gas gun experiments by MPDV system. (**a**) Typical 9-probe MPDV velocity profiles for A36 steel. (**b**) Typical 11-probe MPDV velocity profiles for 304L steel. (**c**) Typical 11-probe MPDV velocity profiles for A36 steel. (**d**) Typical 25-probe MPDV velocity profiles for A36 steel

Acknowledgements This manuscript has been authored by National Security Technologies, LLC, under Contract No. DE-AC52-06NA25946 with the U.S. Department of Energy and supported by the Site-Directed Research and Development Program. The United States Government retains and the publisher by accepting the article for publication acknowledges that the United States Government retains a non-exclusive, paid-up, irrevocable, worldwide license to publish or reproduce the published form of this manuscript, or allow others to do so, for United States Government purposes. The U.S. Department of Energy will provide public access to these results of federally sponsored research in accordance with the DOE Public Access Plan (http://energy.gov/downloads/doe-public-access-plan). DOE/NV/25946-2350.

References

1. Barker, L.M., Hollenbach, R.E.: Laser interferometer for measuring high velocities of any reflecting surface. J. Appl. Phys. **43**, 4669–4675 (1972)
2. Strand, O.T., Goosman, D.R., Martinez, C., Whitworth, T.L., Kuhlow, W.W.: Compact system for high-speed velocimetry using heterodyne techniques. Rev. Sci. Instrum. **77**(083108), 1–9 (2006)
3. Moro, E.A.: New developments in photon Doppler velocimetry. J. Phys. Conf. Ser. **500**, 142023 (2014)
4. Jilek, B.: History of velocimetry technology. In: 7th Annual PDV Workshop. Sandia National Laboratories, Albuquerque (2012)
5. Chau, R., Stolken, J., Kumar, M., Holmes, N.C.: Shock Hugoniot of single crystal copper. J. Appl. Phys. **107**, 023506 (2010)
6. Daykin, E., Perez, C.: Techniques and tools for PDV applications: A work in progress. Presented at the 3rd Annual PDV Workshop, Albuquerque (2008)
7. Daykin, E., Diaz, A., Gallegos, C., Perez, C., Rutkowski, A.: Work in progress: a multiplexed many-point PDV (MPDV) techniques and technologies. In: 5th Annual PDV Workshop, p. 15, Columbus (2010)
8. Daykin, E., Burk, M., Diaz, A., Gallegos, C., Hutchins, M., Pena, M., Perez, C., Rutkowski, A., Teel, M., Theuer, K.: MPDV development, techniques and technologies: a pot load of data. In: 6th Annual PDV Workshop, p. 15, Livermore (2011)
9. Daykin, E.: The limits of multiplexing. In: 7th Annual Photonic Doppler Velocimetry (PDV) Workshop, Albuquerque (2012)
10. Danielson, J.R., Daykin, E.P., Diaz, A.B., Doty, D.L., Frogget, B.C., Furlanetto, M.R., Gallegos, C.H., Gibo, M., Garza, A., Holtkamp, D.B., Hutchins, M.S., Perez, C., Peña, M., Romero, V.T., Shinas, M.A., Teel, M.G., Tabaka, L.J.: Measurement of an explosively driven hemispherical shell using 96 points of optical velocimetry. J. Phys. Conf. Ser. **500**, 142008 (2014)
11. Turley, W.D., Daykin, E., Hixson, R., LaLone, B.M., Perez, C., Stevens, G.D., Veeser, L., Ceretta, E., Gray, G.T., Rigg, P.A., Koller, D.: Copper spall experiments with recompression using HE (copper spall soft recovery experiments). In: 8th Annual PDV Workshop, Las Vegas (2014)

12. Ao, T., Dolan, D.H.: Effect of window reflections on photonic Doppler velocimetry measurements. Rev. Sci. Instrum. **82**, 023907 (2011)
13. Hammond, R.I., Proud, W.G.: Does the pressure-induced α–ε phase transition occur for all low-alloy steels? Proc. R. Soc. London. Ser. A: Math. Phys. Eng. Sci. **460**, 2959–2974 (2004)
14. Jensen, B.J., Rigg, P.A., Knudson, M.D., Hixson, R.S., Gray III, G.T., Sencer, B.H., Cherne, F.J.: Dynamic compression of iron single crystals. AIP Conf. Proc. **845**, 232–235 (2006)
15. Jensen, B.J., Gray, G.T., Hixson, R.S.: Direct measurements of the α-ε transition stress and kinetics for shocked iron. J. Appl. Phys. **105**, 103502 (2009)

Chapter 32
In-service Preload Monitoring of Bolted Joints Subjected to Fatigue Loading Using a Novel 'MoniTorque' Bolt

Anton Khomenko, Ermias G. Koricho, Mahmoodul Haq, and Gary L. Cloud

Abstract Bolted fastening is one of the oldest, most important, and most neglected aspects in engineering design of machines and structures. One of the biggest concerns is the bolt preload control and monitoring during and after the joint assembly. A survey of automobile service managers in the United States has shown that one-quarter of all service problems were traced to loose fasteners. Therefore, fatigue behavior of in-service bolted joints is of special interest for predicting structural integrity and residual lifetime of the structure. In this work, a novel technique 'MoniTorque' (Monitoring Torque) that embeds fiber Bragg-grating (FBG) sensors within the bolt shaft using temporary adhesives is used to monitor pre-load changes during fatigue tests of bolted joints. The inherent small size of the FBG allows precise monitoring without affecting the intrinsic properties of the bolt. Furthermore, temporary adhesives allow easy installation, removal and reuse of FBG sensor, thus converting a simple bolt into a reliable sensing element. Results show good correlation between the number of loading cycles and preload reduction. Overall, the 'MoniTorque' technique demonstrated in this work shows promise in cost-effective and reliable health monitoring of in-service bolted joints.

Keywords 'MoniTorque' bolt • FBG sensor • Preload • Bolted joint • Fatigue

32.1 Introduction

Bolted joints are one of the most important and widely used fastening techniques in manufacturing industries, construction, machine design and many other applications. However, bolted joint assembly and in-service behaviors are a combination of many complex phenomena that are controlled by many different parameters. Fatigue and rotational self-loosening are two of the most widespread reasons for failure of dynamically loaded bolted joints, especially in transverse loading conditions [1]. In service self-loosening of components in lightweight design have been commonly reported [2]. A survey of automobile service managers in the United States reported that one-quarter of all service problems were traced back to loose fasteners [2]. Hence, bolt loosening has become an important research area in mechanical engineering to prevent failures in a variety of applications [3]. Another concern in bolted joints with composite adherends is the effect of stress relaxation due to creep in the through-the-thickness direction of the composite material [4]. Accordingly, a composite panel bolted in the thickness direction is susceptible to loss of preload owing to the viscoelastic nature of the resin, which dominates the response in the through-the-thickness direction. Loss of preload may have a detrimental effect on the strength and the fatigue life of the connection. Thus, it is essential to monitor the bolt preload in service conditions to prevent any catastrophic failure of the bolted joint.

The techniques for bolt tightening control and evaluation can be classified into two main categories: non-contact methods and embedded sensors methods. The first type includes mainly optical inspection [2, 3, 5, 6], whereas the second type consists of acoustic inspection [7–9], smart washers and miniature load cells [4, 10–12], internally embedded strain gages and FBG sensors [13–15], and techniques based on other physical principles. Each of the techniques has its own merits and limitations. Optical inspection is complex and is not suitable for continuous monitoring under field conditions [2, 3]. Acoustic inspection is sensitive to external vibrations [7]. Smart washers have to be incorporated as the part of the structure, and load cells are

A. Khomenko (✉) • E.G. Koricho • G.L. Cloud
Composite Vehicle Research Center, Michigan State University, 2727 Alliance Drive, Lansing, MI 48910, USA
e-mail: khomenka@msu.edu

M. Haq
Composite Vehicle Research Center, Michigan State University, 2727 Alliance Drive, Lansing, MI 48910, USA

Department of Civil and Environmental Engineering, Michigan State University, Engineering Building, East Lansing, MI 48824, USA

A.M. Beese et al. (eds.), *Fracture, Fatigue, Failure and Damage Evolution, Volume 8*, Conference Proceedings of the Society for Experimental Mechanics Series, DOI 10.1007/978-3-319-21611-9_32

quite expensive [11, 12]. Installation of electrical resistance strain gages (RSGs) in the bolt shaft is cumbersome, they are temperature sensitive, and the drilled channel has to be several mm in diameter, which is detrimental to the integrity of the bolt. Moreover, as with A-bolt™ technology [15], RSGs will not work with some types of bolts. FBG sensors are very small in diameter (up to 100 μm) and easy to install in any type of the bolt, but permanent embedding is not practical because the cost of the sensor is much higher than that of traditional RSG sensors. Previously, we developed a novel technique to monitor bolt clamping force (MoniTorque bolt) utilizing removable and reusable embedded FBG sensors, and we validated the technique for static preload measurements in multi-material bolted joint configurations [16]. The MoniTorque technique uses a temporary adhesive for embedding the FBG sensor into the bolt shaft, thereby converting the bolt into a sensor and a reliable tool for evaluating the behavior of the bolted joint. The FBG sensor can be embedded at any given time for any type of bolt, removed and reused later in the same or another bolt. The MoniTorque bolt is resistant to most external environmental effects, except for changes in gross temperature of the bolt, provides real-time precise clamping force measurements and does not require complicated manufacturing or setup preparation.

In this work, a MoniTorque bolt utilizing thermoplastic adhesive as an embedding medium is used for preload monitoring in glass fiber reinforced plastic (GFRP) composite bolted joints subjected to fatigue loading. Initially, the MoniTorque bolt was calibrated using an external load cell, and the calibration was verified prior to every new test. GFRP bolted joints with varying clamping loads were assembled and subjected to static tensile shear testing to obtain the baseline strengths and to establish optimal parameters for fatigue testing. Finally, intermittent fatigue tests were carried out and preload reduction during cyclic loading was continuously monitored. Results show good correlation between the number of loading cycles and preload reduction. Overall, the 'MoniTorque' technique demonstrated in this work shows promise for cost-effective and reliable health monitoring of in-service bolted joints.

32.2 Theory of FBG Sensor Operation

An FBG is a sandwich-like distributed reflector containing a periodic oscillation of refractive index that is embedded into the optical fiber. Such a structure acts as an optical filter that transmits the entire spectrum of the light source and reflects back the resonant Bragg wavelength. The Bragg wavelength (λ_B) is given by the following equation (32.1):

$$\lambda_B = 2n_{eff}\Lambda, \tag{32.1}$$

where n_{eff} is the effective refractive index of the fiber core and Λ is the period of Bragg-grating. The spectrum bandwidth of the back-reflected radiation, which dictates the sensor resolution, also depends on the effective refractive index of the fiber core and the Bragg-grating period. As one can see, any perturbation of the Bragg-grating period results in a shift of Bragg wavelength. The strain response arises from both the physical elongation of the sensor and the change in fiber effective refractive index due to stress-birefringence effects, whereas a thermal response arises from the inherent thermal expansion of the fiber material and the temperature dependence of the refractive index.

The wavelength-encoded nature of the FBG output provides a built-in self-referencing capability for the sensor. Since the wavelength is the fundamental parameter, the output does not depend directly on the total light levels, losses in the connecting fibers and couplers, or source power. Moreover, the fiber and sensor have relatively small dimensions, and only small stresses are required to produce strains over a very broad range, therefore, embedding FBG sensors into all but the most delicate host specimens does not affect their intrinsic properties to any appreciable degree. These advantages of FBG sensors along with their immunity to electromagnetic interference as well as their light weight and high sensitivity make them very appealing for many areas of non-destructive evaluation applications, such as strain, temperature, and vibration measurements.

32.3 Specimen Preparation

32.3.1 MoniTorque Bolt Fabrication

A conventional steel bolt was selected for MoniTorque bolt manufacturing. The bolt was a grade five medium-strength steel hex head (head width: 19.05 mm or 3/4″; head height 7.94 mm or 5/16″) bolt with a right-hand thread having a length of 38.1 mm (1.5″), threaded with a 12.7 mm-13 (0.5″–13) thread. A 1.8 mm diameter hole was drilled along the axis of the bolt through one-half of its length (23 mm), and an FBG sensor was embedded inside the bolt using 3 M™ Scotch-Weld™ Hot Melt Adhesive 3779 PG Amber (thermoplastic). A schematic diagram of the MoniTorque bolt is presented in Fig. 32.1.

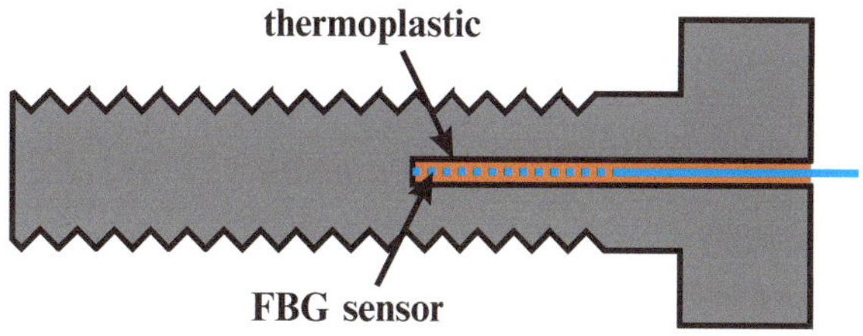

Fig. 32.1 Schematic diagram (not to scale) of the MoniTorque bolt with thermoplastic embedding medium

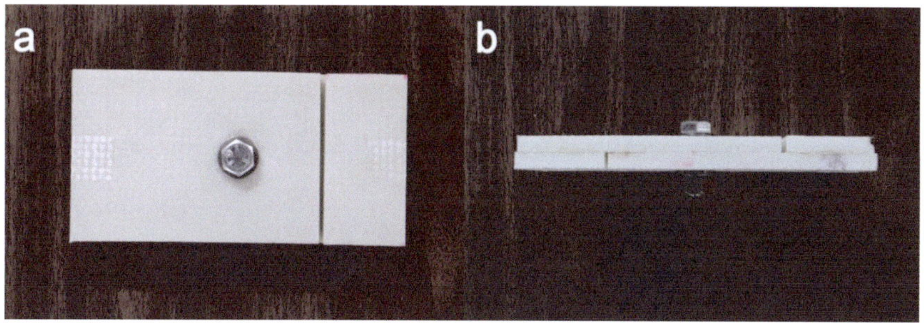

Fig. 32.2 GFRP bolted joint assembly: (**a**) *top view,* (**b**) *side view*

32.3.2 Bolted Joint Assembly

GFRP composite adherends for the bolted joints were manufactured using a liquid molding process. The reinforcement used for the adherend was Owens Corning ShieldStrand® S, S2-glass plain weave fabric with areal weight of 818 g/m^2. Sixteen layers of fabric were used to manufacture the adherends in this study. The distribution medium was Airtech Advanced Materials Group Resinflow 60 LDPE/HDPE blend fabric. The resin used was a two part toughened epoxy, namely SC-15, obtained from Applied Poleramic. A steel mold having dimension of 609.6 × 914.4 mm with point injection and point venting was used to fabricate 508.0 × 609.6 mm plates. After the materials were placed, the mold was sealed using a vacuum bag and sealant tape. The mold was then infused under vacuum with a pressure of 1 atm. The resin-infused panel was cured in a convection oven at 60 °C for 2 h and post cured at 94 °C for 4 h.

The GFRP adherend used in the joint assembly had dimensions of 152 × 101 × 9 mm. An additional tab of dimensions 101 × 50 × 9 mm was bonded to the gripping segments to eliminate any eccentricities during structural testing. A 15.7 mm diameter hole was drilled in the adherends 50.5 mm away from three edges, and a steel sleeve washer was inserted. The sleeve washer had the following dimensions: flange diameter 25.4 mm, sleeve ID of 13.46 mm, sleeve OD of 15.88 mm, flange thickness of 3.18 mm, overall height of 4.75 mm. The resulting assembly is shown in Fig. 32.2. Finally, the MoniTorque bolt was tightened to one of three different torque levels of 20 N-m, 35 N-m and 50 N-m using a digital torque wrench. Three specimens of each assembly were prepared and tested in static and fatigue.

32.4 Experimental Results and Discussion

32.4.1 MoniTorque Bolt Calibration

The MoniTorque bolt was calibrated using a thru-hole load cell model THC-1K-T manufactured by Transducer Techniques® with a maximum load capacity of 4448 N (1000 lb). During the manual tightening of the bolt, the spectral responses of the FBG sensor embedded into the bolt were synchronized with the load cell data and acquired using a Micron Optics sm125 700 Optical Sensing Interrogator. The MoniTorque calibration curve, or typical dependence of the strains measured by FBG on the applied preload, measured by load cell, is demonstrated in Fig. 32.3. The relationship between applied load and strains was found to be linear and the calibration coefficient was found to be 24.1 ± 0.2 nε/N.

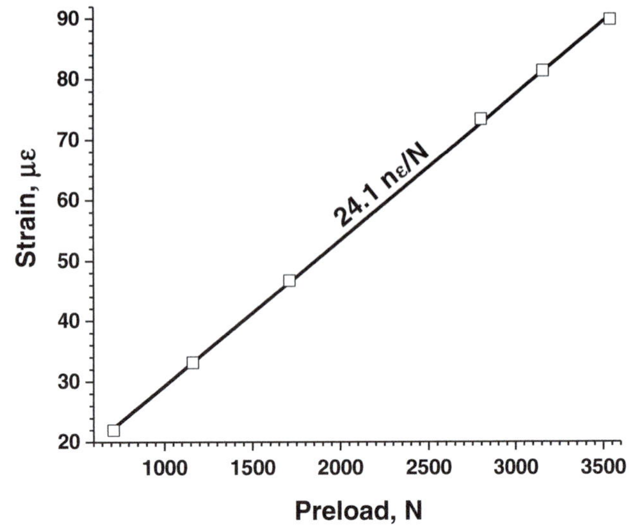

Fig. 32.3 Calibration curve—comparison of strains measured by FBG sensor and bolt preload measured by load cell

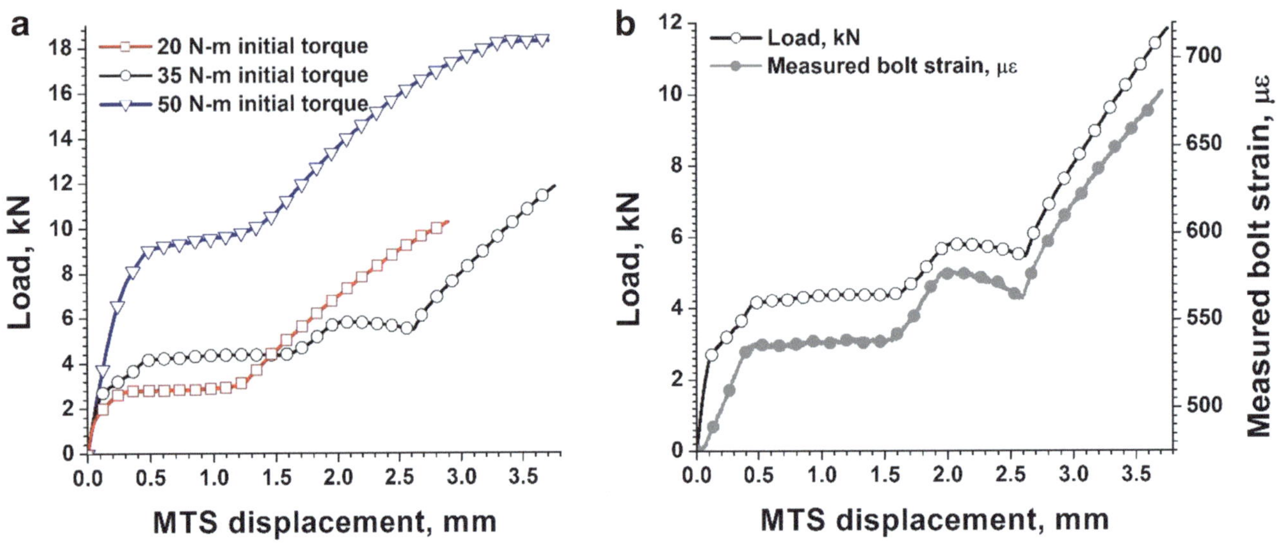

Fig. 32.4 (**a**) Load vs. MTS displacement dependences for GFRP bolted joints with 20 N-m, 35 N-m and 50 N-m initial torques, (**b**) load vs. MTS displacement curve and measured bolt strains vs. MTS displacement for 35 N-m bolted joint assembly

32.4.2 Static Shear Tests

The main purpose of static tensile-shear testing was to obtain ultimate failure load of GFRP bolted joint and identify the slip loads for each tested assembly. The evaluation of the GFRP composite bolted joints in tensile-shear configuration was performed using an MTS 810 servo-hydraulic load frame. All experiments were conducted in displacement control at a rate of 1 mm/min. The displacement and applied load from the MTS were recorded. Three specimens were tested per each bolted joint assembly (i.e. for 20 N-m, 35 N-m and 50 N-m). The initial preload was calculated through use of the MoniTorque calibration result to be 6.8 kN, 11.5 kN and 16 kN for the bolted joints with 20 N-m, 35 N-m and 50 N-m initial torques, respectively. It was found that the ultimate failure load for the bolted joint with 35 N-m initial torque was ~51.5 kN. The load vs. MTS displacement responses prior to GFRP delamination for bolted joints with initial preload/torque levels of 20 N-m, 35 N-m and 50 N-m are illustrated in Fig. 32.4a. Figure 32.4b illustrates the comparison of load vs. MTS displacement curve and the measured strains from the FBG in the MoniTorque bolt vs. MTS displacement for the 35 N-m preloaded joint. Figure 32.4b the measured strains are an indication of the stresses in the bolts along the shaft length due to the applied transverse shear loading. As a result, it follows the same trend as that of the applied load.

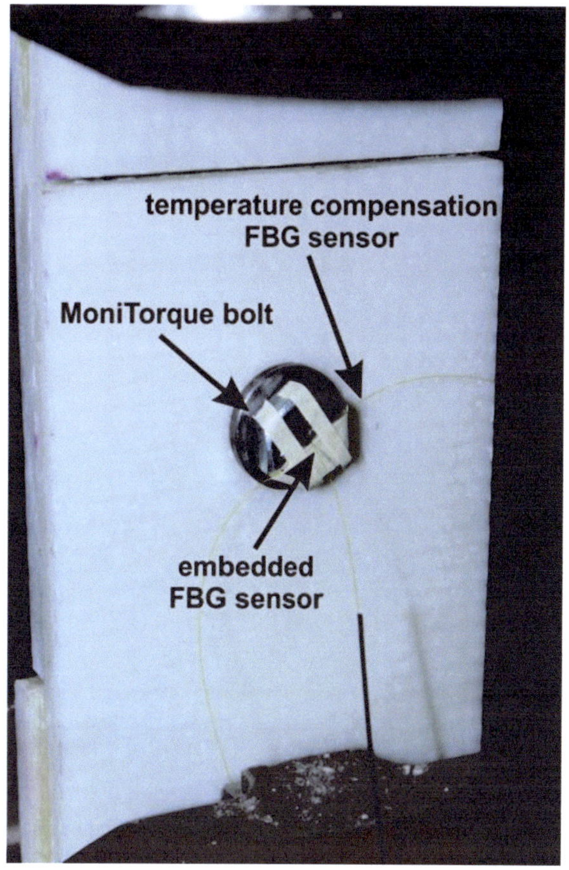

Fig. 32.5 Experimental setup showing the MoniTorque bolt, embedded FBG sensor and temperature compensation FBG sensor

It can be observed that all the bolted joints tested exhibit considerable slip (approximately 1.2 mm) when the applied shear forces exceed the friction forces from the initial clamping preload. The slip loads were found to be approximately 2.8 kN, 4 kN and 9 kN loads for preload levels of 20 N-m, 35 N-m and 50 N-m, respectively. As expected the load capacity that induces slip increases with increasing preload level.

32.4.3 Monitoring of Preload Levels During Fatigue Loading

The experimental evaluation of GFRP composite bolted joints under fatigue loading conditions was performed using an MTS 810 servo-hydraulic load frame. Intermittent fatigue experiments were conducted in load control, using a sinusoidal waveform with a frequency of 1 Hz, and a load ratio $R = 0.1$. The displacement and applied load from the MTS were recorded continuously at a sample rate of 100/s, while the preload was recorded while the testing was paused at 60, 120, 180, 300, 600, 1000, 1500, 2000, and 3000 cycles. In order to evaluate and compensate the temperature dependence, a secondary FBG sensor was used to measure the temperature during the fatigue testing (see Fig. 32.5). It was found that the temperature changes in the bolt were negligible during 1 Hz cyclic loading, thus the temperature effect on the sensor was nil.

Normalized preload reduction as measured during fatigue for the GFRP composite bolted joints with varying initial torques is illustrated in Fig. 32.6.

Notice that, during the first 100 cycles, all bolted joints exhibited rapid reduction in preload level followed by gradual decrease of clamping force. It was found that regardless of initial tightening, the initial preload drop was ~14 %, which is caused by the slippage. Although for 50 N-m initial torque bolted joint the slippage for quasi-static testing was higher than the loading during fatigue, friction degradation between the materials could cause sufficient preload reduction during first 60 cycles. Moreover, it can be noticed that the 20 N-m torque bolted joint showed relatively lower preload drop compared with other bolted joint configurations. This provides an interesting observation that independent of the initial preload level, the drop in normalized preload level is very similar (within statistical variations). This is a valuable information for design engineers as it can considerably reduce testing time for a wide range of torque levels. It should be noted that this observation

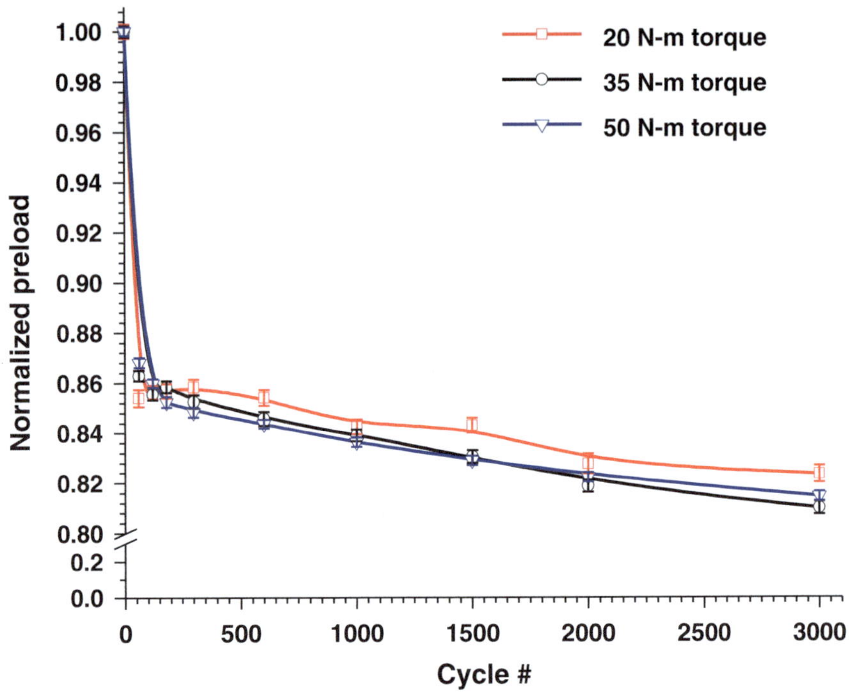

Fig. 32.6 Preload reduction during fatigue loading for GFRP composite bolted joints

applies only to the materials used in this work. Overall, the reduction in preload might be caused by complex interactions of bolted joint components during cycling loading in the case of a loosened bolt. As the preload is getting lower the stress relaxation caused by viscoelastic creep of the composite material becomes smaller, which in turn consequently minimizes the preload drop.

32.5 Conclusions

In this work, a MoniTorque bolt that embeds a fiber Bragg-grating sensor in the bolt shaft and utilizes a thermoplastic embedding medium for bonding the sensor was used for preload monitoring in GFRP composite bolted joints subjected to fatigue testing. The thermoplastic adhesive allows removing and reusing the sensor for multiple joints. Three configurations of GFRP bolted joints with 20 N-m, 35 N-m and 50 N-m initial torques were assembled and tested, and the same MoniTorque bolt was used for preload monitoring, through proper calibration prior to testing of each case. The initial preload was 6.8 kN, 11.5 kN and 16 kN for bolted joints with 20 N-m, 35 N-m and 50 N-m initial torques, respectively. Intermittent fatigue tests were carried out and preload reduction during cyclic loading was continuously monitored using MoniTorque bolt. All GFRP bolted joints exhibited about 14 % initial preload drop followed by monotonic decrease of the clamping force to about 18 % of the initial applied preload. The first preload drop was caused by the slippage between adherends when the applied sliding forces overcome the clamping forces. This was observed in all the joints in the study. After the initial preload drop due to slippage, the rate of reduction in preload was relatively lower for all joints in this study. One possible reason could be due to the viscoelastic creep of the composite material during cycling loading. As the preload gets lower the stress relaxation becomes smaller, which consequently minimizes the preload drop. Overall, the technique shows great potential for applications where precise and inexpensive preload control and monitoring is required.

References

1. Dinger, G., Friedrich, C.: Avoiding self-loosening failure of bolted joints with numerical assessment of local contact state. Eng. Fail. Anal. **18**(8), 2188–2200 (2011)
2. Li, C., Xu, C.-N., Adachi, Y., Ueno, N.: Real-time detection of axial force for reliable tightening control. Proc. SPIE **7522**, 75223G (2010)
3. Huda, F., Kajiwara, I., Hosoya, N., Kawamura, S.: Bolt loosening analysis and diagnosis by non-contact laser excitation vibration tests. Mech. Syst. Signal Process. **40**(2), 589–604 (2013)

4. Caccese, V., Berube, K.A., Fernandez, M., Melo, J.D., Kabche, J.P.: Influence of stress relaxation on clamp-up force in hybrid composite-to-metal bolted joints. Compos. Struct. **89**, 285–293 (2009)
5. Huang, Y.H., Liu, L., Yeung, T.W., Hung, Y.Y.: Real-time monitoring of clamping force of a bolted joint by use of automatic digital image correlation. Opt. Laser Technol. **41**(4), 408–414 (2009)
6. Sims, D.I., Siadat, M.-R., Nassar, S.: Toward real-time clamping load measurement using DSPI. Proc. SPIE **8494**, 84940I (2012)
7. Coelho, C.K., Das, S., Chattopadhyay, A., Papandreou-Suppappola, A., Peralta, P.: Detection of fatigue cracks and torque loss in bolted joints. Proc. SPIE **6532**, 653204 (2007)
8. Doyle, D., Zagrai, A., Arritt, B., Cakan, H.: Damage detection in bolted space structures. J. Intell. Mater. Syst. Struct. **21**(3), 251–264 (2010)
9. Amerini, F., Meo, M.: Structural health monitoring of bolted joints using linear and nonlinear acoustic/ultrasound methods. Struct. Health Monit. **10**(6), 659–672 (2011)
10. Industrial and research news: miniature load cells. Strain. **23**, 196 (1987)
11. Jiang, Y., Zhang, M., Park, T.-W., Lee, C.-H.: An experimental study of self-loosening of bolted joints. J. Mech. Des. **126**(5), 925–931 (2004)
12. Peairs, D.M., Park, G., Inman, D.J.: Practical issues of activating self-repairing bolted joints. Smart Mater. Struct. **13**(6), 1414–1423 (2004)
13. New products: force-sensing bolts. Strain. **20**, 35 (1984)
14. Hay, A.D.: Bolt, stud or fastener having an embedded fiber optic Bragg Grating sensor for sensing tensioning strain. US Patent 5,945,665, 31 Aug 1999
15. Guru, S.M., Fernando, S., Halgamuge, S., Chan, K.: Intelligent fastening with A-BOLT™ technology and sensor networks. Assem. Autom. **24**(4), 386–393 (2004)
16. Khomenko, A., Koricho, E.G., Haq, M.: Clamping load monitoring in bolted joints using innovative 'MoniTorque' bolts. In: Proceedings of Composite and Advanced Materials Expo (CAMX 2014) Orlando, 13–16 Oct 2014

Chapter 33
Fatigue Behavior of Novel Hybrid Fastening System with Adhesive Inserts

Ermias G. Koricho, Anton Khomenko, Mahmoodul Haq, and Gary L. Cloud

Abstract A novel joining technique that incorporates the advantages of both bonded (lightweight) and bolted (easy disassembly) techniques was invented (Provisional Patent 61/658,163) by Dr. Gary Cloud at Michigan State University. The most basic configuration of this invention consists of a bolt that has a channel machined through the bolt-shaft that allows injection of an insert compound that fills the necessary clearance between the bolt and the work-pieces and acts a structural component. More sophisticated versions of the concept incorporate additional sleeves or inserts.

In this paper the fatigue behaviour of composite hybrid bolted joints was studied. Composite plates were bolted with grade-five bolts and preloaded to a torque of 35 N-m. Two types of bolted joint configurations were evaluated. In the first case, a conventional bolted joint was studied. In the second case, pristine SC-15 epoxy resin was used as the structural insert in the hybrid fastening system. The joints were subjected to different fatigue loadings with the maximum loading level up to 80 % of the joint ultimate failure load and stress ratio of $R = 0.1$. Results reveal considerable improvement of fatigue life of the novel fastening system compared with a conventional bolted joint.

Keywords Bolted joint • Hybrid joint • Bolt • Composite • Fatigue • Damage

33.1 Introduction

With increasingly stringent national and international fleet fuel economy regulations, the use of composite materials for primary vehicle and aircraft structures has been expanding considerably for the last decades. Several automotive and aircraft manufactures have developed composite-intensive components not only to reduce the weight of the structure but also to improve the performance of final products. For example BMW has introduced a composite-body for the BMW i3, using carbon fiber rather than traditional steel, in order to eliminate enough weight to compensate for the weight of the battery packs that power the electric motor while still maintaining safety standards [1]. Further, Airbus and Boeing have built the latest generation of composite-intensive commercial aircraft, the A350XWB and Boeing 787, respectively [2].

Even though considerable progress has been made on adhesive and other joining techniques, the joining of composite structures still depends on mechanically fastened bolted joints that allow the disassembly of parts for repair/replacement. However, bolted joints have their own disadvantages, the main one being that drilling and machining bolt holes causes stress concentrations and delaminations that reduce the resistance of the joint construction to applied loads, and thus its efficiency.

Loading conditions, such as type of loading and load configuration, are among the most important factors in designing a joint in a structure. Particularly, the failure behaviour of a bolted joint in a composite structure under fatigue is complex due to the fluctuations of the stress transfer mechanisms through the joint that detrimentally affect the thermo-mechanical behaviour of the composite materials around the hole. Generally, the fatigue damage around bolt holes consists of three types, i.e. hole-wear, damage in the contact surface of the composite, and growth of the delaminations that are induced due to drilling of holes [3]. The material degradation and the hole-wear are caused by the erosion of material around the bolt holeas

E.G. Koricho (✉) • A. Khomenko • G.L. Cloud
Composite Vehicle Research Center, Michigan State University, 2727 Alliance Drive, Lansing, MI 48910, USA
e-mail: ermias.koricho@gmail.com

M. Haq
Composite Vehicle Research Center, Michigan State University, 2727 Alliance Drive, Lansing, MI 48910, USA

Department of Civil and Environmental Engineering, Michigan State University, 2727 Alliance Drive, Lansing, MI 48910, USA

A.M. Beese et al. (eds.), *Fracture, Fatigue, Failure and Damage Evolution, Volume 8*, Conference Proceedings of the Society for Experimental Mechanics Series, DOI 10.1007/978-3-319-21611-9_33

a result of the friction forces. Damage at the contact surfaces is induced by bolt bending and tilting under loading, which result in hole-elongation during fatigue loading. The growth of delamination around bolt holes has been found to decrease the fatigue life of bolted joints [3].

Saunders et al. [3] studied the development of fatigue damage around fastener holes in thick graphite/epoxy composite laminates. Results showed that the dominating failure mode was bolt failure. The static and fatigue behaviour of joints with respect to bolt clamping force and environmental effects was studied in [4]. Herrington and Sabbaghian [5] studied the effects of applied stress level, orientation of the outer layer reinforcing filaments, and the bolt torque level on the fatigue life of a graphite–epoxy composite. They found that the effect of outer ply angle was insignificant and that the fatigue life increased as the clamping torque was increased. Mallick et al. [6] studied the static and fatigue performances of adhesive/bolted (hybrid) joints in a structural injection molded composite. Hybrid joints were shown to have a higher static failure load and longer fatigue life than adhesive joints. Haq et al. [7, 8] have studied the effect of various adhesive insert reinforcements (nano-fillers and glass/carbon sleeves) on the quasi-static behaviour of hybrid bolted joints similar to the joints studied in this work. They have observed zero-slip, considerable improvements in static load carrying capacities, increase strengths and delayed onset of delaminations in resulting joints [7, 8]. The study on fatigue behaviour of such novel hybrid bolted joints does not exist and hence this work aims to evaluate the fatigue behaviour and establish a benchmark for such novel joints.

In this paper, the fatigue behaviours of two types of single lap composite bolted joints, namely, conventional bolted joints and novel hybrid bolted joints with injected resin inserts as originally invented by Gary Cloud (Provisional Patent 61/658,163), were investigated. Additionally, the static tests were also carried out to re-establish the baseline for the materials and processes used in this work. For the sake of convenience and brevity, this bolt system is referred to as the "Hybrid-bolt" in this paper. Load controlled tensile fatigue tests with a stress ratio R of 0.1 were conducted the cyclic loading. The specimens were subjected to varying fatigue loadings with a maximum loading level of up to 80 % of the ultimate strength of the joints. As expected, the life of the joints was found to depend on the level of fatigue loading. The experimental fatigue data and its correlation with failure provide valuable information for better design of composite joints and related components.

33.2 Materials

In this study, the vacuum assisted resin transfer molding (VARTM) technique was used to manufacture the glass fiber reinforced plastic (GFRP) composite laminates used for the single lap composite bolted joints. The reinforcement was S2-glass plain weave fabric (Owens Corning ShieldStrand S) with areal weight of 818 g/m^2. The resin used was a two part toughened epoxy, namely SC-15 (Applied Poleramic Inc., CA). The SC-15/glass-fiber laminates were cured in a convection oven at 60 °C for 2 h followed by a post cure of 4 h at 94 °C. Sixteen layers of plain-weave glass reinforcement were used for each adherend in this study. The thickness of cured laminate was found to be 9.6 mm. The laminate coupons for bolted joint were cut by water cooled, diamond coated-disc cutting machine to achieve a final dimension of 100 mm width by 150 mm length per adherend.

The bolt was a nominal ½ in. grade five, medium-strength steel hex head (head width: 19.05 mm or 3/4 in.; head height 7.94 mm or 5/16 in.) bolt with a right-hand thread having a length of 38.1 mm (one and a half in.), fully threaded with a 12.7 mm-13 (1/2 in.-13) thread.

33.3 Fabrication of Bolted Joints

Two approaches were used to develop the composite single lap bolted joints: conventional and Hybrid-bolt. The schematic diagram of this invention (see Fig. 33.1) basically consists of a bolt that has a channel through the bolt shank that allows the injection of an insert that fills the hole clearance space between the bolt and work-pieces and acts as a structural component. One of major advantages of this invention is that the presence of the adhesive/insert compensates for the variation of hole clearance that typically occurs during conventional machining process. Additionally, in layered composites, the delaminations introduced due to hole-drilling get filled with the adhesive insert, thereby repairing any damage and creating a monolithic part.

The GFRP plates were joined using the bolts described above and flanged sleeve washers. A 15.7 mm diameter hole was drilled in the work pieces 50.5 mm away from three edges. The sleeve washer has the following dimensions: flange diameter 25.4 mm, ID of 13.46 mm, sleeve OD of 15.88 mm, flange thickness of 3.18 mm, overall height of 4.75 mm. Bolts were drilled with 2 mm diameter passageways for resin injection. The clamping force of 35 N-m was applied by digital torque wrench.

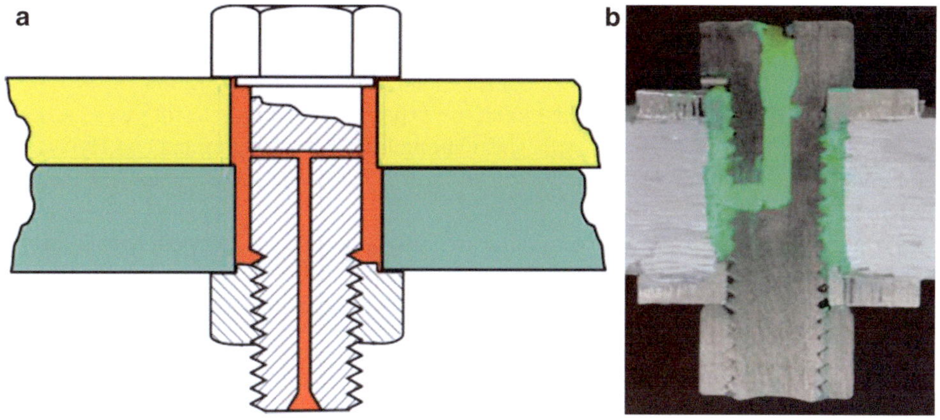

Fig. 33.1 Novel bolted composite joint (hybrid-bolt): (**a**) schematic diagram, (**b**) sectional view of scaled-up channel (*green*)

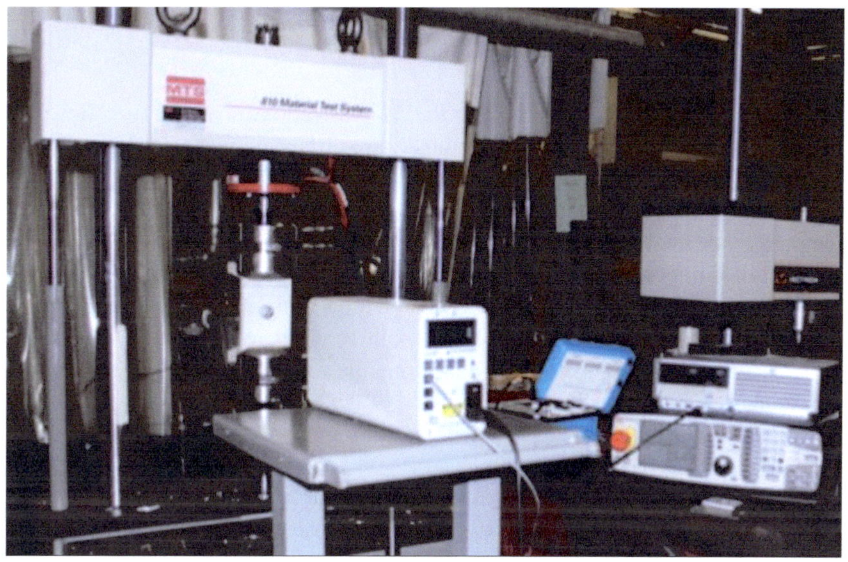

Fig. 33.2 Experimental setup

After assembly, the samples that were selected for the hybrid joint were injected with SC-15 resin through the channels in the bolts so as to fill the clearance space between the threaded bolt shank and the work pieces. The injected inserts were cured by a convection oven at 60 °C for 2 h followed by post curing at 94 °C for 4 h. Fig. 33.1b shows a cross-section of the bolted joint, wherein the channels are exaggerated for better understanding, and the resin is filled with a ultra-violet dye marker. As it could be seen, the resin fills all intricate areas including the threads of the bolts and upon curing creates a monolithic structure.

33.4 Experimental Setup

All quasi-static testing was performed using a 100 kN load cell servo-hydraulic testing machine (MTS 810), and a typical test setup is shown in Fig. 33.2. The machine is equipped with a standard load cell and an internal crosshead displacement measuring extensometer. An external laser extensometer was also used to measure the joint longitudinal displacement, as shown in Fig. 33.2. To acquire the strain gage data, a Vishay Model P3 Strain Indicator and Recorder was used.

Prior to testing the bolted joints, five GFRP specimens were tested according to ASTM D3039 to characterize the material properties. In order to avoid the influence of tab and gripping pressure in the case of GFRP, the experimental setup was modified based on previous experience [9]. As specified in ASTM D3039, the GFRP specimens were subjected to monotonic tensile loading with a stroke rate of 2 mm/min.

Quasi-static joint tests were performed according to ASTM STP1455-EB. In this work, five tests were carried out for each joint configuration. The average shear strength of the joint was calculated by dividing the maximum load reached in each test divided by the cross-sectional area of the bolt.

On the other hand, the fatigue tests were carried out under load control using a 50 kN load cell servo-hydraulic testing machine (MTS 810), with sinusoidal wave, sample recording rate of 100/s, load ratio R = 0.1, and frequency of fatigue test equal to 5 Hz. Three tests were carried out for each fatigue test configuration.

To plot the S–N curve of the bolted joint, the experimental fatigue testing was carried out at various load levels, namely 80, 60, 45, and 30 % of the quasi-static shear strength of both conventional and Hybrid bolted joints. Fatigue tests were run continuously up to 10^6 cycles at the 30 % load level, whereas for 45, 60, and 80 % load levels, the tests were run to material failure.

For the 30 % load level specimens, after the fatigue tests were stopped, tensile tests were performed on the pre-cycled specimens to measure the joint residual shear strength and stiffness. Test procedure for measurement of the residual properties was the same as that used for measurement of the properties of intact specimens.

33.5 Result Discussion

33.5.1 Static Tests

Material characteristics measured for the S2-glass/epoxy material are shown in Table 33.1. As mentioned above prior to the fatigue tests, quasi-static tensile bolted joint tests were performed on conventional and hybrid bolted joints. Figure 33.3 provides a comparison of the obtained tensile shear strength of the two types of joints.

As it can be seen in the Fig. 33.3, the plot for the conventional bolted joint shows a significant slip around ~5 kN, denoted by point 1, where the applied tensile-shear load is sufficient to overcome the clamping forces. At ~13 kN load, onset of delamination and localized fiber buckling occurs around the contact area adjacent to the bolt (see Fig. 33.3, point 2). It is evident that continuous stiffness degradation took place until the applied load reaches ~16 kN. As the applied load increases further, a stable load transfer from the bolt to the laminate was observed. At about 31.5 kN, due to gradual increase of secondary bending and bolt tilting the washer starts excessively damaging the work pieces and causes increasing delamination and out-of plane-deformation of the laminate. The tilting of the bolt is shown as angle θ in the photograph of Fig. 33.4. Owing to this phenomenon, the joint stiffness is significantly reduced, but the load carrying capacity increases monotonically up to the maximum load of ~53 kN.

On the other hand, for the hybrid bolted joint, the load-displacement curve as shown in Fig. 33.3 demonstrates a different type of behavior. In this curve, the slope is initially similar to that of the conventional bolted joint. However, unlike the conventional bolted joint, it exhibits a monotonic increase of load without slippage. Onset of delamination occurs at approximately 32 kN. Once the delamination initiates, the stiffness of the joint reduces gradually until it reaches it's the maximum load of ~51.5 kN. If the stiffness of the conventional bolted joint is calculated between points 1 and 2, the novel

Table 33.1 Material properties

Composite	Weave type	Fiber fraction (wt %)	Tensile strength (MPa)	E_x (GPa)	G_{xy} (GPa)	ν_{xy}
0/90S-glass/epoxy	Plain	0.61	476.264	23.18	4.0	0.08

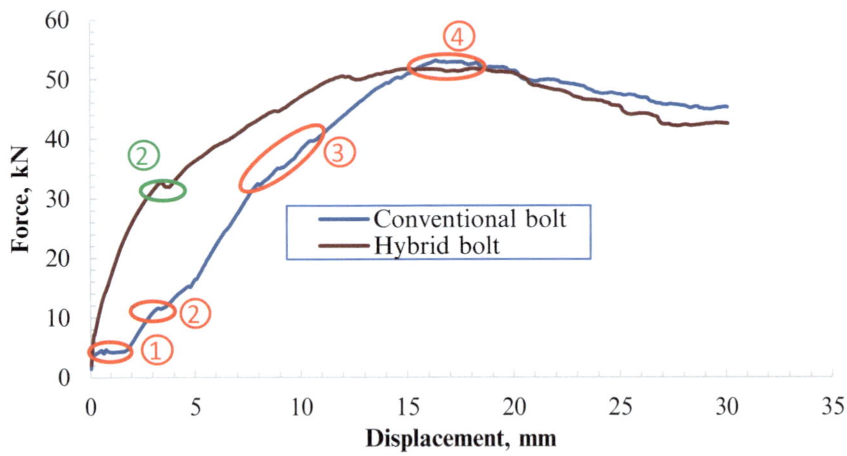

Fig. 33.3 Tensile strength of bolted: conventional (*blue*); hybrid bolt (*brown*)

Fig. 33.4 Conventional bolted joint subjected to tensile-shear loading

hybrid bolted joint improves the stiffness of the joint by ~62 % prior to the onset of delamination. Furthermore, if slip is critical, then the hybrid bolt can be considered infinitely better than the conventional one. If failure is defined as the onset of delamination, then the hybrid bolt is three times better than the conventional one. On the other hand, if ultimate strength is considered as the design criterion (which is not practically adopted by different industries), then, the conventional joint shows a slightly better performance. This phenomenon could be explained by the fact that the presence of large clearance in the joint causes excessive laminate damage due to concentrated load and allows the joint absorb more energy by delamination and localized fiber buckling. As a result, the work done by the plastic deformation of the bolt is delayed, which eventually enhances the ultimate load carrying capacity of the joint. It is worthwhile to mention that the influence of the insert on the ultimate strength of the joint is insignificant, except through alteration of the nature of the interaction between the laminate and the bolt. Similar results for hybrid joints with various types of inserts has been reported in our earlier work [7, 8].

33.5.2 Fatigue Tests

The results of the fatigue test for both conventional and hybrid bolted joints are presented in Fig. 33.5 in the form of S–N diagrams. The best-fit linearized plots were obtained by least-squares curve fitting, which might not be appropriate. It can be observed, that the curve of conventional bolted joint translated to the right with flatter slope, which implies better performance in fatigue life for the same percentage of ultimate failure load (UFL) compared with hybrid bolted joint. In these experiments, the lowest applied load level was restricted to 30 % of UFL at a test frequency of 5 Hz, for which the conventional and Hybrid bolted joints sustained the applied load for one million cycles without significant damage. Hence, the results show that, unlike the quasi-static test results, the hybrid joint exhibits inferior mechanical performance, meaning lower fatigue life, at higher load rates. This could be due to the fact that a large empty clearance between the bolt and the hole results in more concentrated loads on the laminate. These loads, in turn, cause more excessive laminate damage around the bolt, as suggested by the photographs presented in Fig. 33.6. As a result, the work done by the plastic deformation of the bolt/washer is delayed (as shown in Fig. 33.6c), which eventually enhances the load carrying capacity of the joint. Hence, more work is being done by the laminate, which leads to extended fatigue life of the joint and allows the bolt/washer to absorb less energy through plastic deformation.

In Fig. 33.7, standard S–N (stress versus number of cycles) curves of conventional and hybrid bolted joint are presented. As it can be observed, generally for a given number of cycles, the conventional bolt performs considerably better than does the hybrid bolted joint. For a given stress ratio of $R = 0.1$, the life of conventional and hybrid bolted joints can be predicted using (33.1).

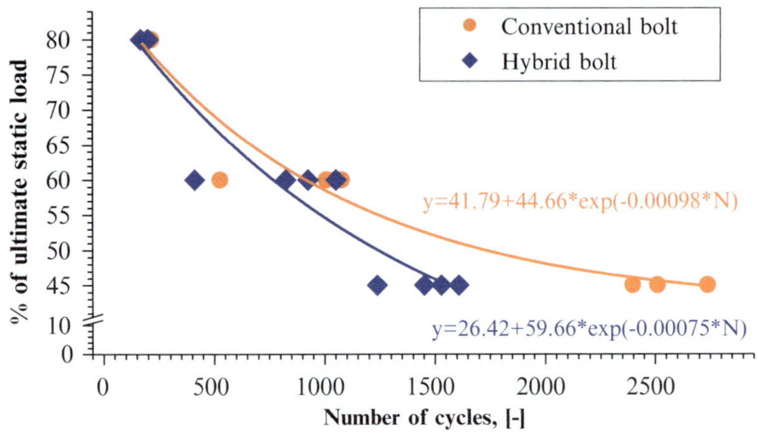

Fig. 33.5 Comparison of logarithmic S–N Curves of conventional and hybrid bolted joints

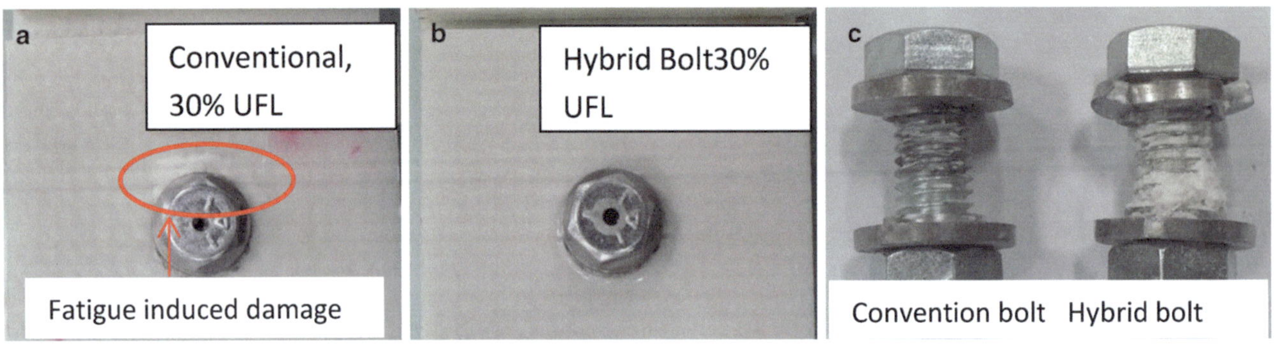

Fig. 33.6 Failure modes: (**a**) pre-cycled conventional bolted joints, (**b**) pre-cycled hybrid GCbolt, (**c**) failure mode of washers at 45 % of UFL

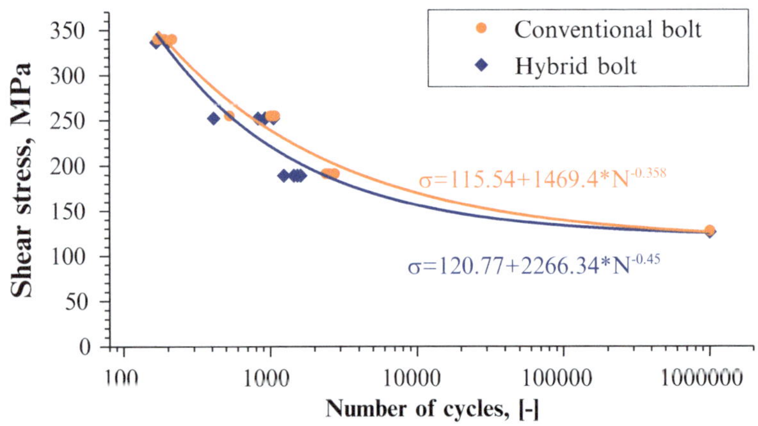

Fig. 33.7 Stress-versus number of cycles diagrams of conventional and hybrid bolted joints

$$\sigma = a * logN + b \quad (33.1)$$

$$N = 10^{(\sigma - b)a} \quad (33.2)$$

Where a and b are constants that can be found from the experimental tests, as shown in Fig. 33.7.

N is number of cycles.

For both the conventional and hybrid bolted joints, which were subjected to 30 % of UFL, the fatigue tests were interrupted at one million cycles and quasi-static tensile tests were performed to obtain the residual strength. Figure 33.8 presents the results of these tests as well as reproducing the results of the quasi-static strength tests on the as-fabricated specimens previously presented in Figure 33.3. As expected, the as-fabricated conventional bolted joint exhibit higher strength than does the pre-cycled specimen. The pre-cycled specimen shows moderate strength reduction, in this particular case by 6 %. Figure 33.6 showed the fatigue induced damage around the hole of the pre-cycled conventional bolted joint, which is evidence

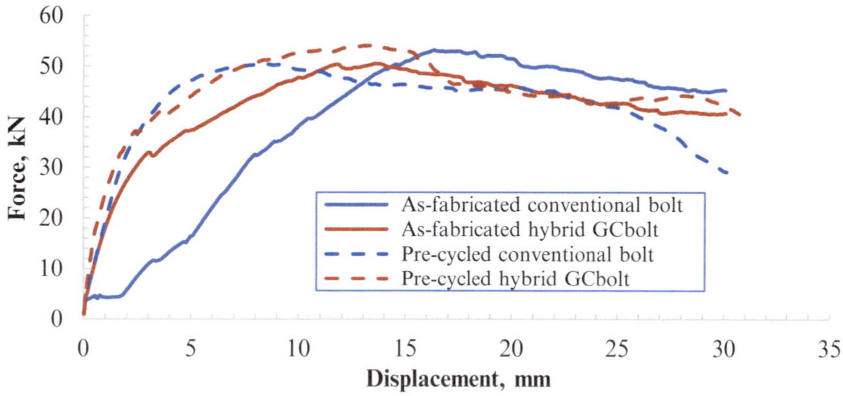

Fig. 33.8 Load-displacement diagram of as-fabricated and pre-cycled conventional and hybrid bolted joints

for possibly having lower residual strength. On the other hand, unlike the conventional joint, the hybrid bolted joint exhibited surprising improvement of residual joint strength after fatigue cycling by about 8 %. This finding could be attributed to many reasons. Firstly, the bolt is tightly assembled with the laminates through the injected resin. This arrangement would help the bolt to carry the applied load uniformly, so that the bolt and the washer could be treated favorably through work hardening and their proof loads/yield strengths increased. Secondly, after one million of cycles, the resin may act as a cushioning media between the laminate and the bolt that would allow the load to transfer smoothly during tensile shear loading. Finally, the residual strength of the pre-cycled hybrid bolted joint is seen to be about 12 % higher than that of the pre-cycled conventional joint. Also, the fiber reinforced laminates used in the adherends could also be contributing to increase in stiffness. The micro-degradation of the matrix within the adherends may cause fibers to slightly re-align along the load-direction. Post-fatigue quasi-static testing would have aligned fibers relative to the non-cycled specimens. This could explain the increase in stiffness for both the conventional and hybrid bolted joint. Similar joints with all metal adherends would confirm this hypothesis and will be reported in our next work. The results are preliminary and statistically significant tests along with insights from numerical simulations need to be performed to fully understand the observed results. Nevertheless, the results show promise in use of the hybrid joints in a wide range of structural applications.

33.6 Conclusion

In this work, the static and fatigue behavior of conventional and the most basic form of novel hybrid bolted joints were investigated. In the hybrid bolted joint, SC-15 epoxy resin was used as the structural insert. This most basic configuration of the hybrid bolt consists of a bolt that has a channel machined through the bolt-shaft that allows injection of an insert compound that fills the hole-clearance of the work-pieces and acts a structural component. The observed results from quasi-static and fatigue tests can be summarized as follows:

- It is well known that the presence of slip between the joined parts leads to rapid loosening, as a result, the reliability of bolted joint is significantly affected. The hybrid joint eliminates the presence of slippage. If failure is defined as the onset of delamination, then the hybrid joint is three times better than the conventional one.
- On the other hand, if the ultimate strength is considered as design criteria (which is not practically adopted by different industries), the conventional joint shows a slightly better performance. This phenomenon could be explained by the fact that the presence of clearance in the joint causes excessive laminate damage by the concentrated load and forces the absorption of more energy by delamination and localized fiber buckling/breakage. As a result, the work done by the plastic deformation of the bolt is delayed, which eventually enhances the load carrying capacity of the joint.
- The hybrid joint exhibits inferior mechanical performance and lower fatigue life at higher load rates. Similar to the quasi-static loading condition, a large clearance between the bolt and the hole results in more concentrated loads on the laminate, causing more excessive laminate damage around the bolt. Hence, more energy is being absorbed by the laminate, which leads to extended life of the joint and allows the bolt to absorb less energy through plastic deformation.
- Residual test results of pre-cycled conventional and hybrid bolted joints show that the as-fabricated conventional bolted joint exhibits higher loads than the pre-cycled specimens.
- The residual strength of the cycled hybrid joint is about 12 % higher than that of the pre-cycled conventional joint.

- On the other hand, a pre-cycled hybrid bolted joint exhibits improvement of joint strength over the as-fabricated hybrid bolted joint. This surprising result could be explained by a combination of few phenomena. Firstly, because of the presence of the resin insert, the bolt carries the applied load uniformly, so that both the bolt and the washer could be favorably stressed through work hardening and their proof loads/yield strengths increase accordingly. Secondly, after one million of cycles, the resin may act as a cushioning media between the laminate and the bolt that would allow the load to transfer smoothly during tensile shear loading. Thirdly, the matrix in the composite adherend may be undergoing micro-degradation allowing the fibers to align in the load direction during the fatigue cycling. Upon post-fatigue quasi-static testing, these aligned fibers may increase the stiffness and strength of resulting joints relative to their non-cycled counterparts. Further experimental work combined with realistic modeling and numerical simulations can aid in understanding the underlying phenomena.

Generally, the conventional and hybrid bolted composite joints under quasi-static and fatigue loadings have shown complex behaviors. However, the observations from the experimental data and their correlations with failure could provide valuable information for better design of innovative composite joints and related components in various applications. Besides, more complex tailored inserts that contain other resins, adhesives, particulates, and fibres could play a significant role in improving the fatigue life of joints and facilitate lightweighting of vehicles and other devices.

References

1. http://www.plasticsnews.com/
2. Heimbs, S., Schmeer, S., Blaurock, J., Steeger, S.: Static and dynamic failure behaviour of bolted joints in carbon fibre composites. Compos. A: Appl. Sci. Manuf. **47**, 91–101 (2013)
3. Saunders, D.S., Galea, S.C., Deirmendjian, G.K.: The development of fatigue damage around fastener holes in thick graphite/epoxy composite laminates. Composites **24**(4), 309–321 (1993)
4. Crews Jr., J.H.: Bolt-bearing fatigue of a graphite/epoxy laminate. In: Kedward, K.T. (ed.) Joining of Composite Materials, ASTM STP 749, pp. 131–144. American Society for Testing and Materials, Baltimore (1981)
5. Herrington, P.D., Sabbaghian, M.: Fatigue failure of composite bolted joints. J. Compos. Mater. **27**(5), 491–512 (1993)
6. Fu, M., Mallick, P.K.: Fatigue of hybrid (adhesive/bolted) joints in SRIM composites. Int. J. Adhes. Adhes. **21**, 145–159 (2001)
7. Haq, M., Khomenko, A., Cloud G.: Novel hybrid fastening system with nano-additive reinforced adhesive inserts. In: SEM 2014 Annual Conference and Exposition on Experimental and Applied Mechanics, Greenville, 2–5 June 2014
8. Haq, M., Cloud, G.: Flexible hybrid fastening system. In: SEM 2013 Annual Conference and Exposition on Experimental and Applied Mechanics, Chicago, 3–5 June 2013
9. Belingardi, G., Paolino, D.S., Koricho, E.G.. Investigation of influence of tab types on tensile strength of E-glass/epoxy fiber reinforced composite materials. Procedia Eng. **10**, 3279–3284 (2011)

MIX
Papier aus verantwortungsvollen Quellen
Paper from responsible sources
FSC® C105338

If you have any concerns about our products,
you can contact us on
ProductSafety@springernature.com

In case Publisher is established outside the EU,
the EU authorized representative is:
Springer Nature Customer Service Center GmbH
Europaplatz 3, 69115 Heidelberg, Germany

Printed by Libri Plureos GmbH
in Hamburg, Germany